FINITE MATHEMATICS

FINITE MATHEMATICS

Daniel P. Maki
Associate Professor of Mathematics
Indiana University

Maynard Thompson
Professor of Mathematics
Indiana University

McGraw-Hill Book Company

New York   St. Louis   San Francisco   Auckland   Bogotá   Düsseldorf
Johannesburg   London   Madrid   Mexico   Montreal   New Delhi   Panama   Paris
São Paulo   Singapore   Sydney   Tokyo   Toronto

FINITE MATHEMATICS

234567890 DODO 78321098

This book was set in Times Roman.
The editors were A. Anthony Arthur, Alice Macnow, and Michael Gardner;
the designer was Jo Jones;
the production supervisor was Leroy A. Young.
The drawings were done by J & R Services, Inc.
The cover was designed by John Hite.
R. R. Donnelley & Sons Company was printer and binder.

Library of Congress Cataloging in Publication Data

Maki, Daniel P.
    Finite mathematics.

      1. Mathematics—1961-   I. Thompson, Maynard,
joint author. II. Title.
QA39.2.M333   510   77-9368
ISBN 0-07-039745-7

CONTENTS

Preface   ix

1 MATHEMATICAL MODELS, SETS, AND FUNCTIONS   1

1-1 Introduction   1
1-2 Mathematical Models   2
1-3 Examples of Model-building Situations   6
1-4 Sets and Set Operations   9
1-5 Venn Diagrams and the Sizes of Sets   15
1-6 Functions   27
    Important Terms and Review Exercises   30

PART ONE: PROBABILITY MODELS

2 PROBABILITY PRELIMINARIES AND COUNTING   35

2-1 Introduction   35
2-2 Probability Preliminaries   36
2-3 Terminology and Notation   41
2-4 Multistage Experiments and Tree Diagrams   46
2-5 Counting Arrangements: Permutations   54
2-6 Counting Partitions: Combinations   59
2-7 Applications of Counting to the Computation of Probabilities   64
    Important Terms and Review Exercises   69

3 PROBABILITY   71

3-1 Introduction   71
3-2 Probability Measures: Basic Rules and Properties   72
3-3 Conditional Probability and Independence   81
3-4 Stochastic Processes and Tree Measures   88
3-5 Independent Trials with Two Outcomes   95
3-6 Bayes Probabilities   102
3-7 Random Variables and Expected Value   113
    Important Terms and Review Exercises   121

4 STATISTICS   123

4-1 Introduction   123
4-2 The Binomial and Normal Distributions   124
4-3 Mean and Standard Deviation   136

4-4 The Normal Approximation to a Binomial Random Variable   147
4-5 Decision Problems   159
    Important Terms and Review Exercises   167

PART TWO: LINEAR MODELS

5 LINEAR PROBLEMS IN TWO VARIABLES AND THEIR GRAPHS   171

5-1 Introduction   171
5-2 The Cartesian Coordinate System   172
5-3 The Equation of a Line   177
5-4 Systems of Lines   186
5-5 Systems of Linear Inequalities in Two Variables   195
5-6 Formulation of Linear Programming Problems   200
5-7 Graphical Solution of Linear Programming Problems   208
    Important Terms and Review Exercises   217

6 SYSTEMS OF LINEAR EQUATIONS AND MATRICES   220

6-1 Introduction   220
6-2 Systems of Linear Equations and Matrix Notation   222
6-3 Solving Systems of Linear Equations   228
6-4 Matrix Algebra   241
6-5 Matrix Inverses   247
6-6 Matrices and the Chain of Command   255
6-7 A Linear Economic Model   260
    Important Terms and Review Exercises   266

7 COMPUTATIONAL METHODS FOR LINEAR PROGRAMMING   268

7-1 Introduction   268
7-2 Formulation of Problems   269
7-3 Slack Variables and Basic Solutions   274
7-4 Tableaus and the Pivot Operation   278
7-5 Optimal Vectors via the Simplex Method   285
7-6 Dual Programming Problems   291
    Important Terms and Review Exercises   302

PART THREE: APPLICATIONS

8 MARKOV CHAINS   307

8-1 Introduction   307
8-2 Basic Characteristics of Markov Chains   310
8-3 Regular Markov Chains   319
8-4 Absorbing Markov Chains   328
    Important Terms and Review Exercises   337

9 TWO-PERSON ZERO-SUM GAMES   **339**

9-1 Introduction   339
9-2 Two Simple Games   340
9-3 Saddle Points and Dominance   343
9-4 Solving $2 \times n$ and $m \times 2$ Games   352
9-5 Solving $m \times n$ Games Using the Simplex Method   359
    Important Terms and Review Exercises   366

10 DIGRAPHS AND NETWORKS   **367**

10-1 Introduction   367
10-2 Definitions and Notation   369
10-3 Networks   379
    Important Terms and Review Exercises   387

11 EVALUATING INVESTMENT OPTIONS   **389**

11-1 Introduction   389
11-2 Interest   390
11-3 Present Value and Annuities   398
11-4 Amortization and Sinking Funds   406
11-5 Stocks and Bonds   410
    Important Terms and Review Exercises   417

Appendix A Areas under the Standard Normal Curve   420

Appendix B Answers to Selected Odd-numbered Exercises   422

Index   449

PREFACE

This book is intended to introduce the basic ideas and techniques of finite mathematics at a freshman-sophomore level. It is written for students who are interested in the uses of mathematics, and consequently the emphasis is on the development of methods to solve problems. We have stressed the use of mathematics in turning vague questions into precise problems, in recognizing the similar features of apparently diverse situations, in organizing information, and in making predictions.

The book is organized as follows:

Introduction: Chapter 1
Part One: probability models (Chapters 2 to 4)
Part Two: linear models (Chapters 5 to 7)
Part Three: applications (Chapters 8 to 11)

Parts One and Two are independent and self-contained. They can be discussed in either order. Each of the first seven chapters contains problems and applications related to the topic of that chapter. In addition, Part Three contains applications which (in general) depend on combinations and extensions of the ideas developed in Parts One and Two. The dependency relations between the chapters are shown in the following diagram. The notation Ⓐ➔Ⓑ (or Ⓐ → Ⓑ) means that there is material in Chapter A which is essential (or useful but not essential) background material for some of the material discussed in Chapter B. In particular, the chapters in Part Three are all independent of each other. The material of Chapter 6 which is required for Chapters 7 to 9 is contained in Sections 6-1 to 6-4. Section 6-5 provides a natural lead into Chapter 10. Section 11-5 requires a knowledge of expected value (Section 3-7), but otherwise Chapter 11 is independent of all other chapters.

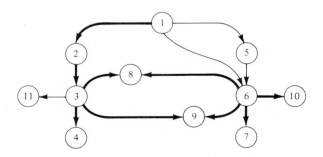

In our view the core material of a course in finite mathematics is contained in Chapters 1 to 3, 5, and 6. In addition to these five basic chapters, we normally cover one additional chapter or parts of two additional chapters in a 15 week semester. Three examples of one-semester courses are given below. The first has a general orientation, the second has an emphasis on probability and its applications, and the third has an emphasis on linear models.

General: Chapters 1 to 4 (Sections 4.1 to 4.3), 5, 6, and one chapter selected from 9, 10, or 11

Probability: Chapters 1 to 6 (Sections 6.1 to 6.4) and 8 (Sections 8.1 to 8.3)

Linear models: Chapters 1 to 3 and 5 to 7

Each chapter is followed by a set of review exercises of varying degrees of difficulty (not necessarily graduated from easy to hard) and related to the topic of the chapter, though not always to a specific section.

The answers to selected odd-numbered exercises are provided in Appendix B.

It is a pleasure to acknowledge the support we have received from our colleagues, several very helpful reviewers, and the McGraw-Hill editors. Our colleagues, Victor Goodman and George Springer, taught from preliminary versions of the manuscript and provided useful comments. We are grateful to the reviewers for their careful evaluation of preliminary versions and their suggestions: Howard T. Bell, Shippensburg State College; Donald C. Cathcart, Salisbury State College; Stephen R. Hilding, Gustavus Adolphus College; James A. Murtha, Marietta College; and Joseph L. Zemmer, University of Missouri, Columbia. Donald Cathcart has been especially helpful, and his perceptive comments have resulted in several improvements. Graduate students at Indiana University, including R. Johns, J. O'Sullivan, J. Weinberger, and T. Whitehurst, have taught from preliminary versions, proofread the manuscript, and aided in preparing the answers to exercises. The support and cooperation of the McGraw-Hill editorial staff, especially Tony Arthur and Alice Macnow, are deeply appreciated.

Daniel P. Maki
Maynard Thompson

FINITE MATHEMATICS

MATHEMATICAL MODELS, SETS, AND FUNCTIONS

1-1 INTRODUCTION

Mathematics is an art, a science, and a practical tool. As an art it can be (and is) pursued and enjoyed entirely for its own sake. As a science it has contributed significantly to our understanding of many biological, social, and physical processes. As a tool it is a standard part of the working knowledge of people in business, science, and technology. Used as a tool, mathematics can help us understand and describe a process or situation, and it serves as a means of quantifying alternatives in decision making. The use of quantitative methods in evaluating, ranking, and selecting various courses of action is a fairly recent (essentially post-1940) and increasingly important application of mathematics in business and the social sciences. The precision of thought, notation, and terminology offered by mathematics is extremely useful in describing our complex environmental, social, and business systems.

The goal of this text is to present several mathematical topics especially useful in describing situations and solving problems which arise in the life and social sciences and in business. The mathematics is organized around a few key ideas and is applied to specific problems through mathematical models. It is the mathematical model which provides the link between the mathematical ideas and the problems, questions, and decisions of the scientific and business world.

Mathematical models are the topic of the first two sections of this chapter. We introduce the basic ideas here and return to them frequently in the chapters which follow. The modeling process is best illustrated through examples, and in a sense the entire book is a collection of models. Thus, it is to be expected that your understanding of Secs. 1-2 and 1-3 will improve as you proceed through the remainder of the text. Sections 1-4 and 1-5 provide the definitions and notation of sets and functions needed to develop and use the mathematics contained in this book.

1-2   MATHEMATICAL MODELS

Attempts to understand situations and to solve problems in business and the sciences often begin with observations and experiments to collect relevant data. After the data are collected, they must be organized and analyzed. Frequently, the result of these activities is a table, chart, diagram, or graph which summarizes the observations in a convenient form. The next step is usually an effort to identify an underlying structure or process which can be used to account for the observations. In some cases a qualitative description is satisfactory or the best one can obtain from the information available. In other cases it may be possible to propose a basic process which accounts for the quantitative features of the observations. When the proposed description is given in terms of mathematical concepts, symbols, equations, or functions, it can be said that a *mathematical model* has been constructed. A mathematical model is an expression in mathematical terms, i.e., involving mathematical concepts and notation, of relationships or processes which have been observed or which have a meaning in a nonmathematical setting. Frequently there is more than one reasonable mathematical model for a set of observations. The model selected for study will depend on the type of information desired, the degree of mathematical complexity acceptable, and the inclinations of the investigator.

EXAMPLE 1-1

A commodity dealer noticed that variations in the sales of a raw material seem to be associated with variations in price, and he sought to determine a relationship between the two. He began by collecting the data provided in Table 1-1.

TABLE 1-1

Price per Unit (in dollars)	Monthly Sales (in millions of units)
8	12.50
9	11.75
10	11.50
12	11.50

There are many possible explanations of the data in this table. Three mathematical models which could serve as working hypotheses in formulating policy are the following:

Model 1:   Sales are independent of price, and the variations noted are due to influences not related to price. The fact that sales are at 11.50 million units per month at both prices 10 and 12 might be cited as evidence to support this model. Thus one proposed model might be

$$Sales = 11.50$$

independent of the price.

Model 2:   Sales decrease as price increases. Specifically, each $1 increase in price per unit results in a decrease in sales of 250,000 units. Let us denote by s the monthly sales (in millions of units) and by p the price per unit (in dollars). Then, for example, the model based on the formula

$$s = 12 - .25(p - 8)$$

accounts for the exact sales recorded with prices 9 and 10. Indeed with $p = 9$ we have

$$s = 12 - .25(9 - 8) = 12 - .25 = 11.75$$

and with $p = 10$ we have

$$s = 12 - .25(10 - 8) = 12 - .50 = 11.50$$

which agree with the data in Table 1-1. However, with $p = 8$ we have

$$s = 12 - .25(8 - 8) = 12.0$$

and with $p = 12$ we have

$$s = 12 - .25(12 - 8) = 12 - 1 = 11.0$$

which are both lower than the values in Table 1-1. An alternative model of the same type is

$$s = 12.5 - .25(p - 8)$$

which accounts for the sales at prices 8 and 12 exactly but gives sales that are too high for prices 9 and 10. The model

$$s = 12.21 - .22(p - 8)$$

TABLE 1-2

Price	Model 1	Model 2 (Third Version)	Model 3
		Sales Predicted by	
8	11.5	12.21	12.5
9	11.5	11.99	12.0
10	11.5	11.77	11.75
12	11.5	11.33	11.5

does not predict sales which agree with any of the data exactly, but in a sense it "splits the difference" and predicts sales which are too low at $p = 8$ and $p = 12$ and sales which are too high at $p = 9$ and $p = 10$. In each of these models the points (price, sales) lie on a straight line, and consequently each is an example of a *linear model*. Models of this sort are studied in detail in Part Two of this book.

Model 3:   Sales decrease as price increases, and sales s and price p are related in the following way

$$s = 11 + \frac{3}{p - 6}$$

Notice that this also gives sales which agree exactly with the data for $p = 8$ and 12 but are too high for $p = 9$ and 10. There are several possible models of this sort, just as there were several linear models above.

The predictions of these three models are tabulated in Table 1-2 and displayed in Fig. 1-1. A selection between these models cannot be based solely on the data in Table 1-2, although each can be analyzed and the results of the analysis used to make predictions. One method of selecting a specific model is discussed in Chap. 5.   ☐

FIGURE 1-1

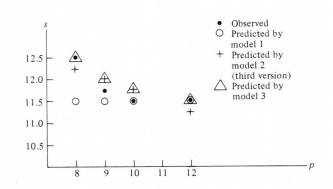

After a problem or situation has been represented in mathematical terms, i.e., after a mathematical model has been constructed, the resulting mathematical problem can be studied using appropriate mathematical techniques. The results of such a study are predictions about the original situation based on that specific mathematical model. For example, a prediction based on the third linear model of Example 1-1 is that for a price of $14 per unit the sales will be 10.89 million units per month. Indeed, the formula $s = 12.21 - .22(p - 8)$ gives $s = 10.89$ for $p = 14$.

The final step, and in some respects the conclusive one, in the model-building process is a comparison of the predictions resulting from the mathematical work with the observations and data of the original situation. The happiest circumstance is that everything actually observed is accounted for in the conclusions of the mathematical study and that other predictions based on a study of the mathematical system are subsequently verified by observation or experiment. Such agreement is not common, at least not on the first attempt. A much more typical situation is that the set of conclusions resulting from the mathematical analysis contains some which seem to agree and some which seem to disagree with the observations. This was the case with each of the models introduced in Example 1-1. This lack of agreement between predictions and observations necessitates a reexamination of the model-building process. One begins this reexamination by asking a number of questions: Does the mathematical model reflect all the important aspects of the original situation, and does it avoid introducing extraneous features not observed there? Is the mathematical work free from error? Have the mathematical results been interpreted correctly in terms of the original situation? Negative answers to any of these questions mean that the modeling process was deficient in some respect. In actual practice it is usually necessary to construct several models, each a refinement of the preceding, until finally an acceptable model is found. For example, it may be determined that while sales of a commodity do depend upon price, they also depend upon a number of other factors, e.g., the season of the year, anticipated production, and interest rates. Therefore predictions based on a model which uses price as the only factor affecting sales are likely to be inaccurate. A model which reflects the dependence of sales on these other factors is needed. The new model might also be linear but more complicated than the linear models of Example 1-1, or it might be completely different from the models introduced in that example. In any case, the new model would also be judged by the accuracy and usefulness of predictions based on it.

Mathematical models were first developed in connection with questions in the physical sciences. Important models in astronomy, chemistry, and physics which are still useful today were formulated in the seventeenth century. Many of these models have been highly developed and in their present forms involve quite complex mathematics. In the last half century the use of mathematical models has rapidly expanded into business and the social and life sciences. In fact, some of the areas of most rapid growth in these fields are those in which mathematical tools have proved to be useful. Frequently it is possible to formulate interesting and useful models for situations arising in business and the social and life sciences using only relatively simple mathematical ideas. It is models of this sort that are emphasized in this text.

1-3　EXAMPLES OF MODEL-BUILDING SITUATIONS

The examples of this section are indicative of the types of problems which can be profitably studied using the ideas and techniques developed in this text. At the moment we are interested only in illustrating the type of situation in which mathematical ideas and techniques have proved useful. Our goal here is not the formulation and solution of precise mathematical problems but the identification of those aspects of a situation which, if pursued, lead to such problems. The initial steps of problem formulation and conversion into mathematical form are crucial ones. If an inappropriate mathematical model is selected, it is very unlikely that the results of the mathematical analysis will be useful.

EXAMPLE 1-2　HOW TO FIND A LIVE VIRUS EFFICIENTLY

The process by which the Downjim Drug Company manufactures a vaccine involves live viruses, and it is very important that no live viruses be contained in the vaccine supplied to the public. The process used to manufacture the vaccine is known to have the property that over the long run about 1 in every 20 batches of vaccine will contain live viruses, but the company has no way of knowing exactly which batches will be contaminated. Since the penalty for releasing vaccine containing live viruses to the public is very severe (loss of production license and a fine), every batch produced must be tested for live viruses. There is a method of testing which requires only a very small sample from a batch in order to determine whether the whole batch contains live viruses. Also, samples can be pooled for testing. That is, samples from several batches can be combined and tested together. If the result of testing a pooled sample for live viruses is negative, none of the batches from which the samples were selected contains any live viruses. However, if the test results are positive, it is known only that some (one or more) of the batches contain live viruses; precisely which batches is unknown. In such a case it is necessary to continue testing according to some scheme until each of the batches containing live viruses is identified.

　　The costs incurred in testing the vaccine consist of extraction costs and laboratory costs. It costs $2 to extract a sample from a batch of vaccine, and it costs $8 to test a sample—pooled or not. Thus, to test a single sample there is a cost of $10. To test a pooled sample involving samples from two batches there is a $12 cost, to test a pooled sample from three batches there is a $14 cost, etc. Moreover, a maximum of 10 samples can be combined into a pooled sample for testing. How should the drug company design a testing program which will guarantee that no vaccine containing live viruses is released and which is as economical as possible in terms of dollar cost?

The situation described here clearly involves the notion of randomness. The company has no way of predicting precisely which batches of vaccine will contain live viruses, but it does have information on the frequency of contaminated batches. We need to agree upon a precise way of translating this information on frequency into information on the likelihood that a specific batch is contaminated. Once this information is available, we need to develop a means of computing or estimating in some appropriate way the likely costs of various testing programs. We then select the testing program with the lowest likely cost.

This problem involves several ideas which will be discussed in detail in the chapters on probability and statistics. A version of the problem is completely solved in Chap. 4. □

EXAMPLE 1-3   DATA AND DOLLARS

The actions of government officials affect everyone, and it is important that these officials have accurate data upon which to base their decisions. In many situations, however, it is both difficult and expensive to obtain reliable data. A typical data-collecting situation is the following.

The firm Public Opinion Survey Enterprises (POSE) has been hired to sample public opinion in Metroburg by conducting a survey. Metroburg consists of an inner city and several suburbs. The contract with POSE specifies the following conditions:

1. At most 2000 people are to be sampled.
2. At least 400 people from the suburbs must be sampled.
3. At least half of those sampled must be from the inner city.
4. POSE is to be compensated with a fee of $2000 plus $10 for each individual surveyed.

Since POSE operates with the intention of making a profit, it is interested in conducting the survey in a manner that produces maximum profits.

From previous experience, POSE knows that it is difficult to conduct a survey of this type entirely by mail. Many addresses are incorrect—especially in the inner city—and many people will not respond to a survey form received in the mail. On the other hand, it is relatively expensive to interview everyone in person, especially in the suburbs where many of the people live far apart. Thus the survey team must decide on some blend of interviews and mailings and a mix of individuals from the suburbs and inner city. The goal of the survey team is to maximize their total profits while meeting the guidelines established.

Problems like the one faced by POSE can often be studied with the aid of mathematics known as linear programming. Linear programming is introduced in Chap. 5 and continued in Chap. 7. A specific instance of a problem of the type faced by POSE is solved in Chap. 5.   □

Examples 1-1 to 1-3 illustrate situations in which basic and important decisions must be made about the mathematical formulation of the problem. There are several ways to proceed in each case, and the conclusions which result from the analysis will vary with the basic assumptions. These examples, while still relatively simple, display many of the features of the typical problems one meets outside of mathematics. There are, however, other problems in which the mathematical model is either given or obvious, but which still require the analysis of the mathematical problem. Since we shall meet many such problems in this text, we conclude this section with one.

EXAMPLE 1-4 SCHEDULING FOR MICE

An experiment in psychology involves observing the interaction of two mice. The experimenter wishes to minimize the effects of the idiosyncrasies of the individual mice, and so he decides to conduct the experiment with every possible pair selected from seven trained mice. He can conduct several experiments each day, but he decides that each mouse should be used at most once each day. How many days should he reserve the laboratory for his tests, and how should the individual mice be scheduled in order to complete the experiment in the shortest possible time?

Example 1-4 has several aspects. First of all, it is a counting problem since the number of pairs of mice must be determined. Such a counting problem can be solved by making a list of all possible pairs of mice. For example, if the mice are identified with the letters $A, B, \ldots, G$, then there must be experiments involving the pairs A and B, A and C, $\ldots$, A and G; B and C, $\ldots$, B and G; and so on. However, instead of making such a list, it is much simpler to apply a general formula which can be used to solve an entire class of such counting problems. We shall develop such a formula in Chap. 2.

Also, this example contains a scheduling problem. To illustrate its nature it is convenient to work with smaller numbers, and we consider a similar situation with five mice labeled A, B, C, D, E. If the experimenter schedules experiments as shown below on the first 3 days,

Day 1	Day 2	Day 3
A and B	A and C	A and D
C and D	B and D	B and C

then he must schedule A and E, B and E, C and E, and D and E as the only experiments on the next 4 days (all other pairs of mice have already been used). This schedule, which you might guess is inefficient because of the underuse of the last 4 days, requires 7 days to complete the experiment. There are more efficient ways of scheduling in which the experiment can be completed in 5 days. It is often the case that although a scheduling problem can be solved by a trial-and-error listing method, such a method may involve many trials and many errors. A direct method of scheduling is preferable. □

1-4 SETS AND SET OPERATIONS

The students in a finite mathematics class form a set. So do the planets in our solar system and the collection of banks which were chartered by the United States government as of January 1, 1978. For our purposes any well-defined collection is a *set*. The phrase *well-defined* means that it is possible to determine by some method exactly which things belong to the collection and which do not. The items in the collection are the *elements* of the set. The set is specified by a rule or other method which makes it possible to determine which things are in the set, i.e., the collection, and which are not. Sets can be described in two ways. The first way is simply to list the elements of the set. By convention, the elements are listed between a pair of braces. Thus, the set P of known planets in our solar system may be denoted by

$$P = \{\text{Mercury, Venus, Earth, Mars, Jupiter, Saturn, Uranus, Neptune, Pluto}\}$$

The second method of denoting a set is by stating a rule which specifies the elements of the set. Again, by convention, the rule is written out between a pair of braces following a symbol denoting a general element of the set and a colon. For example,

$$B = \{b: \quad b \text{ is a bank which was chartered by the United}$$
$$\text{States government as of January 1, 1978}\}$$

In the notation used to describe the set B, the symbols "$B = \{b:$" are read "B is the set of all b such that." With this notation an alternative way of specifying the set P described above is

$$P = \{p: \quad p \text{ is a known planet in the solar system}\}$$

In certain cases a list of its elements provides the simplest and most useful representation of a set, while in other cases specifying a rule is preferable. For sets with a large number of elements the specification of a rule is often the only practical way to define a set.

From one point of view all mathematics can be developed through sets and operations on sets. Whether or not one holds this view, the notation and operations of set theory are very convenient tools for the development of mathematics and its applications. Our goal here is to introduce enough of the basic definitions and notation of set theory to facilitate our work in the remainder of the book.

In general we use uppercase letters such as A and B to denote sets and lowercase letters such as a and b to denote elements of sets. To indicate that s is an element of S we write $s \in S$. To indicate that s is *not* an element of S we write $s \notin S$.

Definition   If every element of a set A is also an element of a set B, then we say that A is *included in* B or that A is a *subset* of B, and we write $A \subset B$. If $A \subset B$ and $B \subset A$, then the sets A and B have exactly the same elements and we say that A and B are *equal*. To indicate that A and B are equal we write $A = B$.

Thus the set $I = \{$Mercury, Venus$\}$ of inner planets is a subset of the set P of all known planets in the solar system; that is, $I \subset P$.

Sets can be combined to form new sets, and one method of combining the sets is similar to the way statements in ordinary English are combined. In English the words *and*, *or*, and *not* are often used to combine and modify statements to form new statements. We now define three symbols $\cap$, $\cup$, and $\sim$, which are used to form new sets from given ones and which are the set-theoretic associates of the words *and*, *or*, and *not*, respectively.

Definition   If A and B are sets, then $A \cap B$ is the set consisting of all elements which are in both A and B. $A \cap B$ is called the *intersection* of A and B.

Definition   If A and B are sets, then $A \cup B$ is the set consisting of all elements which are in A or in B or in both A and B. The set $A \cup B$ is called the *union* of A and B.

These concepts are illustrated in the following examples of sets, subsets, unions, and intersections.

EXAMPLE 1-5

Let the sets S, A, B, and C be defined as follows:

$$S = \{\text{GM, Ford, Chrysler, Goodyear, Goodrich, Firestone, Texaco,}$$
$$\text{Phillips, Exxon}\}$$
$$A = \{\text{GM, Ford, Chrysler}\}$$
$$B = \{\text{GM, Goodyear, Texaco}\}$$
$$C = \{\text{Texaco, Exxon, Marathon}\}$$

Since each element of A is also an element of S, A is a subset of S. Likewise B is a subset of S. On the other hand, since Marathon $\notin S$, the set C is not a subset of S. We have

$$A \cup B = \{\text{GM, Ford, Chrysler, Goodyear, Texaco}\}$$
$$A \cap B = \{\text{GM}\}$$
$$B \cap C = \{\text{Texaco}\}$$
$$B \cup C = \{\text{GM, Goodyear, Texaco, Exxon, Marathon}\}$$
$$(A \cup B) \cap (B \cup C) = \{\text{GM, Goodyear, Texaco}\}$$ □

In Example 1-5 the subsets A and C have no elements in common; i.e., the intersection contains no elements.

Definition   The set which contains no elements is known as the *empty set*, and it is denoted by $\varnothing$.

By convention the empty set is considered to be a subset of every set. The following properties of the empty set are direct consequences of the definitions of intersection and union:

If A is any set, then $A \cap \varnothing = \varnothing$ and $A \cup \varnothing = A$

EXAMPLE 1-6

The receiving clerk at Mul. T. Decibel Music Company has been assigned the task of inspecting a shipment of tapes for defective ones. It has been noted that the quality control of the manufacturer has deteriorated, and the Music Company has had an excessive number of tapes returned for replacement. As a result the receiving clerk has been instructed to sample the incoming shipment for defective tapes and if there are too many to reject the shipment. The shipment consists of 100 cartons of 12 tapes each. The clerk selects 5 cartons at random and checks 2 tapes in each carton. Each tape which is tested is judged to be either acceptable or defective. How can the outcome of this inspection be described?

Solution: From the standpoint of accepting or rejecting the shipment, the only relevant information obtained from the inspection is the number of defective tapes. Thus one can represent the set of possible outcomes of the inspection by the set

$$X = \{0, 1, 2, \ldots, 10\}$$

(Three dots, $\ldots$, indicate that numbers continue in the indicated pattern, i.e., sequentially, up to 10.) The set X gives the set of all possible numbers of defective tapes. Recall that 10 tapes are tested and any number of them could conceivably be defective.

Next, suppose that the shipment is accepted if the inspection results in not more than 1 defective tape and is rejected if more than 2 defective tapes are found. If exactly 2 of the 10 tested tapes prove to be defective, additional tapes will be tested. Let A, R, and C represent the subsets of X which correspond to accepting, rejecting, and continuing to test the shipment, respectively. Then

$$A = \{0, 1\} \qquad R = \{3, 4, \ldots, 10\} \qquad C = \{2\}$$

The sets A, R, and C have the following properties

$$A \cup R \cup C = X$$

[Notice that no parentheses are necessary since $(A \cup R) \cup C = A \cup (R \cup C)$.]

$$A \cap R = \varnothing \qquad A \cap C = \varnothing \qquad \text{and } R \cap C = \varnothing \qquad \qquad \square$$

Definition   Two sets which have the property that their intersection is the empty set are said to be *disjoint*.

Thus sets A and R of Example 1-6 are disjoint. Likewise sets A and C and sets R and C of that example are disjoint. We can say this more concisely by stating that A, R, and C are pairwise disjoint. Moreover, since every one of the possible outcomes is either in A or in R or in C, the union of A, R, and C is the set X of all possible outcomes of the inspection. Sets A, R, and C with these properties are said to form a *partition* of X. Alternatively, we say that X is partitioned into sets A, R, and C. There is a direct generalization of this to an arbitrary number of subsets.

Definition   The sets in a collection are said to be *pairwise disjoint* if every pair in the collection is disjoint.

Definition   A *partition* of a set S is a collection of subsets of S which are pairwise disjoint and whose union is the entire set S.

We have defined the intersection of two sets A and B (the set of elements which are in A *and* in B) and the union of two sets A and B (the set of elements which are in either A *or* B or both). Thus we have set-theoretic operations which associate naturally with *and* and *or*. We consider next a set-theoretic operation which is the natural associate of the term *not*.

Definition | The *complement of the set A with respect to the set B* is the set of all elements in *B* which are not in *A*. The complement of *A* in *B* is written $B \sim A$.

EXAMPLE 1-7

Let $X = \{0, 1, 2, 3, 4, 5\}$, $E = \{0, 2, 4\}$, $O = \{1, 3, 5\}$, and $P = \{2, 3, 5\}$. Then

$$X \sim E = \{1, 3, 5\} = O \qquad X \sim O = \{0, 2, 4\} = E \qquad X \sim P = \{0, 1, 4\}$$

$$E \sim P = \{0, 4\} \qquad P \sim E = \{3, 5\} \qquad \square$$

There is a case in which the notation for the complement of a set with respect to another is shortened.

Definition | A set *U* is said to be a *universal set* for a problem if all sets being considered in the problem are subsets of *U*.

In Example 1-7 the set $X = \{0, 1, 2, 3, 4, 5\}$ could serve as a universal set. Although a problem may have several possible universal sets, frequently there is one which is especially appropriate. This is often the case in probability problems, and it is with such applications in mind that we introduce the concept.

If *U* is a universal set for a problem, and if it is understood that all complements are taken with respect to *U*, then we write $\tilde{A}$ instead of $U \sim A$, and we refer to $\tilde{A}$ as the *complement of A*.

In addition to taking unions, intersections, and complements there are other useful ways of building new sets from given ones. It is frequently helpful to construct a set by pairing the elements of one set with those of another. For example, suppose that a sociologist has sufficient funds to conduct one survey. The survey can be conducted either by mail *M*, or by phone *P* and in one of three cities, Atlanta *A*, Boston *B* or Cleveland *C*. Thus, the choice for the sociologist can be viewed as one of selecting one element from the set

$$\{(M,A), (M,B), (M,C), (P,A), (P,B), (P,C)\}$$

Definition | The *cartesian product* of the sets *A* and *B*, denoted by $A \times B$, is the set of all ordered pairs (a,b) where $a \in A$ and $b \in B$.

EXAMPLE 1-8

If $A = \{a, b, c, 5, 6\}$ and $B = \{d, 7\}$, then

$$A \times B = \{(a,d), (b,d), (c,d), (5,d), (6,d), (a,7), (b,7), (c,7), (5,7), (6,7)\}$$

If A is as above and $C = \{c, d\}$, then

$$A \times C = \{(a,c),\ (b,c),\ (c,c),\ (5,c),\ (6,c),\ (a,d),\ (b,d),\ (c,d),\ (5,d),\ (6,d)\} \qquad \square$$

EXAMPLE 1-9

A football league contains five teams: the Aardvarks A, Bears B, Coyotes C, Dolphins D, and Elephants E. Each game can be represented as an ordered pair of teams in the league in which the first entry represents the home team. With this notation, the set of all games can be represented as a subset of the cartesian product of the set $X = \{A, B, C, D, E\}$ with itself (Exercise 10). $\qquad \square$

It is essential to remember that order is important in the construction of an ordered pair and that (a,d) is to be considered different from (d,a). In Example 1-9 the game (A,B) between A and B, in which A is the home team, is a different game from the game (B,A) between A and B, in which B is the home team.

Exercises for Section 1-4

1. Let the sets M, A, B, C be defined as follows:

$$M = \{\text{Minnesota, Michigan, Montana, Massachusetts}\}$$
$$A = \{\text{Alabama, Arkansas, Michigan}\}$$
$$B = \{\text{Montana, Michigan}\}$$
$$C = \{\text{Alabama, Arkansas}\}$$

Decide which of the following subset relationships are correct:
   a. $B \subset M$ **b.** $B \subset C$ **c.** $C \subset A$
   d. $C \subset B$ **e.** $C \subset M$ **f.** $A \subset (B \cup C)$

2. Let the sets A, B, C, and D be defined by

$$A = \{x:\ x \text{ owns a GM car}\}$$
$$B = \{x:\ x \text{ works for GM}\}$$
$$C = \{x:\ x \text{ is the president of GM}\}$$
$$D = \{x:\ x \text{ owns stock in GM}\}$$

Describe the elements in the sets:
   a. $A \cap B$ **b.** $(A \cup B) \cup D$
   c. $B \sim A$ **d.** $C \cap A$

3. Let U, A, B, and C be defined by

$$U = \{a, b, c, 1, 2, 3\}$$

$$A = \{a, b, c\} \qquad B = \{a, 2, 3\} \qquad C = \{1, 2, 3\}$$

List the elements in each of the following sets.
   a. $A \cup B$ **b.** $B \cap C$ **c.** $(A \cup B) \cap (B \cup C)$
   d. $\tilde{A}$ **e.** $A \sim B$ **f.** $(A \sim B) \cup (B \sim C)$

4. Which pairs of sets in Exercise 1 are disjoint?

5. List all subsets of the sets
  a. {GM, Ford}   **b.** {GM, Ford, Chrysler}

6. Counting the empty set and the set itself, how many subsets does each of the following sets contain?
  a. {GM}
  b. {GM, Ford}
  c. {GM, Ford, Chrsyler}
  d. {GM, Ford, Chrysler, American Motors}
  Is there a pattern? What is it? How many subsets does a set with seven elements contain?

7. Using the sets defined in Exercise 1, decide which of the following set equalities are correct:
  a. $M \cap A = A$ **b.** $A \cup C = A$
  c. $B \sim C = B$ **d.** $B \cap C = B \sim M$

8. For each of the following set equalities give an example of sets A and B which satisfy the stated condition:
  a. $A \cup B = A$ **b.** $A \sim B = A$
  c. $A \cap B = B$ **d.** $A \sim B = B \sim A$

9. Using the sets defined in Exercise 3, decide which of the following pairs of subsets form a partition of U:
  a. A and B **b.** A and C
  c. B and C **d.** $(A \cup B)$ and $(C \sim B)$

10. Let the set $L = \{A, B, C, D, E\}$ represent the five teams in a division of a football league. Suppose that each team plays every other team in a home-and-home series, and suppose these games are represented by ordered pairs where the first member of the pair represents the home team. Thus the pair (B,D) represents the game between teams B and D which is played on B's field. Show that the set of pairs S which represents this home-and-home series is a subset of $L \times L$. How many elements are in S? In $L \times L$?

11. Let $A = \{x, y, z\}$ and $B = \{i, e\}$. List the elements in $A \times B$.

12. Let $A = \{x, y, z, w\}$ and $B = \{y, z, w, e\}$. List the elements in $A \times B$.

1-5 VENN DIAGRAMS AND THE SIZES OF SETS

It is often easier to understand the relations between sets if the sets are represented in pictorial form. A convenient representation, known as a *Venn diagram*, serves this purpose. In a Venn diagram a set U and its subsets are pictured using geometric shapes. By convention the set U is usually a rectangle, and the subsets of U are circles inside the rectangle. For example, if A and B are subsets of U, the Venn diagram for this relationship is that shown in Fig. 1-2.

It is useful to describe each of the subregions of Fig. 1-2 in terms of unions, intersections, and complements. In each of the diagrams of Fig. 1-3 the shaded area is identified below the diagram.

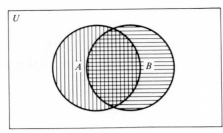

FIGURE 1-2

A comparison of Fig. 1-3c and i illustrates the set equality

$$\widetilde{A \cup B} = \tilde{A} \cap \tilde{B} \tag{1-1}$$

which can be read: "The complement of a union is the intersection of the complements." This relation is true for any pair of subsets A and B of U and can be proved using the definitions of $\cup$, $\cap$, and $\sim$. Likewise, an examination of Fig. 1-3d and j illustrates the set equality

$$\widetilde{A \cap B} = \tilde{A} \cup \tilde{B} \tag{1-2}$$

which can be read: "The complement of an intersection is the union of the complements."

Other useful relations which can be deduced from appropriate Venn diagrams and proved using the definitions are

$$A \cap (B \cup C) = (A \cap B) \cup (A \cap C) \tag{1-3}$$

and
$$A \cup (B \cap C) = (A \cup B) \cap (A \cup C) \tag{1-4}$$

which hold for any three sets A, B, and C. It was noted earlier that $(A \cup B) \cup C = A \cup (B \cup C)$ and consequently the union of three sets can be written simply as $A \cup B \cup C$. The same comment holds for the union of an arbitrary number of sets. Likewise, since $(A \cap B) \cap C = A \cap (B \cap C)$, the intersection of three sets can be unambiguously written as $A \cap B \cap C$. Again the same comment holds for the intersection of an arbitrary number of sets.

Note: It is important to note that, in general, expressions involving both the symbols $\cap$ and $\cup$ require the use of parentheses. Expressions involving both $\cap$ and $\cup$ are not well defined unless parentheses are used to indicate which pair of sets is associated with each intersection and each union operation. For example, $A \cap B \cup G$ is not well defined, but both $A \cap (B \cup G)$ and $(A \cap B) \cup G$ are well defined.

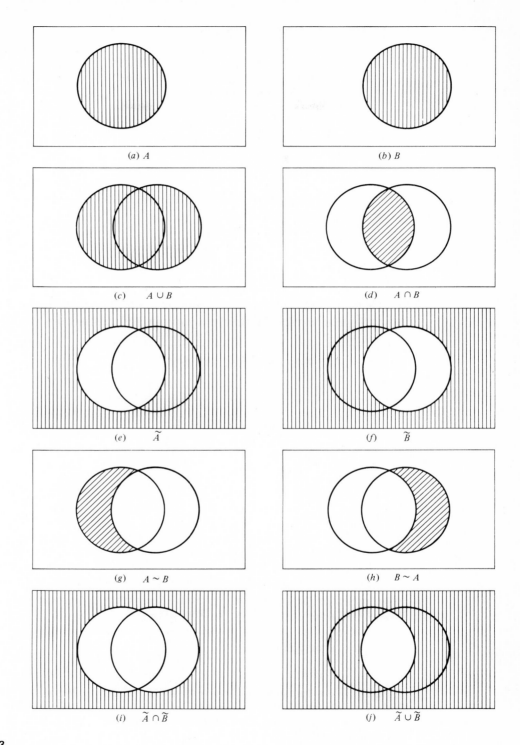

(a) A

(b) B

(c) $A \cup B$

(d) $A \cap B$

(e) $\widetilde{A}$

(f) $\widetilde{B}$

(g) $A \sim B$

(h) $B \sim A$

(i) $\widetilde{A} \cap \widetilde{B}$

(j) $\widetilde{A} \cup \widetilde{B}$

FIGURE 1-3

In our study of probability, where many arguments are based on set theory, we shall need to compute the number of elements in certain sets. We turn to this topic briefly now and continue the discussion in Chap. 2. We begin by introducing notation for the number of elements in a set. Since we consider only sets with a finite number of elements in this book, our definition is given for such sets.

Definition   If A is a set with a finite number of elements, then $n(A)$ denotes the number of elements in the set A.

For example, if X, A, R, and C are the sets of Example 1-6, that is,

$$X = \{0, 1, 2, \ldots, 10\} \qquad A = \{0, 1\} \qquad R = \{3, 4, 5, \ldots, 10\} \qquad C = \{2\}$$

then $n(X) = 11, n(A) = 2, n(R) = 8$, and $n(C) = 1$. The following useful fact follows immediately from the definitions of a partition and the number of elements in a set.

If a set X is partitioned into k subsets $X_1, X_2, \ldots, X_k$, then

$$n(X) = n(X_1) + n(X_2) + \cdots + n(X_k) \qquad (1\text{-}5)$$

In particular, for Example 1-6 we have

$$11 = n(X) = n(A) + n(R) + n(C) = 2 + 8 + 1$$

EXAMPLE 1-10

A survey of 1000 college students was conducted to determine preferences regarding grading systems and withdrawal policies. Of the 1000 students surveyed, 650 preferred a pass-fail grading system to a conventional one, and 800 preferred unrestricted withdrawal from classes to restricted withdrawal. Of the 650 students who preferred pass-fail grading, 500 also preferred unrestricted withdrawal. These 500 were included in the 800 whose preferences for unrestricted withdrawal were noted earlier. How many students preferred neither pass-fail grading nor unrestricted withdrawal?

Solution: We can answer the question by utilizing a Venn diagram and our notation for the number of elements in a set. Let U be the set of 1000 college students whose opinions were sampled; let P be the subset of those who prefer pass-fail grading, and let W be the subset of those who prefer unrestricted

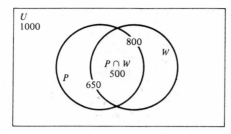

FIGURE 1-4

withdrawal from classes. The information given is $n(U) = 1000$, $n(P) = 650$, $n(W) = 800$, and $n(P \cap W) = 500$. We are asked to determine the number of students in U but in neither P nor W, that is, the number $n(\widetilde{P \cup W})$. A Venn diagram for this situation is given in Fig. 1-4.

We begin by observing that since $P \cup W$ and $\widetilde{P \cup W}$ provide a partition of U,

$$n(P \cup W) + n(\widetilde{P \cup W}) = 1000$$

Thus to compute $n(\widetilde{P \cup W})$ it is sufficient to compute $n(P \cup W)$. Since we know how to compute the number of elements in a set using a knowledge of the number of elements in each subset of a partition of the set, it appears helpful to search for a partition of $P \cup W$. There is a natural partition of $P \cup W$ into the subsets $P \sim W$, $P \cap W$, and $W \sim P$. These are the sets labeled X, Y, and Z, respectively, in Fig. 1-5.

Notice that X and Y are disjoint and $X \cup Y = P$. Therefore, X and Y provide a partition of P, and $n(P) = n(X) + n(Y)$. Since $n(P)$ is given as 650 and

$$n(Y) = n(P \cap W) = 500$$

it follows that

$$n(X) = n(P) - n(Y) = 650 - 500 = 150$$

Likewise Y and Z form a partition of W, and consequently

$$n(W) = n(Y) + n(Z)$$

FIGURE 1-5

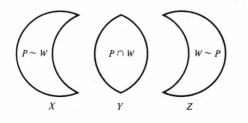

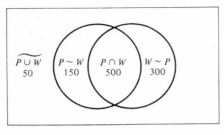

FIGURE 1-6

Since $n(Y) = 500$ and $n(W) = 800$, we conclude that

$$n(Z) = n(W) - n(Y) = 800 - 500 = 300$$

Finally, X, Y, and Z provide a partition of $P \cup W$, and consequently

$$n(P \cup W) = n(X) + n(Y) + n(Z) = 150 + 500 + 300 = 950$$

We conclude that $n(\widetilde{P \cup W}) = 1000 - 950 = 50$. In terms of the original situation, 50 students preferred both a conventional grading system and a restricted withdrawal policy. The number of elements in each of the sets $\widetilde{P \cup W}$, $P \sim W$, $W \sim P$, and $P \cap W$ is shown in Fig. 1-6.   □

Problems similar to that posed in Example 1-10 can obviously be solved using the method developed in analyzing that problem. However, it is more direct to appeal to a general result which holds for all sets and not just for those associated with a specific problem. The general result, which can be proved using an argument entirely similar to that used in the special case of Example 1-10, is as follows:

Let A and B be two sets. Then
$$n(A \cup B) = n(A) + n(B) - n(A \cap B) \tag{1-6}$$

Notice that if $A \cap B = \varnothing$, then A and B provide a partition of $A \cup B$ and Eq. (1-6) is a special case of Eq. (1-5). If we use Eq. (1-6) to solve the problem of Example 1-10, we have immediately

$$n(P \cup W) = n(P) + n(W) - (P \cap W) = 650 + 800 - 500 = 950$$

as we concluded above.

EXAMPLE 1-11

The Flabnomore Exerciser Company requires each of its employees to pass a yearly physical examination. The results of the most recent examination of the 50 employees were that 30 employees were overweight, 25 had high blood pressure, and 20 had a high cholesterol count. Moreover, 15 of the overweight employees also had high blood pressure, and 10 of those with a high cholesterol count were also overweight. Of the 25 with high blood pressure, there were 12 who also had a high cholesterol count. Finally, there were 5 employees who had all three of these undesirable reports. When these reports reached the desk of the president, Jox Chinup, he asked, "Don't we have any completely healthy employees around here?" The answer to his question, and in fact the exact number of healthy employees (in terms of these symptoms) can be obtained using Venn diagrams and relation (1-5).

Solution: We let U, O, B, and C be, respectively, the sets of all employees, employees who are overweight, who have high blood pressure, and who have a high cholesterol count. The information gathered in the tests can be summarized as follows:

$$n(U) = 50 \qquad n(O) = 30 \qquad n(B) = 25 \qquad n(C) = 20$$

$$n(O \cap B) = 15 \qquad n(O \cap C) = 10 \qquad n(B \cap C) = 12 \qquad n(O \cap B \cap C) = 5$$

This problem is concerned with three subsets of U, and the relevant Venn diagram is shown in Fig. 1-7. The intersection $O \cap B \cap C$ is known to have 5 elements. Since $O \cap B$ has 15 elements, it follows that $(O \cap B) \sim (O \cap B \cap C)$ contains 10 elements. Likewise, since $O \cap C$ has 10 elements, $(O \cap C) \sim (O \cap B \cap C)$ contains 5; and since $B \cap C$ contains 12 elements, $(B \cap C) \sim (O \cap B \cap C)$ contains 7 elements. This information is shown in Fig. 1-8a.

Next, since the numbers of elements in O, B, and C are known, we can use the information determined thus far to compute the number in O but not

FIGURE 1-7

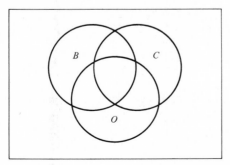

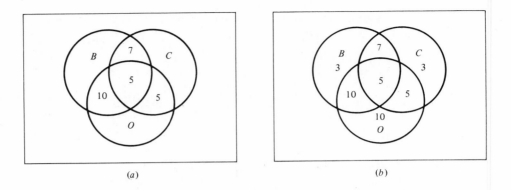

FIGURE 1-8

in B or C, etc. This information is shown in Fig. 1-8b. Note that each number on the Venn diagram gives the number of employees in the set corresponding to that particular portion of the Venn diagram in which the number is printed. Thus, there are 25 employees in the set B but only 3 in the set $B \sim (O \cup C)$, which is that portion of the Venn diagram in which the 3 is printed. The seven regions bounded by arcs of circles in Fig. 1-8b represent a partition of $O \cup B \cup C$, and the number of elements in each of them is known. Using Eq. (1-5), we have

$$n(O \cup B \cup C) = 3 + 10 + 5 + 7 + 10 + 5 + 3 = 43$$

Therefore, since there are 50 elements in U, there are 7 elements in $\overline{O \cup B \cup C}$. That is, 7 employees do not have high blood pressure or high cholesterol and are not overweight.   □

Standard techniques for collecting certain types of sociological data include mail and telephone canvassing. A sociologist is interested in data which can be obtained in 3 different cities. If the sociologist has funds for only one survey, there are 6 different options available. If each survey is represented as an ordered pair in which the first element is mail or phone and the second element is the name of the city, then the set of surveys can be represented as a cartesian product. That is, if M and P denote mail and phone surveys, respectively, and A, B, and C denote the cities, the set of surveys can be represented as

$$S = \{(M,A), (M,B), (M,C), (P,A), (P,B), (P,C)\}$$

It is clear that the cartesian product S can be arranged in the array

$$\begin{array}{ccc} (M,A) & (M,B) & (M,C) \\ (P,A) & (P,B) & (P,C) \end{array}$$

There are two rows in the array, one corresponding to each element in the set $\{M, P\}$, and there are three columns in the array, one corresponding to each element in the set $\{A, B, C\}$. It follows from the definition of ordinary multiplication that there are $n(\{M, P\}) \times n(\{A, B, C\}) = 2 \times 3 = 6$ elements in the array. Therefore $n(S) = 6$.

This technique for counting the number of elements in a cartesian product of two sets is perfectly general, and there is the following result:

If A and B are sets, then

$$n(A \times B) = n(A)n(B) \qquad (1\text{-}7)$$

In our work on probability we shall find it useful to consider cartesian products of more than two sets. For example, the cartesian product $E \times F \times G$ is the set of all ordered triples (e, f, g) with $e \in E$, $f \in F$, and $g \in G$.

Suppose the sociologist has a choice of four months in which to conduct the survey. If these months are September S, October O, November N, and December D, the choices available can be represented as an ordered triple

$$(x, y, z)$$

in which x denotes a month, $x \in \{S, O, N, D\}$; y denotes the method of sampling, $y \in \{M, P\}$; and z denotes the city sampled, $z \in \{A, B, C\}$. The number of surveys available to the sociologist (24) is the number of elements in the cartesian product of the sets $\{S, O, N, D\}$, $\{M, P\}$, and $\{A, B, C\}$.

There is a formula analogous to (1-7) for the number of elements in a cartesian product of more than two sets:

If $E_1, E_2, \ldots, E_k$ are sets, then

$$n(E_1 \times E_2 \times \cdots \times E_k) = n(E_1)n(E_2) \cdots n(E_k) \qquad (1\text{-}8)$$

Exercises for Sec. 1-5

1. Let A, B, and C be subsets of a set U. Draw a Venn diagram to illustrate each of the following sets. In each case shade the area corresponding to the designated set.

a. $\tilde{A} \cap B$ **b.** $A \cup \tilde{B}$ **c.** $A \cup B \cup C$

d. $(A \cup B) \cap \tilde{C}$ **e.** $A \cap B \cap \tilde{C}$ **f.** $\tilde{A} \cap B \cap \tilde{C}$

2. In each case determine which of the points x, y, z, v, w in Fig. 1-9 belong to the designated set.

a. $A \cup C$ **b.** $B \cap A$

c. $\tilde{A} \cup B$ **d.** $\tilde{B} \cup C$

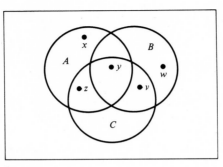

FIGURE 1-9

3. In each case determine which of the points x, y, z, v, w in Fig. 1-10 belong to the designated set.

a. $S \cap C$ **b.** $S \cap C \cap T$ **c.** $C \cup S \cup T$

d. $C \cap (S \cup T)$ **e.** $(C \cup S) \cap (C \cup T)$

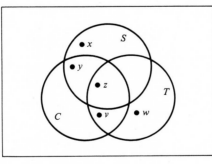

FIGURE 1-10

4. In Example 1-10 how many students are in the set $(P \sim W) \cup (W \sim P)$? What are the preferences of these students?

5. An item is said to be defective if it has a major defect or a minor defect or both. In a batch of 25 defective items, 20 have major defects and 14 have minor defects. How many items in the batch have both major and minor defects?

6. One hundred premedical students who were not admitted to medical school were asked whether they would be interested in careers as medical technicians or registered nurses. Fifty-four of the students indicated an interest in medical technology, 32 in nursing, and 23 in both. How many students were interested in neither of these careers?

7. A construction crew is made up of individuals who have training for operating specific pieces of heavy equipment. The equipment and the number of operators with each skill are shown in Table 1-3. If every crew member can operate one of the machines, how many individuals are in the crew?

TABLE 1-3

Type of Equipment	Number Trained
Backhoe	17
Bulldozer	17
Crane	15
Backhoe & bulldozer	5
Backhoe & crane	8
Bulldozer & crane	7
All three	3

8. A survey of 500 families provided the following data:

> 63 families subscribed to *The Wall Street Journal*
> 41 families subscribed to *Rolling Stone*
> 37 of the families who subscribed to *Rolling Stone* did not subscribe to *The Wall Street Journal*

a. How many families subscribed to both?
b. How many families subscribed to neither?

9. Census data provide the following information. Out of 1000 adult residents of the United States:

> 544 live in a city with population in excess of 100,000.
> 312 are members of families with more than two children.
> 87 live in a city with population above 100,000 and have bought a new car in the last 24 months.

If it is possible to do so from the given information, determine the number of individuals (out of 1000) in each of the following groups:
a. Those who do not live in a city with population above 100,000
b. Those who are not members of families with more than two children
c. Those who have bought a car within the last 24 months and are members of a family with more than two children

d. Those who have not bought a car in the last 24 months and do not live in a city with population above 100,000

10. Express the shaded areas in each of the Venn diagrams of Fig. 1-11 using the operations union, intersection, or complement and the sets A, B, and C.

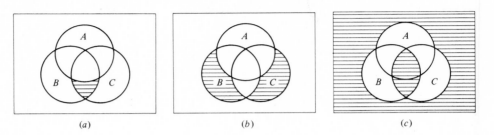

        (a)   (b)   (c)

FIGURE 1-11

11. Let $A = \{1, 7, 3, a, b\}$ and $B = \{3, c\}$. Determine
  a. $n(A \times B)$ **b.** $n(B \times B \times B)$

12. Let U be the set of all real numbers, $A = \{x: \ x$ is a solution of $x^2 = 1\}$, and $B = \{-1, 2\}$. Determine
  a. $\tilde{A}$ **b.** $A \cup B$ **c.** $A \cap B$

1-6   FUNCTIONS

Most applications of mathematics involve attempts to quantify relationships between two ore more sets of objects. For example, in business one is often concerned with the relationship between the sales of an item (or the profits obtained from the sales) and the price of the item. A biologist may be concerned with the relationship between cell growth and factors such as temperature and humidity, and a child psychologist may be concerned with the relationship between a child's ability to read and the number of hours per week that the child watches television. In each of these cases the relationship involved may be quite complex and difficult to determine exactly. Frequently attempts to determine such relationships involve the use of the mathematical concept of a *function* from one set to another. In fact, the relationship is usually expressed as a function.

Definition

A *function from a set A to a set B* is a rule which associates with each element of A an element of B in such a way that to each element of A there is assigned exactly one element of B. The function is said to be *defined on* the set A or to have *domain A* and to *have values* in the set B. The set of elements in B which are assigned to one or more elements of A is the *range* of the function.

Intuitively it is helpful to think of this assignment in the following way. The elements of B are to be used as labels for the elements of A. Each element in A is to be labeled unambiguously, i.e., given exactly one label from B. This does not prevent an element of B from being used as a label for more than one element of A, as the next example illustrates.

EXAMPLE 1-12

Let $A = \{a, b, c, d\}$ and $B = \{2, 4, 6, 8, 10\}$. The rule which associates 2 with a (that is, assigns the element $2 \in B$ to the element $a \in A$), 6 with b, 8 with c, and 8 with d is an example of a function from A to B. If we denote this function by the letter f, we can write the definition of this function as $f(a) = 2$, $f(b) = 6$, $f(c) = 8$, and $f(d) = 8$. We can also describe this function pictorially by using arrows to represent the correspondence (see Fig. 1-12).

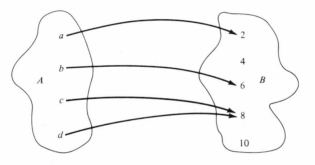

FIGURE 1-12 □

Note that in Example 1-12 the element 8 of the set B is assigned to both elements c and d of the set A. This does not violate the condition given in the definition of a function, since it is required only that to *each* element of A there is assigned exactly one element of B; the same element of B may be assigned to several elements of A. The domain of the function is the set A, and its range is the subset $\{2, 6, 8\}$ of B.

EXAMPLE 1-13

Let A and B be the set of positive natural numbers $\{1, 2, 3, \ldots\}$, and let f be the function which assigns to each number in A the square of that number. Thus $f(1) = 1$, $f(2) = 4$, $f(3) = 9$, etc. As a convenient shorthand notation we write $f(x) = x^2$, for $x \in A$, as an alternative definition of f. □

EXAMPLE 1-14

Let $A = \{0, .2, .4, .6, .8, 1.0\}$, and let B be the set of numbers between 0 and 10. Define a function f by the formula $f(x) = 4x + 1$, for $x \in A$. Then $f(0) = 1.0$ and $f(.6) = 3.4$, for example.

If a function is defined on a set with only a few elements, as in this case, it is often useful to represent it in a table. The function f of this example is represented in Table 1-4.

TABLE 1-4

x	$f(x)$
0.0	1.0
.2	1.8
.4	2.6
.6	3.4
.8	4.2
1.0	5.0

The function of Example 1-14 is a *linear function*. Such functions will be investigated in some detail in Chap. 5.

Exercises for Sec. 1-6

1. Which of the diagrams in Fig. 1-13 are representations of functions from the set A to the set B?

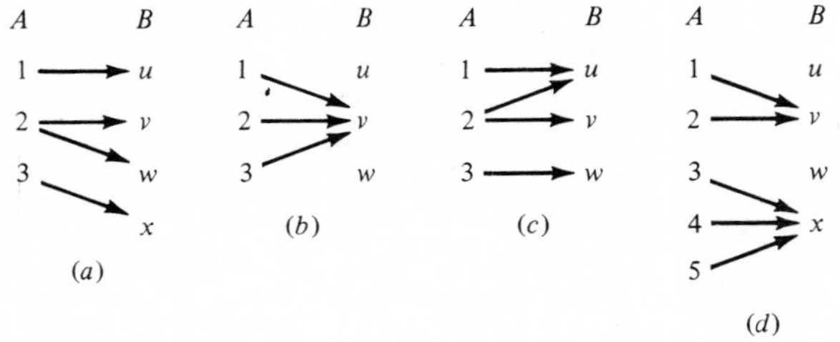

FIGURE 1-13

2. Let A and B be the set of nonnegative numbers. Define a function f from A to B by the formula $f(x) = x^2 - 2x + 1$.
   a. Find the following values of f: $f(1), f(\frac{1}{2}), f(10)$.
   b. For which values of x is $f(x) = 0$?

3. Find a formula for the function f which assigns to each number in the set $A = \{1, 2, 3, 4, 5\}$ a number in the set $B = \{0, \frac{1}{2}, 1, \frac{3}{2}, 2, \frac{5}{2}, 3\}$, as shown in Table 1-5.

TABLE 1-5

$x \in A$	$f(x) \in B$
1	0
2	$\frac{1}{2}$
3	1
4	$\frac{3}{2}$
5	2

TABLE 1-6

x	$f(x)$
1	3
3	2
5	1
7	0
8	$-\frac{1}{2}$

4. Find a formula for the function which assigns values as shown in Table 1-6.

5. Write out a formula for the function which cubes a number, multiplies the result by 3, and adds this number to 5.

6. Give an example of a function from $A = \{a, b, c, d\}$ to $B = \{1, 2\}$.

7. Let P be the set of positive numbers, and let f and g be functions from P to P defined by $f(x) = 2x + 7$ and $g(x) = 4x + 1$.
  a. For which values of x is $f(x) = g(x)$?
  b. What is $g(f(2))$?
  c. What is the formula for the function which assigns the number $g(f(x))$ to x?

8. Sally sells computers and receives a basic salary plus commissions. Her basic salary is $500 per month, and her commission is $100 per sale. Find a formula for a function which assigns her monthly salary to the number of sales she made that month.

9. A farmer wants his crops to receive the equivalent of 1 inch of rainfall per week. He must irrigate for 1 hour to place the equivalent of $\frac{1}{10}$ inch of rainfall on his fields. Write a formula for the number of hours of irrigation needed based on the number of tenths of inches of rainfall which have fallen that week. (Assume at most 1 inch of rain per week.)

10. Let f be as given in Exercise 7. Find a function g such that

$$g(f(x)) = x \qquad \text{for all positive numbers } x$$

IMPORTANT TERMS

You should be able to describe, define, or give examples of each of the following:

Mathematical model

Set

Element

Subset

Set equality

Intersection

Union

Empty set

Disjoint sets

Partition

Complement of A in B

Universal set

Complement of A

Cartesian product

Venn diagram

Function from A to B

REVIEW EXERCISES

1. Let $S = \{1, 2, 3, 4, 5, 6\}$ and $E = \{5\}$. Find a partition of the set S which consists of three sets, one of which is E.

2. Let $S = \{a, b, c, d\}$ and $T = \{1, 2, 3\}$. What is the complement of $R = \{(a,1), (b,2), (c,3)\}$ in $S \times T$?

3. Give an example of sets A, B, and X for which $A \subset X$, $B \subset X$, and $\tilde{A} \subset \tilde{B}$. What is the relationship between A and B?

4. Let A, B, and C be subsets of X. Use Venn diagrams to illustrate the following sets:
  a. $\tilde{A} \cup B$ **b.** $\tilde{B} \cup C$ **c.** $(A \cup \tilde{B}) \cap C$

5. Consider the set equality $(A \cap B) \cup C = (\tilde{A} \cup \tilde{B}) \cap C$.
  a. Give an example of sets A, B, and C to show it is *not always true*.
  b. Give an example of sets A, B, and C to show it is *sometimes true*.

6. Let $U = \{1, 2, 3, \ldots, 20\}$. Assume sets A, B, C, and D are subsets of U.

$$A = \{x: \ x \text{ is a multiple of } 3\}$$
$$B = \{x: \ x < 10\}$$
$$C = \{x: \ x \text{ is an even number}\}$$
$$D = \{5, 7, 12, 19\}$$

Find the elements in each of the following sets:
  a. $A \cap C$ **b.** $C \cap \tilde{B}$
  c. $D \sim A$ **d.** $\tilde{C} \cup D$

Find the following numbers:
  e. $n(B)$ **f.** $n(A \cup C)$
  g. $n(A \cup B \cup C)$ **h.** $n(\tilde{B} \times A \times D)$

7. Refer to the sets in Exercise 6 and fill in these blanks with $\in$, $\notin$, $\subset$, or $\not\subset$. (*Note:* $A \not\subset B$ means that A is *not* a subset of B.)
  a. 9 _____ $A \sim C$ **b.** $\{8, 9, 10\}$ _____ B
  c. D _____ $A \cup \tilde{C}$ **d.** $\varnothing$ _____ $\tilde{D}$
  e. 6 _____ A **f.** $(2,3)$ _____ $A \times C$
  g. B _____ $(C \cap D)$ **h.** $\{10\}$ _____ C

8. Suppose A and B are subsets of X and $n(X) = 50$. If $n(\tilde{A} \cap \tilde{B}) = 10$ and $n(\tilde{A} \cap B) = n(A \cap \tilde{B}) = 15$, what is $n(A \cup B)$? $n(A \cap B)$?

9. Give an example of sets A and B such that

$$n(A \cup B) = n(A \cap B) = 2n(A) = 2n(B)$$

What is the relationship between A and B [in addition to the fact that $n(A) = n(B)$]?

10. Suppose $n(A) = 10$ for some set A.
  a. Find $n(A^k)$. (*Note:* $A^k = A \times A \times \cdots \times A$, k times.)
  b. How many subsets does A have?

11. In a group of 75 students, 43 said they enjoyed tennis and 24 said they enjoyed swimming but not tennis.
  a. How many students do not enjoy either sport?
  b. How many of them enjoy both?

12. Consider the following data concerning 160 English majors:

> 76 take French
> 85 take German
> 33 take French and German
> 35 take German and Russian
> 32 take French as their only foreign language
> 15 take French, Russian, and German

If every student takes at least one foreign language:
  a. How many take Russian?
  b. How many take French and Russian but not German?

13. The following information was gathered from a group of 200 undergraduates:

> 39 own an imported car.
> 32 own a domestically manufactured car.
> 98 own a bike.
> 69 own neither a car nor a bike.

  a. If no student owns two cars, how many have both a car and a bike?
  b. If you also know that 21 owners of imported cars do not own a bike, how many owners of domestically manufactured cars do own a bike?

14. Consider the function $f(x) = 5x^2 + 1$ defined on the set $A = \{-2, -1, 0, 1, 2, 3\}$ with values in the set $B = \{1, 2, 3, 4, 5, \ldots, 50\}$. What values does f take?

15. Let f and g be functions from $\mathbb{Z}$ to $\mathbb{Z}$, where $\mathbb{Z}$ is the set of integers $\mathbb{Z} = \{0, 1, -1, 2, -2, 3, -3, \ldots\}$. Suppose f and g are defined by $f(x) = 20 - 3x$

and $g(x) = 5x + 4$. For what values of x, if any, are the following true?

a. $f(x) = g(x)$ **b.** $f(x) = 0$
c. $g(x) = 39$ **d.** $f(x) = 35$

16. Refer to the functions f and g in Exercise 15.
   a. Find $f(3)$, $g(-6)$, $g(0)$, and $f(-8)$.
   b. Find a formula for $f(x) + g(x)$.
   c. Find a formula for $f(g(x))$.
   d. Find $f(g(0))$ and $g(f(0))$.

17. Mary gets a salary of $60 a week waiting on tables if she does not break any dishes; the manager deducts $1.50 from the $60 for each dish Mary breaks per week. Find a formula for the function which relates Mary's weekly salary to the number of dishes she breaks during the week.

18. To rent a particular car, a firm charges $20 a day plus 18¢ a mile. Dave is renting this car for 5 days. Find a formula for the function which gives Dave's cost of the car in terms of the number of miles he drives.

19. A private school advertises that they have no more than 4 times as many students as teachers. Find a formula for the function which gives the minimum number of teachers in the school in terms of the number of students.

20. The bookstore is selling 3 large notebooks for $2. Find a formula for the function which gives the cost (in dollars) in terms of the number of notebooks. This function will be defined on the set $\{3, 6, 9, \ldots\}$.

PART ONE

PROBABILITY MODELS

PROBABILITY PRELIMINARIES AND COUNTING

CHAPTER

TWO

2-1   INTRODUCTION

Statements involving direct references to probabilities are a common part of ordinary conversation. For example, in listening to a weather report one may hear, "The chance of precipitation in the next 24 hours is 30 percent" and in the production department of a drug manufacturer one may hear, "There is probability of one-tenth that this process will produce a batch of vaccine containing live viruses." In this chapter we discuss several meanings of the term *probability*, and we begin our development of the mathematical techniques needed to study probability. First, however, some general comments about the nature of the subject are appropriate.

Applied probability theory concerns the assignment of numbers to events in such a way that the numbers assigned reflect the likelihood of the events' occurring. To be useful the numbers must be assigned in such a way that a specific number always represents the same likelihood of occurrence of an event. Also, for the numbers to be useful it should be possible to readily convert the number assigned to an event into a statement about the occurrence of the event. As we shall see later (Sec. 2-3), there is a commonly accepted set of rules which are followed in assigning numbers to events. These rules reflect some of the characteristics of measurements of likelihood which seem to be desirable on intuitive grounds (Sec. 2-2).

Probability theory was developed to meet the need for a precise method of analyzing situations which involve uncertainty. Examples of such situations, and historically the first ones seriously studied, are games of chance such as roulette, blackjack, and craps. Now, however, applications of probability theory can be found in almost all fields of study; they are especially prevalent in the social and life sciences and in business. The example of Chap. 1 concerned with the testing of vaccines illustrates the importance of analyzing situations involving uncertainty. Problems of this sort can be solved directly by using the methods of probability theory; however, it is quite difficult to solve them without these methods. Other examples are given in the sections which follow.

In Sec. 2-2 we discuss some of the possible points of view toward the meaning of probability, and we introduce some aspects of applied probability theory. In Sec. 2-3 we describe the settings and present the notation and terminology used in this and subsequent chapters. These two sections are informal, and issues are raised which are considered in a more precise way in Chap. 3, where the development of probability theory and its applications is continued. In Secs. 2-4 to 2-7 we study the counting principles which are necessary for many applications of probability.

2-2 PROBABILITY PRELIMINARIES

The study of probability has two parts: (1) the theoretical aspect of the development of a mathematical system and (2) the practical aspect of using the ideas and techniques to solve problems encountered outside of mathematics. Mathematical probability theory (the theoretical aspect) is a well-defined, active, and rich area of current mathematical interest. Like such other areas as geometry and algebra, there are axioms and definitions on which probability theory is based, and the theory is developed by studying the consequences, i.e., the theorems, which can be logically deduced from the axioms. Such a development is not our intention here. Our goal in this chapter is to present the basic ideas of probability in an informal way. This discussion is intended for those with no background in the subject; readers already acquainted with the basic ideas of probability may proceed directly to Sec. 2-4. A more precise development will be given in Chap. 3.

The practical aspect of probability, i.e., the study of its applications, is not as well defined as the theoretical aspect, and, in fact, there is by no means complete agreement on what constitutes a legitimate application of probability theory. (Even the formulation of basic definitions in applied probability is difficult. In order to be precise enough to formulate a definition that is useful, clear, and unambiguous, it is necessary to be very careful, and usually this care is reflected in long and elaborate statements. This difficulty is one of the reasons behind the informality of these sections.) One school of thought holds that probability theory should be applied only to the study of experiments which can be repeated any number of times under essentially the same conditions. Under these circumstances an estimate of the probability of the occurrence of a specific outcome or result of the experiment, i.e., the number assigned to that outcome, is provided by the ratio of the number of times the outcome occurs to the number of repetitions of the experiment. If an experiment is repeated 100 times, and if a specific outcome occurs 62 times, then $\frac{62}{100}$ would provide an estimate of the probability of that outcome. Another example is provided by the drug company introduced in Example 1-2. In that example the probability that any specific batch of vaccine contains live viruses is $\frac{1}{20}$ because this is the known ratio of the number of batches with live viruses to the total number of batches. Similarly, the probability that a batch of vaccine will not contain live viruses is $\frac{19}{20}$ since this is the ratio of the number of batches which do not contain live viruses to the total number of batches. We shall call this method of specifying the probability of an outcome the *sample-frequency method*.

Many situations for which a probability model is appropriate are such that it is not practical to repeat an experiment thousands of times. In these cases one cannot legitimately use the sample-frequency method to define the probabilities of various outcomes. In some such situations it is possible to deduce what fraction of the time a certain outcome should occur by simply analyzing the experiment being conducted. This is the method traditionally used with games of chance that involve activities such as flipping coins, rolling dice, spinning spinners, and dealing cards.* For example, consider an ordinary coin selected from the change in your possession. The coin has two faces (commonly called *heads* and *tails*), and the different designs on the two faces do not seem to alter the symmetry of the coin significantly. Thus, a reasonable assumption is that if the coin is flipped thousands (or millions) of times, each face will come up about the same number of times. Thus, the outcomes "a head comes up" and "a tail comes up" should each occur about half the time. From this symmetry argument it is reasonable to assign each of these events the probability $\frac{1}{2}$. Therefore, even though a particular coin may not have been flipped a great many times to verify that heads and tails occur equally often, most people will accept that this is a legitimate assumption. Of course, as a practical matter, the coin may actually be biased (as opposed to being fair), and thus the sample frequencies of heads and tails may not be $\frac{1}{2}$. To a mathematician, the concept of a fair coin or a fair die provides an abstract notion which can be conveniently used in examples. The abstraction is an idealization of actual coins and dice, and for these idealizations it is natural to use probabilities which are especially chosen to exhibit certain properties. Thus, in an example one might well consider a biased coin for which the probability of a head is $\frac{1}{6}$ and the probability of a tail is $\frac{5}{6}$. This coin would be an idealization of an actual coin which is somehow weighted so that in the long run tails come up 5 times as often

---

* A *die* (plural *dice*) is a cube with the numbers 1, 2, 3, 4, 5, 6 showing on the six faces. A *deck of cards* has 52 cards divided as follows. There are four suits of 13 cards each. The suits are called clubs, diamonds, hearts, and spades. Two suits are black (clubs and spades), and two are red (hearts and diamonds). In each suit the 13 cards are labeled A (ace), 2, 3, 4, 5, 6, 7, 8, 9, 10, J (jack), Q (queen), and K (king). The cards J, Q, K, A are called the *face cards*.

as heads. Similarly we use the notion of a fair die, which corresponds to a completely symmetric cube with 6 equally weighted faces. Accordingly, in the experiment of rolling the die, each possible outcome is assigned the probability $\frac{1}{6}$. Here, and in many similar situations, symmetry plays a role in an assignment. One assigns equal probabilities to several outcomes because of the symmetry of some real or idealized object. In many practical situations it is reasonable to assume that some object or experiment is sufficiently like an idealized object or experiment to permit the probabilities used in the idealized case (the numbers deduced as being appropriate) to be used in the actual case as well. We call this method of assigning probabilities the *deductive method*. We shall use both the sample-frequency method and the deductive method in our examples and exercises. In this book the deductive method will always be used except in cases in which sample data are provided. In many examples and problems probabilities will be given which in practice are likely to arise from application of the sample-frequency method, but the origin of these probabilities is ordinarily of no concern to us. Of course, if in actual practice an experiment with an object results in sample frequencies which are incompatible with those deduced from an idealization, say 950 heads in 1000 tosses of a presumed fair coin, one is forced to reconsider the appropriateness of the idealization. One of the questions to be considered in the chapter on statistics (Chap. 4) is that of deciding when an assumption about an object is incompatible with experimental results.

A third method of assigning probabilities to outcomes corresponds to the use of the term *probability* in the sentence: "The probability of a manned flight to Mars by the year 2000 is $\frac{1}{100}$". In this case the assigned probability is neither a sample frequency nor a probability deduced in the sense described above. Instead, it is a *measure of the belief* of the speaker that a certain event will take place. Although this is a common use of the term probability, and one we recognize as valid, for our purposes, we investigate only the first two methods of assigning probabilities and not the measure-of-belief method.

We now turn to the specifics of the process we shall use for assigning numbers (probabilities) to outcomes. We begin by considering an experiment which has a fixed, finite number of possible outcomes. For example, the experiment might consist of testing audio tapes with outcomes "acceptable" and "defective." Alternatively, the experiment might consist of rolling a die and noting the number of dots on the uppermost face. We refer to the outcomes of the latter experiment as 1, 2, 3, 4, 5, 6 (the number of dots on the six faces of a die). In general we assume that there are n possible outcomes, which we denote by $\mathcal{O}_1, \mathcal{O}_2, \ldots, \mathcal{O}_n$.* For example, if the experiment consists of testing audio tapes, the outcomes acceptable and defective might be denoted by $\mathcal{O}_1$ and $\mathcal{O}_2$, respectively. After identifying the outcomes, we assign a probability to each outcome. The

---

* Frequently in this book we use subscripts to distinguish between symbols. We could use x, y, z to represent three outcomes of an experiment; the symbols $\mathcal{O}_1$, $\mathcal{O}_2$, $\mathcal{O}_3$ serve the same purpose. Subscripts are especially useful when the number of elements in a set is either very large or not known exactly.

probability, i.e., the number, assigned to outcome $\mathcal{O}_i$ is called the *weight* of $\mathcal{O}_i$, and it is denoted by w_i. We suppose w_i to be defined for $i = 1, 2, 3, \ldots, n$. The w_i may be sample frequencies, if they are available; alternatively they may be obtained by the deductive method. For example, suppose that in an experiment consisting of testing audio tapes the sample frequency of acceptable tapes is $\frac{9}{10}$ and of defective tapes is $\frac{1}{10}$. Then weights $w_1 = .9$ and $w_2 = .1$ should be assigned. If we consider w_i to be the fraction of times outcome $\mathcal{O}_i$ occurs, it follows that *the weights w_i are never negative*. Continuing with this interpretation, we use the fact that one of the outcomes must occur to conclude that the sum of the w_i's must be 1. Indeed, if w_i is the fraction of the total number of outcomes that $\mathcal{O}_i$ occurs, then

$$w_1 + w_2 + \cdots + w_n$$

is the fraction of the time that some outcome occurs and it must be 1. These two properties can be summarized as

$$w_i \geq 0 \qquad i = 1, 2, \ldots, n$$

$$w_1 + w_2 + \cdots + w_n = 1$$

(2-1)

and they are fundamental to applied probability theory.

EXAMPLE 2-1

An experiment consists of rolling a die and noting the number of dots on the face which lands uppermost. The outcomes can be denoted by $\mathcal{O}_1 = 1$, $\mathcal{O}_2 = 2$, $\mathcal{O}_3 = 3$, $\ldots$, $\mathcal{O}_6 = 6$. If we assume that it is a fair die, i.e., it behaves like an idealized die in which each outcome is equally likely, then the weights are (by deduction) $w_1 = w_2 = \cdots = w_6 = \frac{1}{6}$. $\qquad\qquad\square$

Example 2-1 is typical of many in that all weights are deduced to be the same number. In such a case, since the sum of the weights is 1, it follows that the weight assigned to each outcome is $1/n$, where n is the number of possible outcomes. Rolling a die is an experiment with 6 outcomes; and if the die is fair, each outcome should be assigned weight $\frac{1}{6}$. Flipping a coin is an experiment with 2 outcomes; and if the coin is fair, each outcome has weight $\frac{1}{2}$. Experiments like this are said to be experiments with *equally likely outcomes*.

For experiments with equally likely outcomes, all weights are the same, and they are determined by finding the total number of outcomes and then taking the reciprocal of that number. This brings us to the main topic of this chapter, namely, the study of techniques for determining the total number of possible outcomes of an experiment. At first glance this would seem to be a simple problem since it is solved by listing all outcomes and counting them. However, as the examples and exercises in the sections which follow will indicate, it is often much easier to talk about listing all outcomes than it is to do it. Also, even if one knows a method for listing all the outcomes of an experiment, the list may be long and it is often useful to be able to determine the number of outcomes without actually using a list to do so.

Exercises for Sec. 2-2

1. In inspecting widgets it is found that about 5 per 1000 are too long, 10 per 1000 are too short, and the others are of the proper length. Consider an experiment which consists of selecting a widget and measuring it. From the data given above what are the possible outcomes for this experiment, and what weights should be assigned to these outcomes?

2. Suppose an experiment has three possible outcomes, $\mathcal{O}_1$, $\mathcal{O}_2$, and $\mathcal{O}_3$. Also suppose that $\mathcal{O}_1$ occurs with frequency $1/a$, $\mathcal{O}_2$ with frequency $2/a$, and $\mathcal{O}_3$ with frequency $3/a$. What value must be assigned to a so that these frequencies can be used for the weights of the outcomes?

3. Suppose that a deck of cards is shuffled, a card is drawn by a blindfolded person, and its suit and number are noted. How many possible outcomes are there for this experiment? If each outcome is assumed to be equally likely, what weights should be assigned to the outcomes?

4. Suppose, as in Exercise 3, that a deck of cards is shuffled and then a single card is drawn and the color and number are noted. Thus typical outcomes would be "a red 4 was drawn," or "a black jack was drawn." How many possible outcomes are there for this experiment, and how should these outcomes be weighted to represent the assumption that they are all equally likely?

5. Consider a coin weighted so that when the coin is flipped, heads lands uppermost 3 times as often as tails. What weights should be assigned to the outcomes "heads" and "tails" to reflect this weighting?

6. Consider a die constructed so that all even numbers are equally likely and all odd numbers are equally likely but the even numbers occur twice as often as the odd numbers. What weights should be assigned to the outcomes to reflect this construction?

7. Consider an experiment which consists of planting 10,000 pine seeds. Suppose that it is known that about 90 percent of these seeds will germinate and of those which germinate 90 percent will live and grow into seedlings. The outcomes of the experiment of planting a single one of these seeds are as follows:
   a. The seed does not germinate.
   b. The seed germinates but dies.
   c. The seed germinates and grows into a seedling.
   What weights should be assigned to each of these outcomes?

8. A large company requires yearly physical exams for its employees. Over a period of many years it is found that each year about 12 employees out of every 500 examined will have emphysema and about 3 employees out of every 100 will have lung cancer. Moreover, about 1 employee per 1000 examined will have both emphysema

and lung cancer. Consider an experiment which consists of examining employees and which has the possible outcomes:

a. Emphysema is found but no lung cancer.

b. Lung cancer is found but no emphysema.

c. Both emphysema and lung cancer are found.

d. Neither emphysema nor lung cancer is found.

From the data given, what weights should be assigned to each of the four outcomes of this experiment?

2-3 TERMINOLOGY AND NOTATION

The primary goal of this and the following two chapters is to provide ideas and techniques which will enable you to discuss random phenomena in quantitative terms. The first step in achieving this goal is to introduce some terminology and notation to give us a standard vocabulary.

In probability and statistics the term *experiment* is used in a much more general sense than it is in the physical sciences. Testing a batch of vaccine for the presence of live viruses, checking a shipment of audio tapes for defective tapes, and sampling a production run of widgets for items that are too large or too small are all examples of experiments. We shall also find it convenient for purposes of illustration to discuss experiments such as flipping a coin, drawing a ball from an urn which contains several balls of different colors, and rolling a die. Examples like these are useful because of their simplicity even though they are artificial and not obviously connected to applications. All six examples just mentioned have two characteristics in common: (1) the operation (testing, checking, sampling, flipping, selecting, rolling) has several possible outcomes; and (2) the outcome which results in a specific performance of the operation is unknown in advance. Any operation with these characteristics will be referred to as an *experiment*. As a practical matter we also include as experiments certain operations whose outcome is known in advance, e.g., flipping a two-headed coin or drawing a ball from an urn containing three red balls.

A *performance* of an experiment is one completion of the operation which defines the experiment. An experiment might consist of testing five batches of vaccine. In this case a performance of the experiment consists of testing five batches of vaccine. Three *repetitions* of this experiment would consist of testing three sets of five batches each. Likewise an experiment might consist of checking 20 audio tapes or sampling 100 widgets. The operation which specifies the experiment may consist of several suboperations. Also one experiment might consist of flipping a coin twice, and another experiment might consist of flipping the same coin 3 times. In both of these cases one flip of the coin is *not* a performance of the experiment.

Definition   The *sample space of an experiment* is the set of all possible outcomes of one performance of the experiment.

For example, if an experiment consists of testing five batches of vaccine, then the elements of the sample space for this experiment can be listed as follows (a batch containing no live viruses will be termed *pure*):

$\mathcal{O}_1$:   All 5 batches are pure.
$\mathcal{O}_2$:   4 batches are pure, and 1 batch is impure.
$\mathcal{O}_3$:   3 batches are pure, and 2 batches are impure.
$\mathcal{O}_4$:   2 batches are pure, and 3 batches are impure.
$\mathcal{O}_5$:   1 batch is pure, and 4 batches are impure.
$\mathcal{O}_6$:   All 5 batches are impure.

There are six elements in the sample space of this experiment, one corresponding to each of the possible outcomes. Normally, it is useful to adopt a condensed notation for sample spaces. In this example we might represent each element in the sample space as an ordered pair of whole numbers, the first of the pair denoting the number of pure batches and the second number in the pair denoting the number of impure batches. Thus, the first item in the above list would be written (5,0), and the second (4,1). Using this notation, the sample space of the experiment can be written as the set

$$\{(5,0), (4,1), (3,2), (2,3), (1,4), (0,5)\}$$

Definition   An *event* is a subset of the sample space of an experiment.

In the above example the event that there are more pure batches than impure ones is $\{(5,0), (4,1), (3,2)\}$. The event that all the batches are the same, either all pure or all impure, is the set $\{(5,0), (0,5)\}$.

EXAMPLE 2-2

Suppose that widgets are to be sampled during a production run to determine whether they are oversized, undersized, or normal. Also, suppose that the samples are to be taken during two different shifts so that it is important to keep track of when each widget was sampled. If two widgets are to be sampled, one each shift, and if we use the letters O, U, and N to denote oversized, undersized, and normal widgets, respectively, the sample space for the experiment can be denoted by the set

$$S = \{(O,O), (O,U), (O,N), (U,O), (U,U), (U,N), (N,O), (N,U), (N,N)\}$$

In this list (N,U) means, for example, that a normal widget is sampled during the first shift and an undersized widget is sampled during the second shift.

Examine this listing until you are sure that you understand the pattern that was used to be sure that all outcomes were listed. (All outcomes which resulted from the selection of an oversized widget on the first shift were written first, etc.)

The event that at least one normal widget is sampled is

$$E = \{(O,N), (U,N), (N,O), (N,U), (N,N)\}$$

The event that there are no normal sized widgets in the sample is

$$F = \{(O,O), (O,U), (U,O), (U,U)\} \qquad \square$$

Examples like 2-2 illustrate that even in relatively simple experiments the sample space can be surprisingly large. (For example, if 1 widget is sampled in each of 3 shifts, the space has 27 elements). We noted at the end of Sec. 2-2 that it is often important to know exactly how many elements are in a sample space. Thus, it is useful to develop methods (other than listing and counting) for determining the size of a sample space.

EXAMPLE 2-3

Suppose that an urn (Fig. 2-1) contains two red balls, one blue ball and 1 white ball. An experiment consists of drawing three balls in succession without replacement. What is the sample space of the experiment? What is the event that balls of all three colors are selected? What is the event that both red balls are selected? What is the event that a red ball is selected first?

Solution: The first decision we need to make is to decide how to represent the outcomes of the experiment. The balls are selected in succession, so there is a first, a second, and a third ball selected on each performance of the experiment. The concept of an ordered triple is ideally suited to represent such outcomes. For example, we represent the result of drawing a red ball on the first draw, a blue ball on the second draw, and a white ball on the third draw by the ordered triple (R,B,W). With this notation the sample space can be represented as the set

$$S = \{(R,W,R), (R,W,B), (R,B,R), (R,B,W), (R,R,W), (R,R,B),$$
$$(B,W,R), (B,R,W), (B,R,R), (W,B,R), (W,R,B), (W,R,R)\}$$

FIGURE 2-1

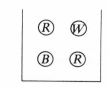

The event that balls of all three colors are selected is

$$E = \{(R,W,B), (R,B,W), (B,W,R), (B,R,W), (W,B,R), (W,R,B)\}$$

The event that both red balls are selected is

$$F = \{(R,W,R), (R,B,R), (R,R,W), (R,R,B), (B,R,R), (W,R,R)\}$$

The event that a red ball is selected first is

$$G = \{(R,W,R), (R,W,B), (R,B,R), (R,B,W), (R,R,W), (R,R,B)\} \qquad \square$$

Remark: In Example 2-3 the balls are selected *without replacement*. This means that the first ball is not returned to the urn before the second ball is drawn; likewise, neither the first nor the second ball is returned to the urn before the third ball is drawn. An alternative experiment would consist of drawing three balls in succession *with replacement*; i.e., each ball is returned to the urn as soon as its color has been noted. The sample space for experiments with and without replacement are usually different. See, for example, Exercise 9 at the end of this section.

EXAMPLE 2-4

Suppose one has the same urn and balls as in Example 2-3 and an experiment is defined as follows: select three balls simultaneously from the urn and note their colors. In this case there is no order associated with the three balls; all are drawn simultaneously. Describe the sample space S for this experiment, and the events E, "balls of all three colors are selected," and F, "two red balls are selected."

Solution: The sample space of this experiment can be represented by a set of ordered triples of whole numbers:

(Number of red balls, number of blue balls, number of white balls)

Thus, (2,0,1) would indicate that 2 red balls, 0 blue balls, and 1 white ball were drawn. With this notation the sample space is

$$S = \{(1,1,1), (2,1,0), (2,0,1)\}$$

In this experiment the event E that balls of all three colors are selected is $E = \{(1,1,1)\}$, and the event F that two red balls are selected is $F = \{(2,1,0), (2,0,1)\}$.

$\square$

Exercises for Sec. 2-3

1. Using the sample space of Example 2-2, list the elements in the following events:

G: An oversized widget is found on the first shift.
H: An oversized widget is found on at least one shift.
I: No oversized widgets are found.

2. Using the sample space of Example 2-3, list the elements in the following events:

J: A red ball is drawn second.
K: No blue ball is drawn.
L: No red ball is drawn.

3. Consider an experiment which consists of testing three batches of vaccine in the following manner:

1. Each batch is tested and labeled pure or impure according to the test results.
2. Each batch labeled pure is tested a second time, 24 hours later, and again is labeled pure or impure according to the test results.

Describe the sample space S for this experiment and list the elements of the event E "all batches tested are eventually found to be impure."

4. Two batches of vaccine are tested as in Exercise 3, but now the batches labeled pure on the second day are tested for a third time 24 hours later. Describe the sample space S for this experiment and list the elements in the event E "no impure batches are found on the first or second day."

5. A nickel and quarter are flipped simultaneously and each lands with either heads H or tails T uppermost. What is the sample space for this experiment? What is the event that the quarter lands with tails uppermost?

6. A red and a green die are rolled simultaneously. Each die has 6 faces with 1 dot on one face, 2 dots on another face, etc. The number of dots on the uppermost face of each die is noted. What is the sample space of this experiment? What is the event that at least one die has 4 dots uppermost? What is the event that the sum of the numbers of dots uppermost on the two dice is 7?

7. Four balls are drawn simultaneously from the urn shown in Fig. 2-2 and their colors noted. What is the sample space for this experiment? What is the event that two white balls are drawn?

FIGURE 2-2

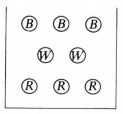

8. An experiment consists of rolling a die and flipping a coin. The number of dots on the uppermost face of the die and whether the coin lands heads H or tails T are noted.
   a. List the elements in the sample space of this experiment.
   b. List the elements in the event E_1 that the number of dots on the die is even.
   c. List the elements in the event E_2 that the coin lands with tails uppermost.
   d. List the elements in the events $E_1 \cup E_2$ and $E_1 \cap E_2$.

9. Work Example 2-3 with the condition without replacement changed to with replacement. As before answer the questions:
   a. What is the sample space of the experiment?
   b. What is the event that balls of all three colors are selected?
   c. What is the event that a red ball is selected first?

2-4   MULTISTAGE EXPERIMENTS AND TREE DIAGRAMS

Many experiments are of the sort that can reasonably be described as *multistage*. For example, if a shipment of audio tapes is to be checked for defectives, and if two tapes are to be selected and examined, then the diagram in Fig. 2-3 provides a method of representing all outcomes. Such a diagram is called a *tree diagram*. In Fig. 2-3 A denotes acceptable, and D denotes defective (not acceptable).

In this diagram the box at the far left (the first fork) represents the beginning of the experiment, i.e., the examination of the first tape. If it is acceptable, the first stage of the experiment corresponds to the upper branch from the box $\boxed{\text{Begin}}$, and the result of this examination is noted at the end of this branch in the circle $\textcircled{A}$. If the first tape is defective, this stage corresponds to the lower branch from the

FIGURE 2-3

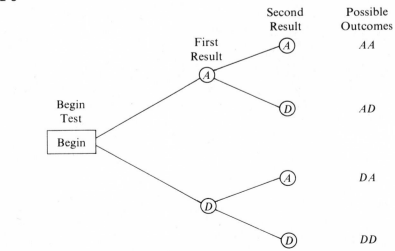

box, which terminates in the circle $\textcircled{D}$. In either case the result of the examination of the first tape is shown under the heading First Result. The second stage of the experiment, the examination of the second tape, is represented by the branches from the second forks (two of them). Again in each case the result can be A (acceptable tape, upper branch) or D (defective tape, lower branch). The result of the examination of the second tape is shown under the heading Second Result. The tree diagram leads at once to the sample space of the experiment

$$\{AA,\ AD,\ DA,\ DD\}$$

Here we have used AD, for example, as shorthand for the ordered pair (A,D), which in turn stands for first tape acceptable, second tape defective. We make use of similar conventions frequently in the future.

EXAMPLE 2-5

Suppose that an experiment consists of checking audio tapes until either two consecutive defective tapes have been examined or until four tapes have been checked. Draw a tree diagram for this experiment and find the sample space.

Solution: The tree diagram for this experiment is shown in Fig. 2-4 (see p. 48). The sample space for this experiment is

$$S = \{AAAA,\ AAAD,\ AADA,\ AADD,\ ADAA,\ ADAD,\ ADD,$$
$$DAAA,\ DAAD,\ DADA,\ DADD,\ DD\}$$

where we have adopted the shorthand notation introduced above.

Notice that in this example, even more than in the example of Fig. 2-3, the use of a tree diagram provides a systematic means of determining the sample space of the experiment.   □

EXAMPLE 2-6

Suppose that one is confronted with the situation depicted in Fig. 2-5 and an experiment which consists of first selecting a box, then an urn in the box, and finally a ball from the urn. Draw the tree diagram for this experiment.

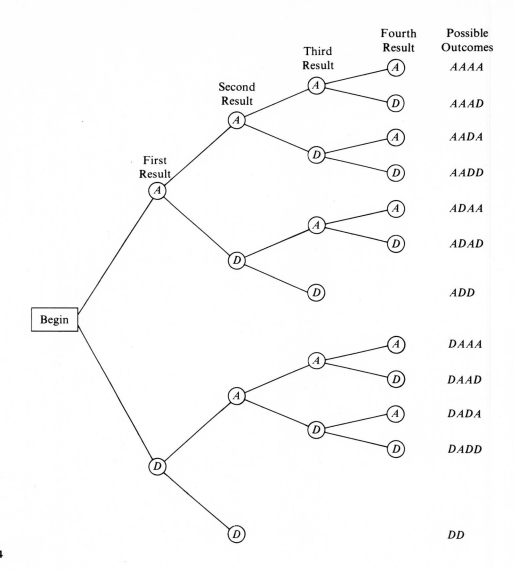

FIGURE 2-4

FIGURE 2-5

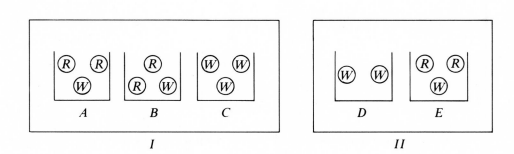

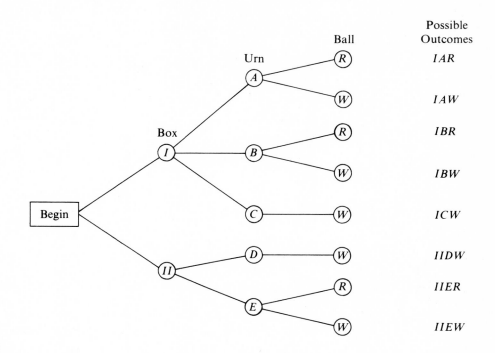

FIGURE 2-6

Solution: The solution is given in Fig. 2-6. □

So far we have used tree diagrams only as an aid in determining and listing the outcomes of a multistage experiment. However, for some of these experiments the tree diagram has symmetry properties which allow us to apply a simple formula and compute the number of outcomes of the experiment directly without listing them. We illustrate this type of experiment with an example.

EXAMPLE 2-7

An experiment consists of flipping a coin, noting whether it lands heads or tails, then drawing a ball from an urn which contains one red, one blue, and one white ball and noting its color. Draw the tree diagram for this experiment and determine the sample space.

Solution: The tree diagram is shown in Fig. 2-7.

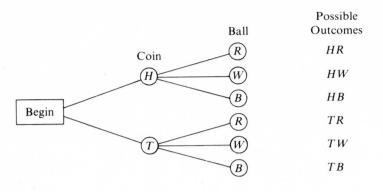

FIGURE 2-7

We note that this tree diagram has the property that at the second stage (drawing the ball) the number of outcomes (3) is independent of the outcome at the first stage. Thus if the coin comes up heads, there are 3 possible outcomes at the second stage, and if the coin comes up tails, there are still 3 possible outcomes at the second stage. Since there are 2 possible results at the first stage (H or T), and after the first stage is completed there are 3 possible results at the second stage (R, W, or B), we see that the tree for this experiment has 2 initial branches, each of which splits into 3 branches. Thus the tree represents the $2 \times 3 = 6$ possible outcomes for the experiment. The sample space is the set

$$S = \{HR, HW, HB, TR, TW, TB\} \qquad \square$$

Example 2-7 is a simple one in that the tree is easy to draw and it is clear that there are 6 possible outcomes for the experiment. The example can be made more complicated by increasing the number of results possible at each stage; however, if the structure of the experiment remains the same, it is still a simple matter to determine the number of possible outcomes of the experiment. For example, suppose that at the first stage of a 2-stage experiment, any of 8 possible results can occur. Also, suppose that at the second stage, no matter what occurs at the first stage, there are 12 possible results. Then the associated tree diagram has 8 branches from the initial fork and each of these 8 branches splits into 12 branches (one such split is shown in Fig. 2-8.) Since there are 8 initial branches and each splits into 12 branches, the full tree represents $8 \times 12 = 96$ possible outcomes for the experiment.

In the general case of a two-stage experiment like that of Example 2-7, if the first stage has n_1 possible results and the second has n_2 possible results (no matter which occurred at the first stage), then the experiment has $n_1 \times n_2$ possible outcomes. Notice that this is the number of elements in the cartesian product of the outcome sets of the first and second stages of the experiment. Indeed, an outcome of a multistage experiment is an ordered set of outcomes of the

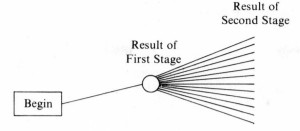

Result of
Second Stage

Result of
First Stage

Begin

FIGURE 2-8

component stages, i.e., an element of the cartesian product of the outcome sets of the individual stages. Thus counting the number of outcomes is the same as counting the number of elements in a cartesian product. The general result can be formulated as the following principle.

Multiplication Principle

Consider a multistage experiment consisting of k stages. If the first stage has n_1 possible outcomes, if the second stage has n_2 possible outcomes regardless of the outcome of stage 1, if the third stage has n_3 outcomes regardless of the outcomes of the first two stages, . . . , if the kth stage has n_k outcomes regardless of the outcomes of the earlier stages, then there are

$$n_1 \times n_2 \times \cdots \times n_k$$

elements in the sample space of the experiment.

EXAMPLE 2-8

An experiment consists of flipping a coin and noting which side lands uppermost, then rolling a die twice and noting in succession the numbers of dots on the uppermost faces. How many elements are in the sample space of this experiment?

Solution: This experiment consists of 3 stages: flip a coin, then roll a die, then roll a die again. The first stage has 2 outcomes, so $n_1 = 2$. The second and third stages have 6 outcomes each, and thus $n_2 = 6$, and $n_3 = 6$. It follows from the multiplication principle that there are $2 \times 6 \times 6 = 72$ elements in the sample space. ☐

The multiplication principle can also be phrased in terms of tree diagrams. To develop such a formulation let us refer to the beginning point of the tree, i.e., the fork at the box ⌐Begin⌐ at the left, as the *first fork*. The branches of the tree emanating from this fork will be called *first branches*. If the first branches terminate in forks, they are known as *second forks*, and so on. Now consider a multistage experiment for which the tree diagram has k sets of forks. If n_1 branches emanate from the first fork and each of them terminates in a fork, n_2 branches emanate from each of the second forks and each of them terminates in a fork, . . . , and n_k branches emanating from each of the kth forks, then there are $n_1 \times n_2 \times \cdots \times n_k$ elements in the sample space of the experiment.

As an example in which the formulation in terms of tree diagrams is helpful, consider the tree diagram in Fig. 2-3. There are 2 sets of forks, (that is, $k = 2$ in the above principle) and at each fork there are two branches (that is, $n_1 = 2$ and $n_2 = 2$). Therefore, there are $2 \times 2 = 4$ elements in the sample space of the experiment. These 4 elements are listed to the right in Fig. 2-3.

Notice that the principle does not apply to either Example 2-5 or 2-6. In Example 2-5 there are some branches from second forks which terminate in forks and some which do not. Therefore, the hypotheses of the statement of the principle are not fulfilled, and we cannot conclude that the result holds. Likewise, the hypotheses of the assertion are not fulfilled for the tree diagram of Example 2-6.

Let us use the term *path in a tree diagram* to denote an ordered set of connected branches, the first originating at the first fork, the second originating at the second fork, etc. Each path in the tree diagram of an experiment corresponds to one element in the sample space of the experiment, and conversely, each element in the sample space corresponds to one path in the tree diagram. It follows that if the multiplication principle can be applied, i.e., if the hypotheses are satisfied, then it gives us a means of computing the number of paths in the tree diagram of an experiment. Indeed, the number of paths is equal to the number of outcomes of the experiment.

EXAMPLE 2-9

A wildlife biologist plans an experiment to determine the effect of a growth inhibitor on freshwater algae. There are 3 different ways in which the tests can be performed, the biologist can choose from 4 different lakes, and the tests can be performed at any 1 of the lakes during any month from March through September, inclusive. From how many plans must the biologist make a choice?

Solution: A plan for the experiment, i.e., a choice of test, lake, and month, is equivalent to a path in the tree diagram of the experiment. The multiplication principle can be applied in this case, and we conclude that there are $3 \times 4 \times 7 = 84$ outcomes and consequently 84 paths in the tree diagram. The biologist must choose among 84 plans.   □

Exercises for Sec. 2-4

1. In Example 2-5 list the elements in the event E "at least two acceptable tapes are inspected."

2. Suppose that in Example 2-6 urn B is *removed* from box I and a red ball is added to urn D. Draw the tree diagram for the experiment of selecting a box, selecting an urn, and drawing a ball.

3. Suppose that in Example 2-5 the testing procedure is changed so that testing stops as soon as:

   1. Three acceptable tapes are found or,
   2. Three defective tapes are found or,
   3. Four tapes have been checked in all

   Draw a tree diagram to represent this experiment and list the elements in the sample space.

4. A trial of an experiment consists of drawing three balls in succession and *without replacement* from the urn shown in Fig. 2-9 and noting their colors. Draw a tree diagram for the experiment and list the elements in the sample space.

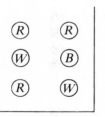

FIGURE 2-9

5. A trial of an experiment consists of drawing 3 balls *with replacement* from the urn shown in Fig. 2-9. That is, a ball is drawn, its color is noted, and it is replaced in the urn, etc. Draw a tree diagram for this experiment. How many elements are in the sample space of this experiment?

6. A trial of an experiment consists of rolling a die 4 times and noting whether the uppermost face is even or odd. How many elements are in the sample space of this experiment?

7. A shipper has 3 routes from New York to Chicago, 4 routes from Chicago to Denver, and 3 routes from Denver to San Francisco. How many different routes are there from New York to San Francisco?

8. A utility company must lay lines from the point labeled A to the point labeled B on the map shown in Fig. 2-10. The company is constrained to lay lines only in streets (shown by line segments), and it is interested in keeping its lines as short as possible. How many paths must be considered if all blocks are the same length? (*Hint:* Label all vertices and use a tree diagram.)

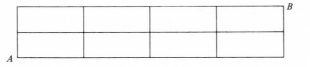

FIGURE 2-10

9. A chair manufacturer has 4 types of upholstery, 2 types of wood, and 4 designs to choose from. If 2 designs allow any choice of upholstery and wood and the other 2 designs allow only a choice of upholstery, how many different chairs can he make? (*Hint:* Use a tree diagram where the first choice is the design.)

10. A trial of an experiment consists of rolling a die until either an odd-numbered face lands uppermost 3 consecutive times or 6 rolls have been made. Draw a tree diagram for the experiment and find the sample space.

11. A product carries an identification number which consists of a letter, a single-digit number, and a four-digit number, for example, B-3-1018. The letter corresponds to the inspector, the single-digit number to the place of manufacture, and the four-digit number to the date (month and day) of manufacture. If there are 8 plants with 15 inspectors for each plant, and if in each month there are 20 working days, how many possible identification numbers are there?

12. Work Exercise 11 if the plants work every day of the month (assume it is not a leap year).

2-5 COUNTING ARRANGEMENTS: PERMUTATIONS

The multiplication principle provides us with a method of solving problems of the following sort. An experiment consists of selecting a letter from a set of five letters, say A, B, C, D, E, and a number from a set of three numbers, say 1, 2, 3. How many elements are in the sample space of this experiment? Using the multiplication principle, we know that the experiment has $5 \times 3 = 15$ possible outcomes.

There is another useful way of thinking of this experiment. Imagine two boxes as shown in Fig. 2-11. One of the 5 letters is to be selected and put in box I and one of the 3 numbers is to be selected and put into box II. The answer to the question "In how many ways can this be done?" is 15.

Now suppose that our experiment is slightly different. Instead of a set of letters and a set of numbers, suppose that we have only a set of 5 distinct letters and we are to put one letter into box I and another into box II. In how many

FIGURE 2-11

ways can this be done? A tree diagram for this experiment has 2 forks (corresponding to the 2 decisions which must be made) with 5 branches at the first fork and 4 branches at each of the second forks. By the multiplication principle, we conclude that there are $5 \times 4 = 20$ elements in the sample space. That is, there are 20 ways of placing 1 of 5 distinct letters in box I and another of them into box II. Note that in this experiment the sample space element which has A in box I and B in box II is distinct from the element which has B in box I and A in box II. That is,

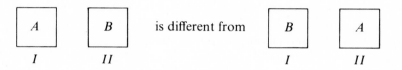

If we had three boxes, I, II and III, and five letters, A, B, C, D, E, there would be 60 elements in the sample space of the experiment which assigns one letter to box I, another to box II, and a third to box III. The justification of this uses the multiplication principle in the same way as it was used in the argument just above.

The two examples just discussed provide instances of an application of the multiplication principle in a way which is sufficiently common to merit its own name. We state the following principle in terms of boxes for clarity even though in most applications the boxes are an artificial construction which exists only in our imagination as an aid to understanding.

Permutation Principle

Given n distinct objects and r distinguishable boxes, $r \leq n$, there are

$$n \times (n - 1) \times \cdots \times (n - r + 1)$$

ways to put r of the objects into the boxes so that there is one object in each box.

EXAMPLE 2-10

Three districts of a city are to be surveyed for opinions on municipal services, and there are 5 experienced survey takers available to do the job. How many different assignments of people to districts can be made if no one surveys more than 1 district?

Solution: There are 5 individuals to be assigned to 3 districts. Using the permutation principle, we conclude that this can be done in

$$5 \times 4 \times 3 = 60$$

ways.   □

It is helpful to give two alternative phrasings of the permutation principle. In the first we suppose the boxes to be labeled 1, 2, ..., r.

1. An experiment is defined as follows:
Put one of the n objects in box 1
Put another in box 2
. . . .
Put another in box r.
The sample space of this experiment contains $n \times (n - 1) \times \cdots \times (n - r + 1)$ elements.
2. There are $n \times (n - 1) \times \cdots \times (n - r + 1)$ ordered arrangements of n distinct objects taken r at a time.

The second alternative formulation results from the identification of a box with a position in the arrangement. For example, instead of writing

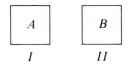

as we did above, we could simply write AB with the implicit understanding that AB means that A goes into box I and B goes into box II. It is clear that AB is different from BA; the arrangement is ordered. This convention is entirely consistent with the shorthand use of AB for (A,B) introduced in Sec. 2-4. In both cases the A can be thought of as the outcome of the first stage of a two-stage experiment and the B as the outcome of the second stage.

EXAMPLE 2-11

A map showing 4 countries is to be colored so that each country is a different color. If there are 8 colors available, in how many different ways can the map be colored? We consider 2 colorings to be different unless the color of all 4 countries is identical in both colorings.

Solution: We apply the permutation principle to conclude that the map can be colored in

$$8 \times 7 \times 6 \times 5 = 1680$$

different ways. ☐

It is convenient to introduce the notation $n!$ (read "n factorial") to denote $n \times (n-1) \times \cdots \times 2 \times 1$. Thus, $5! = 5 \times 4 \times 3 \times 2 \times 1 = 120$ and $10! = 10 \times 9 \times 8 \times \cdots \times 2 \times 1 = 3,628,800$. When this notation is used, the expression $n \times (n-1) \times \cdots \times (n-r+1)$, which occurs in the permutation principle, can be written

$$n \times (n-1) \times \cdots \times (n-r+1)$$

$$= \frac{n \times (n-1) \times \cdots \times (n-r+1) \times (n-r) \times \cdots \times 2 \times 1}{(n-r) \times \cdots \times 2 \times 1} = \frac{n!}{(n-r)!}$$

There is one case which requires special comment. If $r = n$, the term $(n-n)! = 0!$ occurs in the denominator of this fraction. The symbol $0!$ (which is not defined by the definition of $!$ given above) is defined to be 1. With this definition the formula given above matches the original permutation principle in the case $r = n$.

EXAMPLE 2-12

A signal consisting of 5 flags is to be raised on a vertical pole. If there are 7 different flags available, how many distinct signals can be displayed? We assume that the order of the flags is important and that 2 signals are distinct unless they have exactly the same flags in the same order.

Solution: Applying the permutation principle (thinking of the top flag as in box 1, the second from the top as in box 2, etc.), we conclude that

$$7 \times 6 \times 5 \times 4 \times 3 = \frac{7!}{(7-5)!} = \frac{7!}{2!} = 2520$$

distinct signals can be displayed. ☐

EXAMPLE 2-13

A director of a community theater is conducting auditions for a play which has parts for 4 female and 2 male players. There are 7 females and 8 males trying out for the play. How many possible casts are there?

Solution: This problem cannot be solved by a direct application of the permutation principle; instead one has to use both the permutation and multiplication principles. If we assume that the parts are all distinguishable (certainly a reasonable assumption), then there are $7 \times 6 \times 5 \times 4 = 840$ ways of selecting 4 females to play the female parts and $8 \times 7 = 56$ ways of selecting 2 males to play the male parts. Consider the task of selecting a cast as a two-stage experiment: select the female players and then select the male players. The first stage has 840 outcomes. For each of these outcomes, there are 56 possible outcomes of the second stage. Consequently there are $840 \times 56 = 47{,}040$ outcomes to the experiment; i.e., there are 47,040 possible casts. (Probably a good deal more than the director even cares to contemplate—let alone try out.) □

Exercises for Sec. 2-5

In each of the following exercises, decide whether to use the more general multiplication principle or the special permutation principle and then solve the problem.

1. A reviewer has 5 different books to review, and he reviews them 1 at a time. In how many different ways can he organize his work?

2. Suppose that in Example 2-10 there are 4 districts in the city and 6 individuals available to carry out the surveys. How many different assignments of survey takers to districts can be made, assuming no one surveys more than 1 district?

3. Evaluate the following numbers expressed in terms of the factorial symbol.

   a. $6!$ **b.** $\dfrac{8!}{4!}$ **c.** $\dfrac{8!}{4!\,4!}$ **d.** $\dfrac{52!}{48!\,4!}$ **e.** $\dfrac{100!}{97!\,3!}$

4. Suppose that each of the 5 positions on a basketball team is assumed to be different. How many different teams can be formed from 10 individuals? (Note that if 2 players change positions on a team, a new team is formed.)

5. Suppose two basketball teams of 5 players each are formed from 10 players. Assume as above that the same 5 players can form several different teams as they switch positions. How many different game lineups can be formed from the 10 players?

6. The time in a television variety show is utilized as indicated below, where C denotes commercial and S denotes skit:

C   S   S   C   S   S   C

If the producer has 5 skits and 3 commercials, in how many ways can he create a television show? We assume here that a skit is never repeated but (unfortunately) commercials can appear any number of times.

7. In Example 2-13 suppose that only 6 females and 5 males are trying out for the 4 female and 2 male parts, respectively. How many possible casts are there?

8. An experiment consists of flipping a coin 6 times and recording the sequence of heads and tails which is obtained. How many outcomes are possible for this experiment?

9. An economics exam consists of 10 statements which are to be labeled true or false. How many different answer sheets can be submitted?

10. A license plate consists of a 2-digit number followed by a letter, followed by a 3-digit number, for example, 35-D-621. How many different license plates can be formed?

2-6 COUNTING PARTITIONS: COMBINATIONS

In the preceding section we developed a method for determining the number of ways that r objects can be selected from n distinct objects and placed in r labeled boxes. Suppose now that we are interested simply in selecting r objects from n and not in placing these r objects into labeled boxes. To study this new operation it is convenient to view the process discussed in the preceding section as a two-step process consisting of *selection* and *insertion*. Then by the multiplication principle.

| Number of ways that r objects can be selected from n objects and put into r labeled boxes | = | number of ways r objects can be selected from n distinct objects | × | number of ways r objects can be put into r labeled boxes |

The permutation principle can be used to evaluate the numbers represented by the boxes on the extreme right and left. We have

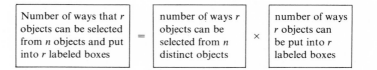

$$\frac{n!}{(n-r)!} = \boxed{\begin{array}{c}\text{number of ways } r \\ \text{objects can be} \\ \text{selected from } n \\ \text{distinct objects}\end{array}} \times r!$$

and this expression leads immediately to a formula for the number of ways r objects can be selected from n distinct objects. The result is summarized as follows:

Combination Principle

A subset of r objects can be selected from a set of n distinct objects in

$$\frac{n!}{(n-r)!r!}$$

different ways.

EXAMPLE 2-14

A sample of 3 audio tapes is to be selected from a package of 12 for testing. How many such samples are there?

Solution: In this example we argue along lines parallel to the derivation of the combination principle instead of simply applying the result. In later examples we shall apply the result directly.

The number of ways that 3 tapes can be selected from 12 and put into 3 labeled boxes is $12 \times 11 \times 10$. Moreover, for each choice of 3 tapes, these 3 can be put into the boxes in $3 \times 2 \times 1$ different ways. Therefore, viewing the total process as selection and insertion, we have

$$12 \times 11 \times 10 = \boxed{\begin{array}{l}\text{number of ways 3}\\\text{tapes can be}\\\text{selected from 12}\\\text{tapes}\end{array}} \times (3 \times 2 \times 1)$$

Consequently there are

$$\frac{12 \times 11 \times 10}{3 \times 2 \times 1} = 220$$

ways of selecting 3 tapes from a set of 12 tapes.   □

EXAMPLE 2-15

There are 8 engineers capable of serving on a design team. If the team is to consist of 5 engineers, how many possible teams can be created?

Solution: This is a problem of selecting subsets of 5 elements each from a set of 8 elements. An application of the combination principle yields

$$\frac{8!}{(8-5)!\,5!} = \frac{8 \times 7 \times 6}{1 \times 2 \times 3} = 56$$

as the number of possible teams.   □

The expression

$$\frac{n!}{(n-r)!\,r!}$$

will occur frequently in our study of probability, and it is useful to introduce a shorter notation for it. We define

$$\binom{n}{r} = \frac{n!}{(n-r)!\,r!} \tag{2-2}$$

That is, $\binom{n}{r}$ denotes the number of different ways of selecting r-element subsets from a set of n elements. The numbers $\binom{n}{r}$ are known as the *binomial coefficients*, and they have many interesting properties. For example, the fact

$$\binom{n}{r} = \binom{n}{n-r}$$

is often useful in working problems. This fact follows directly from Eq. (2-2). Indeed, if we use this equation to evaluate $\binom{n}{n-r}$, we must replace r in Eq. (2-2) by $n-r$. We have

$$\binom{n}{n-r} = \frac{n!}{[n-(n-r)]!\,(n-r)!} = \frac{n!}{(r)!\,(n-r)!}$$

which is [by Eq. (2-2) again] just $\binom{n}{r}$. This fact also follows from the observation that to each r-element subset which is selected from a set of n elements there is associated an $(n-r)$-element subset consisting of those elements which are *not* selected.

In many counting problems it is necessary to combine the principles with each other and with special arguments created for individual problems. The next examples illustrate the use of the combination and multiplication principles together.

EXAMPLE 2-16

A candle manufacturer produces 5 types of round candles and 4 types of square candles. If a gift package consists of 2 different round candles and 2 different square candles, how many different gift packages can be prepared?

Solution: We apply the combination principle to conclude that it is possible to select the 2 round candles in $\binom{5}{2}$ ways and the 2 square candles in $\binom{4}{2}$ ways. The process of pairing 2 round candles with 2 square ones to form a gift package can be understood as a two-stage experiment. The first stage consists of the selection of 2 different round candles. There are $\binom{5}{2}$ outcomes at the first stage. The second stage consists of the selection of 2 different square candles. For each outcome at the first stage, there are $\binom{4}{2}$ outcomes at the second stage. Thus, using the multiplication principle, we conclude that there are

$$\binom{5}{2}\binom{4}{2} = \frac{5 \times 4}{1 \times 2} \ \ \frac{4 \times 3}{1 \times 2} = 10 \times 6 = 60$$

ways of selecting 2 round candles and 2 square ones. That is, there are 60 different gift packages. □

EXAMPLE 2-17

Suppose that the candle manufacturer of Example 2-16 prepares gift packages containing 2 different candles which are either both round or both square. How many different gift packages can be prepared?

Solution: A direct application of the combination principle shows that there are $\binom{5}{2}$ gift packages containing 2 round candles and $\binom{4}{2}$ gift packages containing 2 square candles. Since no gift package contains both 2 round and 2 square candles, Eq. (1-6) of set theory can be applied to conclude that there are $\binom{5}{2} + \binom{4}{2} = 10 + 6 = 16$ different gift packages. □

EXAMPLE 2-18

A design team of 3 engineers, 2 marketing specialists, and 1 finance expert is to be created. There are 6 engineers, 5 marketing specialists and 3 finance experts qualified for the jobs. How many different design teams can be created?

Solution: We begin by applying the combination principle to determine the number of ways the subteams can be created. The 3 engineers can be selected in $\binom{6}{3}$ ways, the 2 marketing specialists can be selected in $\binom{5}{2}$ ways, and the finance expert can be selected in $\binom{3}{1}$ ways. Next we consider how the teams can be formed. It is useful to think of the formation of the teams as a three-stage experiment; the first stage is the selection of 3 engineers, which has $\binom{6}{3}$ outcomes; the second stage is the selection of 2 marketing specialists, which has $\binom{5}{2}$ outcomes; and the third stage is the selection of a finance expert, which has $\binom{3}{1}$ outcomes. We apply the multiplication principle to conclude that the team can be created in

$$\binom{6}{3}\binom{5}{2}\binom{3}{1} = 20 \times 10 \times 3 = 600$$

ways. □

EXAMPLE 2-19

A psychologist plans an experiment in group dynamics which can be conducted with either 3, 4, or 5 subjects. There are 7 subjects available for the experiment. With how many different groups can the experiment be conducted?

Solution: This problem should be approached by first considering separately the cases of groups with 3, 4, or 5 subjects. There are $\binom{7}{3}$ ways to select 3 subjects, $\binom{7}{4}$ ways to select 4 subjects, and $\binom{7}{5}$ ways to select 5 subjects. The entire set of groups can be partitioned into subsets consisting of those groups with 3, 4, and 5 members, respectively. It follows from Eq. (1-5) that the total number of groups is

$$\text{Number of groups with 3 members}$$
$$+ \text{ number of groups with 4 members}$$
$$+ \text{ number of groups with 5 members}$$
$$= \binom{7}{3} + \binom{7}{4} + \binom{7}{5}$$
$$= 35 + 35 + 21 = 91$$

The experiment can be conducted with 91 different groups. ☐

Exercises for Sec. 2-6

1. Compute the following numbers:

   a. $\dfrac{7!}{3!}$ **b.** $\binom{9}{4}$ **c.** $\dfrac{6!}{2^6}$

   d. $\binom{6}{0} + \binom{6}{1} + \binom{6}{2} + \binom{6}{3} + \binom{6}{4} + \binom{6}{5} + \binom{6}{6}$

2. Determine the number of ways that a sample of 5 items can be selected from a set of 20.

3. A fruit packager prepares packages consisting of 3 varieties of fresh fruit and 2 types of preserved fruit. He has apples, grapefruit, oranges, pears, and plums as fresh fruit and apricots, dates, and prunes as preserved fruit. How many different packages must he stock if he is to have a sample of each on hand?

4. A student is to take 1 mathematics course, 1 history course, 1 course in business, and 2 electives. There are 3 courses in mathematics, 3 in history, and 5 in business for which she is prepared. Also, she is interested in 5 elective courses. From how many different sets of courses must she make a choice?

5. A candle manufacturer produces 5 types of round candles and 4 types of square candles. Gift packages containing 3 different candles are to be prepared. How many different gift packages are there? Note that a gift package may contain candles which are all of one shape or 2 candles of one shape and 1 of the other shape.

6. An investment portfolio is to be formed by investing in *1 or more* of the following 4 types of investments: stocks, corporate bonds, utility bonds, and government securities. If 2 portfolios have the same types of investments, they are considered to be the same kind of portfolio. How many different kinds of portfolios are there?

7. A shipment of 100 audio tapes is known to contain 5 defective tapes.
   a. In how many ways can a sample of 5 tapes be selected from the 100?
   b. In how many ways can a sample of 5 defective tapes be selected?
   c. In how many ways can a sample consisting of 3 defective tapes and 2 nondefective tapes be selected?

8. In selecting books to take on vacation a student decides to take 2 novels, 2 biographies, and 1 travel book. If she has 8 novels, 4 biographies, and 5 travel books, in how many different ways can she select 5 books to take on vacation?

9. A psychologist plans an experiment in group dynamics that requires subjects to play specific roles. There are 2 roles that must be played by females and 2 roles that can be played by either males or females. There are 5 female and 3 male subjects available. How many possible ways can the experiment be set up?

10. A political party has rules to the effect that the officers of every party committee, i.e., the chairperson, vice-chairperson, and secretary, must always include at least 1 woman and 1 black. A certain committee of 6 members consists of 2 black males, 1 black female, 2 white males, and 1 white female. In how many different ways can the officers of the committee be selected to obey party rules? Note that if members change offices, a different selection exists.

11. An economics exam has 2 parts. The first part consists of 10 true-false questions, and the second part consists of 5 multiple-choice questions where 1 answer in 4 possibilities is correct. How many different answer sheets can be submitted assuming an answer is given for each question?

12. Suppose that in Example 2-18 one of the engineers dislikes one of the marketing specialists and also dislikes one of the finance experts. How many design teams can be formed so that this engineer is never on a team with one of these people?

13. See Example 1-4. A psychologist wishes to conduct an experiment with each different pair of mice which can be selected from 7 trained mice. How many experiments must be scheduled?

2-7 APPLICATIONS OF COUNTING TO THE COMPUTATION OF PROBABILITIES

Although counting problems frequently arise directly in the applications of mathematics to business and the social sciences, the material on counting, or *combinatorics* as it is called in mathematics, presented here is included to provide a means of computing probabilities. In particular, as indicated in Sec. 2-2, we need counting techniques to determine the number of elements in the sample space of an experiment. In that case, if we assume that every outcome occurs with the same likelihood, an assumption which might be based on sample frequencies or on symmetry, then we can compute the probability of each outcome. In fact, with this assumption the weight w_i assigned to the ith outcome should be $1/n$, where n is the number of outcomes of the experiment. Once the weights are assigned to individual outcomes, we can assign a probability to each event associated with the experiment. In cases where the weights are all the same, we say the outcomes are *equally likely*. Thus the term equally likely refers to our assumption about the likelihood of outcomes.

The following definition provides an assignment of probabilities to events in the special case of equally likely outcomes.

Definition   If we assume that all outcomes of an experiment are equally likely, then the *probability* assigned to event E, denoted by $\Pr[E]$, is

$$\Pr[E] = \frac{n[E]}{n[S]}$$

where $n[E]$ denotes the number of elements in E and $n[S]$ the number of elements in the sample space S.

This assignment of probabilities to events provides an example of a *probability measure*. Since it assigns an equal weight to each outcome, it is known as the *equiprobable measure*. Probability measures in general will be defined and studied in the next chapter. The equiprobable measure is frequently adopted in applications, and it gives the assignment of probabilities which corresponds most closely to an intuitive idea of randomness. Indeed, when we use a phrase like "3 tapes selected at random from a set of 10 tapes" without further comment regarding the likelihood of certain outcomes, it is to be understood that we assume all outcomes, i.e., all selections of 3, to be equally likely. That is, the phrase *selected at random* implies that the equiprobable measure is to be used.

EXAMPLE 2-20

If 2 balls are selected simultaneously and at random from an urn containing 5 balls numbered 1, 2, 3, 4, and 5, what is the probability that 2 balls with odd numbers are selected?

Solution: Since there are 5 balls in the urn, there are $\binom{5}{2} = 10$ ways of selecting 2 balls. Since there are 3 balls in the urn with odd numbers, there are $\binom{3}{2} = 3$ ways of selecting 2 balls with odd numbers. Therefore, if S denotes the sample space of the experiment and E denotes the event that 2 balls with odd numbers are selected, then $n[S] = 10$ and $n[E] = 3$. Consequently, $\Pr[E] = \frac{3}{10}$. The probability of selecting 2 balls with odd numbers is .3.   □

EXAMPLE 2-21

Suppose that you know there are 3 defective and 9 acceptable tapes in a package of 12 audio tapes. If you select a sample of 3 at random, what is the probability that all 3 are acceptable?

Solution: In Example 2-14 it was shown that there are 220 ways of selecting 3 audio tapes from a package of 12. Since there are 9 acceptable tapes, there are $\binom{9}{3} = 84$ ways of selecting 3 acceptable tapes. Therefore, the probability of selecting 3 acceptable tapes is $\frac{84}{220} = \frac{21}{55}$.   □

EXAMPLE 2-22

A candle manufacturer produces 4 types of round candles and 3 types of square candles. All possible gift packages are prepared which contain 3 different types of candles. What is the probability that a randomly selected gift package contains 3 round candles?

Solution: There are $\binom{4}{3} = 4$ ways of preparing gift packages containing 3 round candles and $\binom{3}{3} = 1$ way of preparing gift packages containing 3 square candles. Also, there are $\binom{4}{2}\binom{3}{1} = 6 \times 3 = 18$ ways of preparing gift packages containing 2 round candles and 1 square one and $\binom{4}{1}\binom{3}{2} = 4 \times 3 = 12$ ways of preparing a package with 1 round and 2 square candles. Therefore, there are

$$4 + 1 + 18 + 12 = 35$$

ways of preparing gift packages containing 3 candles. We assume that the sample space of the experiment consists of these 35 outcomes and that each is equally likely. Since there are 4 outcomes which result in a gift package with 3 round candles, the probability of this event is $\frac{4}{35}$.

Pursuing this example a little further, we see that the probability of selecting a gift package at random with exactly 2 round candles is $\frac{18}{35}$, and the probability of selecting a gift package at random with at least 2 round candles is

$$\frac{4 + 18}{35} = \frac{22}{35}$$ □

EXAMPLE 2-23

Sam and Sally are members of a committee studying the effect of government regulations on business. The committee consists of 4 men and 3 women, and a subcommittee of 3 is to be chosen to study paperwork. The subcommittee must include at least 1 woman, but it cannot be all women. If the sub-committee is selected at random from a list of all subcommittees which meet these conditions, what is the probability that the subcommittee chosen will include Sam? Sally?

Solution: The subcommittee may consist of 1 man and 2 women or 2 men and 1 woman. The number of subcommittees with 1 man and 2 women is

$$\binom{4}{1}\binom{3}{2} = 4 \times 3 = 12$$

The number of subcommittees with 2 men and 1 woman is

$$\binom{4}{2}\binom{3}{1} = 6 \times 3 = 18$$

Thus there are $18 + 12 = 30$ possible subcommittees.

To answer the questions about Sam and Sally we need to know the number of subcommittees which include them. If Sam is on a subcommittee, the remaining 2 members could be both women or 1 man and 1 woman. The number of such subcommittees is

$$\binom{3}{2} + \binom{3}{1}\binom{3}{1} = 3 + 3 \times 3 = 12$$

Thus the probability that the subcommittee selected will include Sam is

$$\frac{12}{30} = \frac{2}{5}$$

The analysis is similar for Sally, but if she is on the subcommittee, the other 2 members could be both men or 1 man and 1 woman. The number of such subcommittees is

$$\binom{4}{2} + \binom{4}{1}\binom{2}{1} = 6 + 8 = 14$$

and the probability that the subcommittee selected will include Sally is

$$\frac{14}{30} = \frac{7}{15}$$

□

Exercises for Sec. 2-7

1. For Example 2-20 give the probability of choosing:
   a. 2 balls with even numbers
   b. 1 odd and 1 even

2. Suppose that a bag containing 10 basketballs has 4 with defective valves. If 2 balls are selected at random from the bag, what is the probability that at least 1 of them will have a defective valve?

3. A green die and a red die are rolled, and the numbers of dots on the uppermost faces are noted. Assume both dice are fair.
   a. What is the probability the numbers are the same on the two dice?
   b. What is the probability that the sum of the numbers is even?
   c. What is the probability that the sum of the numbers is 7?

4. It is known that there are 3 defective tapes in a package of 10 audio tapes. A random selection of 3 tapes is made from the package.
   a. What is the probability that at most 2 are defective?
   b. What is the probability that at least 1 is defective?

5. A coach has 6 stop watches, 3 of which are accurate, 1 is fast, and 2 are equally slow. The coach makes a random selection of 2 watches to be used to time the first place finisher in a race.
   a. What is the probability that the first-place finisher is timed accurately by both watches?
   b. What is the probability that the first-place finisher is timed accurately by at least 1 watch?
   c. What is the probability that the watches used to time the first-place finisher both show the same time?

6. A shipment of 100 widgets contains 90 normal widgets, 6 oversized ones, and 4 undersized ones. If a random sample of 10 widgets is selected, what is the probability that exactly 1 oversized and 1 undersized widget are selected?

7. There are 5 black and 5 white mice available for an experiment which requires four mice. If a random selection of 4 mice is made from the set of 10, what is the probability that 2 black and 2 white mice are selected?

8. A fair coin is flipped 4 times. What is the probability that both heads and tails occur?

9. A lot of 20 items is known to contain 5 overweight items. If a sample of 10 items is selected at random:
   a. What is the probability that no overweight items are selected?
   b. What is the probability that all 5 overweight items are selected?

10. A telephone repairman knows that 2 circuits are working properly and 2 circuits are working improperly. He tests the circuits 1 by 1 and at random until both defective circuits are located. What is the probability that the 2 defective circuits are identified after 2 tests?

11. Six men enter a restaurant and check their coats. The checker puts all 6 coats on the same hook and gives one of the men a tag. When the men leave the restaurant, the checker hands each of the men a coat. If the checker hands the coats out at random, what is the probability that each man receives the correct coat?

12. In Example 2-23 what is the probability that a subcommittee will be selected which includes both Sam and Sally?

IMPORTANT TERMS

You should be able to describe, define, or give examples of each of the following:

Sample frequency
Deductive method
Weights
Equally likely outcomes
Experiment
Sample space
Event
Selection with (without) replacement
Multistage experiment
Tree diagram

Multiplication principle
Branch
Fork
Path
Permutation principle
Combination principle
Binomial coefficients
Probability measure
Equiprobable measure
Random selection

REVIEW EXERCISES

1. A coin is flipped 3 times and the face which lands uppermost each time is noted. What is the sample space for this experiment? What is the event that at least 2 heads land uppermost?

2. An experiment consists of flipping a coin, rolling a die, and randomly selecting a letter from the word CHOCOLATE and noting the face of the coin which lands uppermost, the number of dots on the top of the die, and the letter selected. How many elements are in the sample space of this experiment?

3. Referring to Exercise 2, list the elements in the event E, "the coin lands with heads uppermost, the die lands with an even number of dots on the top, and the letter selected is a vowel."

4. A pack of cards is made up of all the hearts, diamonds, and clubs of a regular deck of cards. An experiment consists of drawing a card, noting its *color* and *rank*, and replacing it, then drawing a second card and noting its color and rank. How many outcomes are there for this experiment?

5. Referring to Exercise 4, how many outcomes would there be if the first card were not replaced before the second is chosen?

6. Referring again to Exercise 4, determine the probability of the event of drawing at least 1 black card? Drawing exactly 1 black card? Drawing at least 1 ace?

7. A quiz which is graded A, P (Pass), F (Fail) is given to a class each Friday for 3 weeks. A student who receives an A grade is exempt from future quizzes. If a student views the quiz taking as an experiment, how many outcomes are there? Draw a tree diagram for the experiment.

8. A student takes a quiz on which the possible grades are A, P, and F. In the student's view a grade of A is twice as likely as an F and a P is 3 times as likely as an F. What weights should be assigned to the outcomes of the experiment of taking a quiz to reflect the student's assumptions?

9. There are 4 adjacent seats in a row in a theater. In how many different ways can 4 people be seated?

10. First, second, and third prizes are to be awarded at a science fair in which 16 exhibits have been entered. In how many different ways can the prizes be awarded?

11. Three married couples attend a football game. In how many ways can they be seated in six seats in a row if:
   a. There are no restrictions?
   b. All men are seated together, and all women are seated together?
   c. All women are seated together?
   d. Each married couple sits together?
   e. No 2 women sit together?

12. The chief designer for a large automobile company is considering 4 different radiator grills, 2 different styles of headlights, and 5 different rear fender designs. With respect to these items alone, how many different styles of automobile can be designed?

13. On a 10-question exam each question is worth 10 points (no part credit). In how many ways can you make 70? In how many ways can you make at least 80?

14. There is a list of 8 books for an English class. Seniors are required to read 5 of them, and sophomores are required to read 3 of them. Juniors are required to read 4 of them, but they can not get credit for reading the eighth book since they have already read it in a previous course. Which group has the hardest decision to make, i.e., which group has the most choices, and why?

15. A fair die is rolled 3 times. What is the probability that 1 or 2 dots land on top all 3 times?

16. Of 20 unmarked cans in a box, 3 are vegetables and the rest are fruits. In how many ways can you select:
    a. 3 cans from the box?
    b. 3 cans of fruit?
    c. 2 cans of vegetables?
    d. 3 cans of which at least 2 are cans of fruit?

17. Refer to Exercise 16. Find the probability of selecting 3 cans from the box such that:

    a. All are fruit.
    b. All are vegetable.
    c. At least 2 are fruit.
    d. There is at least 1 fruit and at least 1 vegetable.

18. There are 6 red balls and 4 blue balls in an urn. If a random selection of 4 balls is made from the urn, what is the probability that 2 red balls and 2 blue balls are selected?

19. There are 2 urns with 3 colored balls in each urn. Urn *I* contains 2 red and 1 white ball and urn *II* contains 2 white and 1 blue ball. An experiment consists of randomly selecting a ball from urn *I* noting its color and placing it in urn *II* and then randomly selecting a ball from urn *II* and noting its color. Draw the tree diagram for the experiment and find the sample space.

20. a. How many 3 digit numbers can be formed with the digits 1, 2, 4, 6, 9 if each digit is used at most once?
    b. How many of the numbers in part a are less than 400?
    c. How many 3 digit numbers can be formed with the digits 1, 2, 4, 6, 9 if each digit can be used any number of times?

PROBABILITY

CHAPTER

THREE

3-1 INTRODUCTION

The ideas and techniques of Chap. 2 were introduced to provide a background for dealing with problems involving randomness and uncertainty. In this chapter we continue our study of probability, becoming both more precise and more general. One goal is to formulate precise rules which govern the behavior of probabilities. We shall use the various properties of probabilities in many different ways, and it is essential to know exactly which operations are legitimate and which are not. A second goal is to consider more general settings.

The probability applications of Chap. 2 were exclusively concerned with equally likely events and the use of the equiprobable measure. A knowledge of these topics is adequate only for a very restricted class of problems. For instance, consider the following situation. A supply of widgets contains 90 percent acceptable widgets and 10 percent defective ones. Two widgets are selected, and one of them is tested. If this widget is defective, the testing stops; if it is acceptable, the second widget is tested. What is the probability that a defective widget turns up in this testing procedure? There is no simple definition of an outcome for this experiment for which the equiprobable measure provides the appropriate model. Another probability measure must be introduced. In this chapter we shall consider rather general probability measures and their properties. We continue, however, to restrict ourselves to finite sample spaces.

3-2 PROBABILITY MEASURES: BASIC RULES AND PROPERTIES

We have noted that there are interesting and important problems for which a model utilizing the equiprobable measure is not appropriate. In such cases one must modify the model and introduce probabilities which cannot be obtained by arguments based on the assumption that all outcomes are equally likely. Of course, the new model should be consistent with the intuitive concepts of probability discussed in Sec. 2-2. Thus the probability of an event, i.e., of a subset of a sample space, should still correspond to the likelihood of that event's occurring. To ensure this correspondence in the general case we begin by reexamining the equiprobable measure and identifying more of its properties. We then formulate our basic definition as an abstraction of the relevant properties of this simple example.

Recall that the equiprobable measure is obtained by assigning weight $1/n$ to each outcome in a sample space containing n outcomes and by assigning probability m/n to an event, i.e., a subset, containing m outcomes. Thus, the probability of each outcome is a nonnegative number, the probability of the entire sample space (which contains n outcomes) is 1, and if we have two disjoint events E and E' containing m and m' outcomes, respectively, then there are $m + m'$ outcomes in the event $E \cup E'$, and consequently

$$\Pr[E \cup E'] = \frac{m + m'}{n} = \frac{m}{n} + \frac{m'}{n} = \Pr[E] + \Pr[E']$$

The properties we require of a general probability measure are abstractions of these.

Definition

A *probability measure* is a function Pr which assigns a number to each event of a sample space S and which satisfies the following conditions.

   i. If E is any event of S, then $\Pr[E] \geq 0$.
  ii. If E_1 and E_2 are disjoint events, then $\Pr[E_1 \cup E_2] = \Pr[E_1] + \Pr[E_2]$.
 iii. $\Pr[S] = 1$.

These conditions are commonly referred to as the *axioms of probability theory*.

EXAMPLE 3-1

A biologist collects and classifies snails of 2 varieties, striped and plain, and of both sexes of each variety. An experiment consists of searching a specified area until a snail is found, and recording its sex and variety. We assume that the sex and variety of each snail can be determined. Based on experience, probabilities of various outcomes are assigned as shown in Table 3-1.

TABLE 3-1

Outcome	Probability
Striped female (*SF*)	.34
Striped male (*SM*)	.28
Plain female (*PF*)	.18
Plain male (*PM*)	.20

It is clear that the probability assigned to each outcome is nonnegative. As before, we can define the probability of an event E to be the sum of the probabilities of the outcomes in E. Thus, for example, we define

$$\text{Pr[female]} = \text{Pr}[\{SF, PF\}] = \text{Pr}[\{SF\}] + \text{Pr}[\{PF\}] = .34 + .18 = .52$$

With this definition conditions 1 to 3 are satisfied. Also, one has

$$\text{Pr[not } SF] = \text{Pr}[\{SM, PF, PM\}]$$
$$= \text{Pr}[\{SM\}] + \text{Pr}[\{PF\}] + \text{Pr}[\{PM\}]$$
$$= .28 + .18 + .20 = .66$$

and since $\text{Pr}[\{SF\}] = .34$, we see that in this instance

$$\text{Pr}[\{\widetilde{SF}\}] = 1 - \text{Pr}[\{SF\}]$$

In addition,

$$\text{Pr[striped or female]} = \text{Pr}[\{SF, SM, PF\}]$$
$$= \text{Pr}[\{SF\}] + \text{Pr}[\{SM\}] + \text{Pr}[\{PF\}]$$
$$= .34 + .28 + .18 = .80$$

and

$$\text{Pr[striped]} = \text{Pr}[\{SF, SM\}] = .62$$
$$\text{Pr[female]} = \text{Pr}[\{SF, PF\}] = .52$$
$$\text{Pr[striped and female]} = \text{Pr}[\{SF\}] = .34$$

so that

$$\text{Pr[striped]} + \text{Pr[female]} - \text{Pr[striped and female]} = .62 + .52 - .34 = .80$$

Consequently, in this instance, we have

$$\Pr[\text{striped or female}] = \Pr[\text{striped}] + \Pr[\text{female}] - \Pr[\text{striped and female}] \quad \square$$

The relations between probabilities illustrated in this example are special cases of general results. It is to these results that we now turn. We begin by noting that since an event is a subset of S, a probability measure must be defined on the set of all subsets of S. Some additional properties of probability measures which can be deduced from the axioms (conditions i to iii) and which will be useful in applications are listed here for easy reference:

1. For any event E, $0 \le \Pr[E] \le 1$.
2. For any collection of pairwise disjoint events $E_1, E_2, \ldots, E_k$,

$$\Pr[E_1 \cup E_2 \cup \cdots \cup E_k] = \Pr[E_1] + \Pr[E_2] + \cdots + \Pr[E_k]$$

3. For any event E, $\Pr[\tilde{E}] = 1 - \Pr[E]$.
4. For any events E and F,

$$\Pr[E \cup F] = \Pr[E] + \Pr[F] - \Pr[E \cap F]$$

In property 3 the complement is taken with respect to the sample space S, which serves as a universal set for probability problems.

Notice that if E and F are events, then so are $\tilde{E}$, $E \cup F$, and $E \cap F$; thus properties 3 and 4 are meaningful statements. Also, the union of any collection of events is an event, and hence property 2 is also meaningful.

Other properties can be deduced easily from these four. For example, since the empty set $\varnothing$ is a subset of S, $\Pr[\varnothing]$ is defined. Also, $\tilde{S} = \varnothing$. It follows from condition iii and property 3 that

$$\Pr[\varnothing] = \Pr[\tilde{S}] = 1 - \Pr[S] = 1 - 1 = 0$$

We shall find that the solution of many of the problems in applied probability considered in this book rests on properties 1 to 4. Also, even though these properties are obviously related to simple features of the equiprobable measure, it is not obvious that they must hold for any function Pr for which conditions i to iii hold. Consequently, it is useful to indicate briefly how each of the properties 1 to 4 can be deduced from the axioms of probability theory.

In order to justify property 3 we note that E and $\tilde{E}$ are disjoint sets and $E \cup \tilde{E} = S$. Therefore, by condition ii we have

$$\Pr[S] = \Pr[E] + \Pr[\tilde{E}]$$

Next, using condition iii to replace $\Pr[S]$ by 1, we obtain

$$\Pr[E] + \Pr[\tilde{E}] = 1$$

which can be rewritten as property 3.

Property 1 follows from property 3 and condition i. Indeed, if the event E were such that $\Pr[E] > 1$, then $\Pr[\tilde{E}] = 1 - \Pr[E]$ would be negative. But by condition i this is impossible. Consequently, property 1 holds.

Property 2 follows immediately from condition ii. We give the details for three sets; the justification of the general case is similar. If E_1, E_2, and E_3 are pairwise disjoint, i.e., each pair is disjoint, then E_1 and $E_2 \cup E_3$ are disjoint. Since $E_1 \cup E_2 \cup E_3 = E_1 \cup (E_2 \cup E_3)$, we have from condition ii that

$$\Pr[E_1 \cup E_2 \cup E_3] = \Pr[E_1] + \Pr[E_2 \cup E_3]$$

Since E_2 and E_3 are disjoint we have, again from condition ii, that

$$\Pr[E_2 \cup E_3] = \Pr[E_2] + \Pr[E_3]$$

Combining these two results, we obtain

$$\Pr[E_1 \cup E_2 \cup E_3] = \Pr[E_1] + \Pr[E_2] + \Pr[E_3]$$

which is property 2 for $k = 3$. The general case can be justified by an argument which is similar but which uses mathematical induction.

Finally, we turn to property 4. The justification of this property is a little more involved and rests on an elementary result of set theory. The result is that if E and F are any sets, then $(E \cap \tilde{F}) \cup (E \cap F) \cup (\tilde{E} \cap F) = E \cup F$, and indeed $E \cap \tilde{F}$, $E \cap F$, and $\tilde{E} \cap F$ from a partition of $E \cup F$. The Venn diagram for this relationship is shown in Fig. 3-1, where E is represented by the left-hand circle (horizontally shaded) and F is represented by the right-hand circle (vertically shaded). In Fig. 3-1 $E \cap \tilde{F}$ is the portion which is shaded *only* horizontally,

FIGURE 3-1

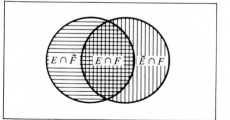

$\tilde{E} \cap F$ is the part shaded *only* vertically, and $E \cap F$ is the part that is shaded both horizontally and vertically. We use this fact together with

$$E = (E \cap F) \cup (E \cap \tilde{F}) \qquad F = (F \cap E) \cup (F \cap \tilde{E})$$

and property 2 to conclude that

$$\Pr[E] = \Pr[E \cap F] + \Pr[E \cap \tilde{F}] \tag{3-1}$$

$$\Pr[F] = \Pr[F \cap E] + \Pr[F \cap \tilde{E}] \tag{3-2}$$

and $\quad \Pr[E \cup F] = \Pr[E \cap \tilde{F}] + \Pr[E \cap F] + \Pr[\tilde{E} \cap F] \tag{3-3}$

From Eqs. (3-1) and (3-2) we have, respectively, $\Pr[E \cap \tilde{F}] = \Pr[E] - \Pr[F \cap E]$ and $\Pr[F \cap \tilde{E}] = \Pr[F] - \Pr[E \cap F]$. Using these in Eq. (3-3) we have, finally,

$$\begin{aligned} \Pr[E \cup F] &= \Pr[E \cap \tilde{F}] + \Pr[E \cap F] + \Pr[\tilde{E} \cap F] \\ &= (\Pr[E] - \Pr[E \cap F]) + \Pr[E \cap F] + (\Pr[F] - \Pr[E \cap F]) \\ &= \Pr[E] + \Pr[F] - \Pr[E \cap F] \end{aligned}$$

which is the desired result.

EXAMPLE 3-2

Suppose that a randomly selected widget can be normal, mismanufactured (too large or too small), or improperly labeled (mislabeled or not labeled at all). An analysis of data leads to the assignment of the probabilities in Table 3-2.

TABLE 3-2

Event	Probability
Too large (F_1)	.12
Too small (F_2)	.08
Mislabeled (G_1)	.15
Unlabeled (G_2)	.02
Normal (H)	.72

Find the probability of each of the following events:

(a) A randomly selected widget is mismanufactured.
(b) A randomly selected widget is not mismanufactured.
(c) A randomly selected widget is improperly labeled.
(d) A randomly selected widget is either mismanufactured or improperly labeled.
(e) A randomly selected widget is both mismanufactured and improperly labeled.

Solution: (a) The event that a randomly selected widget is mismanufactured is $F_1 \cup F_2$. Call this event E_1. Then since $F_1 \cap F_2 = \varnothing$, we have

$$\Pr[E_1] = .12 + .08 = .20$$

(b) This event is $\tilde{E}_1$, and

$$\Pr[\tilde{E}_1] = 1 - \Pr[E_1] = 1 - .20 = .80$$

(c) This event is $G_1 \cup G_2$, and we shall denote it by E_2. Since $G_1 \cap G_2 = \varnothing$, we have

$$\Pr[E_2] = .15 + .02 = .17$$

(d) Since $E_1 \cup E_2 \cup H = S$, and since H is disjoint from $E_1 \cup E_2$, we conclude that $E_1 \cup E_2 = \tilde{H}$. Therefore,

$$\Pr[E_1 \cup E_2] = \Pr[\tilde{H}] = 1 - \Pr[H] = 1 - .72 = .28$$

(e) This event is $E_1 \cap E_2$, and we can determine $\Pr[E_1 \cap E_2]$ by using property 4. Indeed, property 4 applied to E_1 and E_2 gives

$$\Pr[E_1 \cup E_2] = \Pr[E_1] + \Pr[E_2] - \Pr[E_1 \cap E_2]$$

and since we know the values of all the probabilities in this expression except $\Pr[E_1 \cap E_2]$, we can use them to evaluate $\Pr[E_1 \cap E_2]$. We have $\Pr[E_1 \cap E_2] = \Pr[E_1] + \Pr[E_2] - \Pr[E_1 \cup E_2] = .09$. $\square$

Probability measures can be obtained in a number of ways. One of the most common is through the use of weights. Suppose that an experiment has a finite sample space consisting of n outcomes, $S = \{\mathcal{O}_1, \ldots, \mathcal{O}_n\}$. We say that we have an *assignment of weights* for this experiment if to each outcome $\mathcal{O}_i \in S$ there is associated a number w_i such that

$$0 \leq w_i \leq 1 \qquad i = 1, 2, \ldots, n$$
$$w_1 + w_2 + \cdots + w_n = 1$$

$$(3\text{-}4)$$

We remark that the assignment of weight $1/n$ to each outcome of an experiment with n distinct outcomes meets these conditions.

An assignment of weights can be used to define a probability measure on the set of events.

Definition   *The probability measure defined by an assignment of weights.*   For any assignment of weights satisfying (3-4) a probability measure Pr is defined by setting, for any event E,

$$\Pr[E] = \text{sum of weights of outcomes in } E$$

If $E = \varnothing$, this sum is to be interpreted as zero. This interpretation is necessary since we have shown that $\Pr[\varnothing] = 0$.

Note that if we have an event consisting of a single outcome, the probability assigned to that event is simply the weight of that outcome. In the intuitive discussion of Chap. 2 (Sec. 2-2) we referred to "the probability of an outcome." In order to be consistent, it is preferable to refer to the probability of the event consisting of that outcome. The distinction (between an element and a set consisting of that element) is not a crucial one for us, and we shall not belabor the point. Henceforth if, for example, the sample space contains an element e, then we shall freely refer to $\Pr[e]$ as shorthand for $\Pr[\{e\}]$. For example, in flipping a coin, we write $\Pr[\text{head}]$ as shorthand for $\Pr[\{\text{head}\}]$.

We should, of course, verify that the probability measure defined by an assignment of weights to the outcomes of an experiment is in fact a probability measure, i.e., that conditions i to iii are fulfilled.

Condition i follows from the fact that $0 \leq w_i$ for $i = 1, 2, \ldots, n$. Since $S = \{\mathscr{O}_1, \ldots, \mathscr{O}_n\}$, the measure defined by the weights assigns probability $w_1 + w_2 + \cdots + w_n$ to S. But $w_1 + w_2 + \cdots + w_n = 1$ and, consequently, condition iii holds. Finally, for condition ii let $\mathscr{O}_1, \ldots, \mathscr{O}_k$ be the outcomes in E and $\mathscr{O}'_1, \ldots, \mathscr{O}'_m$ be the outcomes in F. Then since $E \cap F = \varnothing$, we have $E \cup F = \{\mathscr{O}_1, \ldots, \mathscr{O}_k, \mathscr{O}'_1, \ldots, \mathscr{O}'_m\}$. It follows that if the weight associated with $\mathscr{O}_i$ is w_i for $i = 1, 2, \ldots, k$ and the weight associated with $\mathscr{O}'_i$ is w'_i for $i = 1, 2, \ldots, m$, then by the definition of the probability of an event

$$\Pr[E] = w_1 + \cdots + w_k \qquad \Pr[F] = w'_1 + \cdots + w'_m$$

and
$$\Pr[E \cup F] = w_1 + \cdots + w_k + w'_1 + \cdots + w'_m$$

Combining these last three expressions, we obtain

$$\Pr[E \cup F] = (w_1 + \cdots + w_k) + (w'_1 + \cdots + w'_m) = \Pr[E] + \Pr[F]$$

which verifies that condition ii holds.

EXAMPLE 3-3

Suppose that the sample space of an experiment is $S = \{\mathcal{O}_1, \mathcal{O}_2, \mathcal{O}_3, \mathcal{O}_4, \mathcal{O}_5\}$ and that weights are assigned as follows:

$$w_1 = \tfrac{1}{8} \qquad w_2 = \tfrac{1}{8} \qquad w_3 = \tfrac{1}{4} \qquad w_4 = \tfrac{1}{3} \qquad w_5 = \tfrac{1}{6}$$

This is clearly an assignment of weights since each w_i is between 0 and 1 and $w_1 + w_2 + w_3 + w_4 + w_5 = \tfrac{1}{8} + \tfrac{1}{8} + \tfrac{1}{4} + \tfrac{1}{3} + \tfrac{1}{6} = 1$. The probability measure defined by these weights assigns probabilities to the events $E_1 = \{\mathcal{O}_1, \mathcal{O}_3, \mathcal{O}_5\}$, $E_2 = \{\mathcal{O}_2, \mathcal{O}_4\}$ and $E_3 = \{\mathcal{O}_1, \mathcal{O}_2, \mathcal{O}_3\}$. What are these probabilities?

Solution: Using the definition of the probability measure defined by an assignment of weights, we have

$$\Pr[E_1] = w_1 + w_3 + w_5 = \tfrac{1}{8} + \tfrac{1}{4} + \tfrac{1}{6} = \tfrac{13}{24}$$

$$\Pr[E_2] = w_2 + w_4 = \tfrac{1}{8} + \tfrac{1}{3} = \tfrac{11}{24}$$

and $$\Pr[E_3] = w_1 + w_2 + w_3 = \tfrac{1}{8} + \tfrac{1}{8} + \tfrac{1}{4} = \tfrac{1}{2}$$ $\square$

In assigning probabilities to events, we assign probability 1 to any event which always occurs and probability 0 to any event which never occurs. Thus, in Example 3-2 we assign probability 1 to the event that a randomly selected widget is either normal or not normal, and we assign probability 0 to the event that it is both normal and not normal. In particular, in any experiment which always has an outcome, $\Pr[\varnothing] = 0$. It is possible for there to be an event G in a sample space, $G \neq \varnothing$, and $\Pr[G] = 0$.

Definition   If E and F are events such that $\Pr[E \cap F] = 0$, then E and F are said to be *mutually exclusive*.

It follows from this definition that disjoint events are always mutually exclusive since $\Pr[E \cap F] = \Pr[\varnothing] = 0$. Moreover, for mutually exclusive events we have $\Pr[E \cup F] = \Pr[E] + \Pr[F]$. We note that in Example 3-2 the events F_1 and F_2 are mutually exclusive and the events G_1 and G_2 are mutually exclusive.

Exercises for Sec. 3-2

1. Which of the following assignments of real numbers to the outcomes in a sample space $S = \{\mathcal{O}_1, \mathcal{O}_2, \mathcal{O}_3, \mathcal{O}_4\}$ gives an assignment of weights and therefore can be used to construct a probability measure?

a. $w_1 = \frac{1}{2}$, $w_2 = \frac{1}{3}$, $w_3 = \frac{1}{4}$, $w_4 = \frac{1}{5}$

b. $w_1 = \frac{1}{4}$, $w_2 = \frac{1}{2}$, $w_3 = \frac{1}{4}$, $w_4 = 0$

c. $w_1 = \frac{1}{3}$, $w_2 = \frac{1}{2}$, $w_3 = \frac{1}{3}$, $w_4 = -\frac{1}{6}$

d. $w_1 = 1$, $w_2 = 0$, $w_3 = 0$, $w_4 = 0$

2. A sample space $S = \{\mathcal{O}_1, \mathcal{O}_2, \mathcal{O}_3, \mathcal{O}_4, \mathcal{O}_5\}$ has the associated weights $w_1 = .20$, $w_2 = .10$, $w_3 = .15$, $w_4 = .45$, and $w_5 = .10$, which define a probability measure. Determine the probability of each of the following events:

a. $\{\mathcal{O}_1, \mathcal{O}_3\}$ **b.** $\{\widetilde{\mathcal{O}_1}\}$ **c.** $\{\mathcal{O}_2\} \cup \{\mathcal{O}_3\}$

d. S **e.** $\{\mathcal{O}_1, \mathcal{O}_2, \mathcal{O}_3\} \cup \{\mathcal{O}_2, \mathcal{O}_4\}$ **f.** $\{\mathcal{O}_1\} \cap \{\mathcal{O}_4\}$

3. In Example 3-1 what is the probability of the event "plain or male"?

4. In Example 3-3 what is the probability of the event $E_1 \cup E_3$? $E_2 \cap E_3$?

5. An inspector on an assembly line of a refrigerator plant classifies each refrigerator according to the quality of its enamel. From his data the inspector assigns the probabilities listed below:

Event	Probability
E_1, too much enamel	.04
E_2, too little enamel	.12
E_3, uneven application	.09
E_4, no defects noted	.82

Assume that no refrigerator has both too much and too little enamel and that the defects associated with E_1, E_2, E_3 are the only ones of concern to the inspector. What is the probability that a randomly selected refrigerator has:

a. A paint defect?

b. A paint defect which includes an improper amount of paint?

c. A paint defect which results from the proper amount of paint but uneven application?

d. A paint defect which results from an improper amount of paint and uneven application?

6. Suppose an experiment has sample space S with outcomes $\mathcal{O}_1$, $\mathcal{O}_2$, $\mathcal{O}_3$, $\mathcal{O}_4$, $\mathcal{O}_5$ and weights $w_1 = .37$, $w_2 = .02$, $w_3 = .13$, $w_4 = .21$, and $w_5 = .27$. If $E_1 = \{\mathcal{O}_1, \mathcal{O}_2, \mathcal{O}_3, \mathcal{O}_4\}$, $E_2 = \{\mathcal{O}_1, \mathcal{O}_4\}$, and $E_3 = \{\mathcal{O}_2, \mathcal{O}_4, \mathcal{O}_5\}$, determine:

a. $\Pr[\tilde{E}_1]$ **b.** $\Pr[E_1 \cup E_2 \cup E_3]$

c. $\Pr[E_1 \cap E_2]$ **d.** $\Pr[(E_1 \cap E_2) \cup (E_2 \cap E_3)]$

7. Suppose an experiment has a sample space $S = \{\mathcal{O}_1, \mathcal{O}_2, \mathcal{O}_3, \mathcal{O}_4\}$ and events $E_1 = \{\mathcal{O}_1, \mathcal{O}_2\}$, $E_2 = \{\mathcal{O}_2, \mathcal{O}_3\}$, $E_3 = \{\mathcal{O}_3, \mathcal{O}_4\}$, and $E_4 = \{\mathcal{O}_4\}$. Give an assignment of weights to the outcomes in S in such a way that

$$\Pr[E_1] = \tfrac{3}{4} \qquad \Pr[E_2] = \tfrac{3}{8} \qquad \Pr[E_3] = \tfrac{2}{8} \qquad \Pr[E_4] = \tfrac{1}{8}$$

8. In Example 3-2 there are not enough data given to determine the probabilities of all events. For which of the following events are there not enough data to determine the probability of the event?

    a. $F_1 \cup G_1$ **b.** $F_1 \cup H$

    c. $F_2 \cap H$ **d.** $F_2 \cap G_2$

9. In Example 3-3 which pair(s) of the events E_1, E_2, and E_3 are mutually exclusive?

3-3 CONDITIONAL PROBABILITY AND INDEPENDENCE

It can happen that with a certain amount of information we would determine the probability of an event to be one number but with additional information we would determine it to be a different number.

EXAMPLE 3-4

Suppose that we have an urn containing 3 red balls, 2 blue balls, and 4 white balls (Fig. 3-2) and that a ball is selected at random and its color noted. Assuming that each ball is equally likely to be selected, i.e., each ball has probability $\frac{1}{9}$ of being selected, we conclude that

$$\Pr[\text{red ball}] = \tfrac{3}{9} \qquad \Pr[\text{blue ball}] = \tfrac{2}{9} \qquad \Pr[\text{white ball}] = \tfrac{4}{9}$$

If we have the additional information that the ball selected is not white, these probabilities change. They become

$$\Pr[\text{red ball given that ball selected is not white}] = \tfrac{3}{5}$$

$$\Pr[\text{blue ball given that ball selected is not white}] = \tfrac{2}{5}$$

$$\Pr[\text{white ball given that ball selected is not white}] = 0$$

FIGURE 3-2

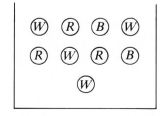

Indeed, in the latter case there are only 5 balls, 3 red and 2 blue, which satisfy the condition of being not white, and the probabilities assigned to the outcomes are those determined by the equiprobable measure with 5 possible outcomes.

□

Example 3-4 illustrates the concept of conditional probability; i.e., the probability assigned to an event (in this case drawing a red ball) with the knowledge that another event has occurred (the ball drawn is not white).

The precise definition is as follows:

Definition

Let A and B be events in the sample space of an experiment with $\Pr[B] \neq 0$. The *conditional probability of A given B*, written $\Pr[A|B]$, is

$$\Pr[A|B] = \frac{\Pr[A \cap B]}{\Pr[B]}$$

Referring back to Example 3-4, we see that the definition of the conditional probability of an event is just the probability of that event in the smaller sample space specified by the "condition" B.

In that example, if A is the event "a red ball is selected" and B is the event "a white ball is not selected," then

$$\Pr[A|B] = \frac{3/9}{5/9} = \frac{3}{5}$$

This ratio is simply the result of using the equiprobable measure in the sample space consisting of those outcomes in which the ball selected is not white.

In the general case we can illustrate the concept of conditional probability by using Venn diagrams. A sample space S and events A and B are shown in the Venn diagram of Fig. 3-3a. The probability of A is the sum of the weights of all outcomes in A. Equivalently, since the weight of all outcomes in S is 1,

FIGURE 3-3

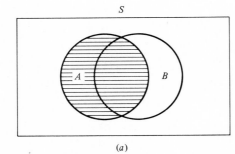

(a)

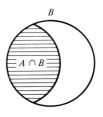

(b)

this sum is the fraction of the total weight due to outcomes in A. The conditional probability of A given B corresponds to the situation in which B represents the sample space and $Pr[A|B]$ is the weight of the outcomes in $A \cap B$ compared to the total weight of B. This is shown in Fig. 3-3b.

EXAMPLE 3-5

Let A and B be events of an experiment whose sample space is S. Suppose $Pr[A] = \frac{1}{3}$, $Pr[B] = \frac{1}{4}$, and $Pr[A \cap B] = \frac{1}{6}$. Find the conditional probability $Pr[A|B]$.

Solution:

$$Pr[A|B] = \frac{Pr[A \cap B]}{Pr[B]} = \frac{1/6}{1/4} = \frac{4}{6} = \frac{2}{3}$$

☐

EXAMPLE 3-6

An inspector on an assembly line checks the size of each widget produced by a new improved widget-making machine. The data collected yield the probabilities shown in Table 3-3. Therefore, the probability that a randomly selected widget is too big is .15. Suppose now that the inspector knows that a randomly selected widget is not the correct size; what is the probability that it is too big?

TABLE 3-3

Event	Probability
Too big (E_1)	.15
Too small (E_2)	.25
Correct size (E_3)	.60

Solution: Let A be the event that it is too big, $A = E_1$, and let B be the event that it is not the correct size, $B = E_1 \cup E_2$. Then, since $E_1 \cap E_2 = \varnothing$, $Pr[B] = Pr[E_1] + Pr[E_2] = .40$. Using the definition of conditional probability, we have

$$Pr[A|B] = \frac{Pr[A \cap B]}{Pr[B]} = \frac{.15}{.40} = .375$$

The conditional probability that it is too big given that it is not the correct size is .375.

☐

EXAMPLE 3-7

Suppose that an urn contains red balls marked 1, 2, 3, a blue ball marked 4, and white balls marked 5, 6, 7, 8. A ball is selected at random, and its color and number are noted. What is the probability that it is red? If the ball is known to have an even number, what is the probability that it is red?

Solution: Let A be the event that the ball is red, and let B be the event that the ball has an even number. Then, since $n(A) = 3$ and $n(S) = 8$, we have $\Pr[A] = \frac{3}{8}$. This gives an answer to the first question. Next, $n(B) = 4$ and $n(A \cap B) = 1$. Therefore,

$$\Pr[A|B] = \frac{\Pr[A \cap B]}{\Pr[B]} = \frac{n(A \cap B)/n(S)}{n(B)/n(S)} = \frac{1/8}{4/8} = \frac{1}{4}$$

and the answer to the second question is $\frac{1}{4}$. Clearly, in this case, $\Pr[A] \neq \Pr[A|B]$. Thus the knowledge that B has occurred does affect the probability assigned to the occurrence of A. ☐

It can also happen that knowledge that event B occurs does not affect the probability that A occurs. Events for which this is the case are said to be *independent*. It is convenient to define the concept in slightly different terms and then to connect the definition with this interpretation. First the definition.

Definition

Events A and B are said to be *independent* if

$$\Pr[A \cap B] = \Pr[A]\Pr[B]$$

The connection between this definition and the intuitive interpretation of the word independent is provided by the following observation. If A and B are independent events and $\Pr[B] \neq 0$, then

$$\Pr[A|B] = \frac{\Pr[A \cap B]}{\Pr[B]} = \frac{\Pr[A]\Pr[B]}{\Pr[B]} = \Pr[A]$$

Thus, for independent events A and B with $\Pr[B] \neq 0$ we have $\Pr[A|B] = \Pr[A]$. Hence a knowledge of the occurrence of B does not affect the probability that A occurs. Likewise, if A and B are independent and $\Pr[A] \neq 0$, then

$$\Pr[B|A] = \frac{\Pr[B \cap A]}{\Pr[A]} = \frac{\Pr[B]\Pr[A]}{\Pr[A]} = \Pr[B]$$

Thus, a knowledge of the occurrence of A does not affect the probability that B occurs. This interpretation justifies the use of the term independent. We have adopted the above definition because it is completely symmetric with respect to the events A and B.

It is important to recognize that independence (as we have defined the concept) has a precise mathematical meaning, and it may not be clear from the description of an experiment whether events are or are not independent. For instance, in Example 3-7 we have $\Pr[A] = \frac{3}{8}$, $\Pr[B] = \frac{4}{8}$, and $\Pr[A \cap B] = \frac{1}{8}$. Consequently $\Pr[A] \cdot \Pr[B] = \frac{3}{16} \neq \Pr[A \cap B]$, and we see that the events A and B are *not* independent. Note, however, that in the same example the situation changes if we consider other events. For example, if E is the event that a white ball is drawn, we have $\Pr[E] = \frac{4}{8}$, $\Pr[B] = \frac{4}{8}$, and $\Pr[E \cap B] = \frac{2}{8}$. Consequently $\Pr[E \cap B] = \Pr[E] \cdot \Pr[B]$, and the events E and B are independent.

Notice that the concepts of independence and mutual exclusiveness both relate to pairs of events A and B and to $\Pr[A \cap B]$:

If A and B are mutually exclusive, then $\Pr[A \cap B] = 0$.
If A and B are independent, then $\Pr[A \cap B] = \Pr[A] \cdot \Pr[B]$.

In general, events which are independent are not also mutually exclusive. In fact, two events A and B which are independent are also mutually exclusive only if $\Pr[A] = 0$ or $\Pr[B] = 0$ (Exercise 10).

EXAMPLE 3-8

Suppose that a fair die is rolled, the uppermost number is noted, and a card is drawn at random from a completely shuffled deck and its color noted. Assume that any event which involves only the die is independent of any event which involves only the cards. That is, if A is any event in which every outcome has the same color card and B is any event in which every outcome has the same number on the die, then A and B are independent. What is the probability of the event in which a number less than 3 occurs on the die and a black card is drawn?

Solution: We represent the sample space for this experiment by a set of ordered pairs. The first entry in each pair is the number obtained on the die, and the second entry is the color of the card which is drawn. Since there are 6 numbers which can occur on the die and 2 colors which can occur for the card drawn, there are 12 ordered pairs in our sample space (we have used the multiplication principle of Sec. 2-4). Next, since the die is fair and the card is drawn at random, and since we assume the results on the die and card are independent, we assign each entry in the sample space a weight of $\frac{1}{12}$; that is we use the equiprobable measure.

Let E be the event "a number less than 3 is obtained on the die," and let F be the event "a black card is drawn." Then

$$E = \{(1,B), (2,B), (1,R), (2,R)\},$$
$$F = \{(1,B), (2,B), (3,B), (4,B), (5,B), (6,B)\}$$

where the symbols B and R denote the draw of a black and red card, respectively. Thus, using the equiprobable measure, we have

$$\Pr[E] = \tfrac{4}{12} = \tfrac{1}{3} \quad \text{and} \quad \Pr[F] = \tfrac{6}{12} = \tfrac{1}{2}.$$

The event of interest in this example is $E \cap F$; "both a number less than 3 occurs and a black card is drawn." We have $E \cap F = \{(1,B), (2,B)\}$, and hence, $\Pr[E \cap F] = \tfrac{2}{12} = \tfrac{1}{6}$. We note that $\Pr[E \cap F] = \tfrac{1}{6} = \tfrac{1}{3} \cdot \tfrac{1}{2} = \Pr[E] \cdot \Pr[F]$, and hence E and F are indeed independent events. $\qquad\square$

FIGURE 3-4

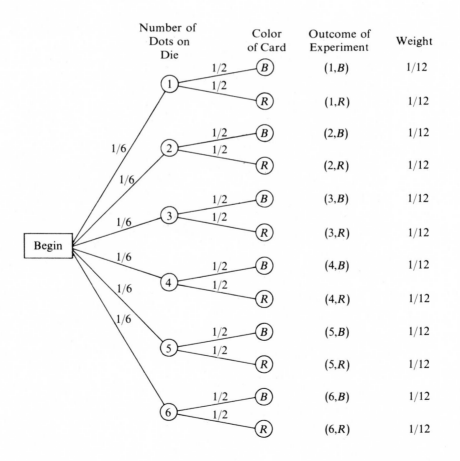

The experiment of Example 3-8 can be viewed as a multistage experiment, and hence it can be represented by a tree diagram (see Sec. 2-4). Such a tree diagram is shown in Fig. 3-4, where the number on a branch is the probability of the result which corresponds to that branch. These numbers are called *branch weights*. Since the branches at each fork correspond to all possible results at that stage of the experiment, the branch weights sum to 1 at each fork in the tree.

Notice that in Fig. 3-4, the probabilities assigned to the colors (*B* and *R*) are the same for every result on the die. This reflects the assumption that the results of rolling the die and drawing the card are independent events. In similar experiments where the results at the various stages are not independent, the branch weights must reflect this fact by being conditional probabilities. Examples of this sort are the topic of Sec. 3-4, where we shall develop a general technique for assigning weights to trees. We shall preserve a key property of the tree in Fig. 3-4, namely that the probability of an outcome is always the product of the branch weights on the path which corresponds to that outcome.

EXAMPLE 3-9

The Mul T. Decibel Music Company purchases its audio tapes from Wraparound, Inc. and its speakers from Tone Down, Inc. Approximately 2 percent of the audio tapes are defective, and the failure rate of speakers is known to be 3 out of 500. If the results of the tests of tapes and speakers are assumed to be independent, what is the probability that a random customer who purchases both an audio tape and a speaker receives two defective items?

Solution: The probability that a randomly selected audio tape will be defective is .02, and the probability of failure of a randomly selected speaker is .006. Since these events are assumed to be independent, we have

$$\text{Pr[defective tape and defective speaker]} = .02 \times .006 = .00012 \qquad \square$$

Exercises for Sec. 3-3

1. Let A and B be events such that $\Pr[A] = .8$, $\Pr[B] = .5$, and $\Pr[A \cap B] = .35$. Determine $\Pr[A|B]$ and $\Pr[B|A]$.
2. In Example 3-4, what is $\Pr[\text{red ball}|\text{not a blue ball}]$?
3. Two fair dice are rolled, and the numbers on the uppermost faces are noted.
   a. What is the probability that exactly one die shows a 4 given that the sum of the numbers is 7?
   b. What is the probability that the sum of the numbers is 7 given that exactly one die shows a 4?
   c. What is the probability that the sum of the numbers is 7 given that at least one die shows a 4?

4. In Example 3-5, find $\Pr[B|A]$. Are A and B independent events in this example?

5. A package of 10 audio tapes contains 2 defective ones. If 2 tapes are selected at random and tested, one after the other, what is the probability that the second is defective?

6. Let A and B be events such that $\Pr[A \cup B] = .8$ and $\Pr[A] = .6$. What is $\Pr[B]$ if:
   a. A and B are independent?
   b. A and B are mutually exclusive?

7. A fair coin is flipped 3 times. What is the probability that there are 2 heads and 1 tail? What is the probability that there are 2 heads and 1 tail given that both heads and tails occur?

8. There are two boxes labeled I and II. Box I contains urns A, B, and C, and box II contains urns D and E. Colored balls are distributed in the urns as follows:

   A: 2 white and 1 red
   B: 2 red
   C: 1 white and 1 red
   D: 2 white and 1 red
   E: 1 white and 1 red

   An experiment consists of selecting a box, an urn from that box, and a ball from that urn. Draw a tree diagram for this experiment and assign branch weights to the branches of the tree. Assume all selections are random.

9. In Example 3-7 are the following two events independent?
   F, "a blue ball is drawn."
   B, "an even-numbered ball is drawn."

10. Let A and B be independent events. If A and B are also mutually exclusive, show that either $\Pr[A] = 0$ or $\Pr[B] = 0$.

11. Let E and F be events in a sample space S. Suppose $\Pr[E] = \frac{1}{2}$, $\Pr[F] = \frac{3}{8}$, and $\Pr[\widetilde{E \cup F}] = \frac{1}{4}$. Find $\Pr[E|F]$.

12. Using the data in Exercise 11, decide whether the events $\tilde{E}$ and $\tilde{F}$ are independent.

13. The problem in Example 3-8 can also be solved by considering the outcomes of the experiment to be ordered pairs where the first entry in the pair is "less than 3" or "not less than 3" and the second entry is "black card" or "not black card." Work the problem using this format and draw the tree diagram which corresponds to the tree in Fig. 3-4.

3-4 STOCHASTIC PROCESSES AND TREE MEASURES

Many experiments (the term is used as described in Sec. 2-2) consist of sub-experiments which are performed sequentially.

Definition   A sequence of experiments is a *stochastic process*.

The tree diagrams introduced in Chap. 2 are especially helpful in describing and analyzing stochastic processes. This was illustrated in Fig. 3-4 and the accompanying discussion. As another example of how tree diagrams are used we consider a three-stage process, i.e., a process with three experiments, whose tree diagram is shown in Fig. 3-5. In addition to indicating the possible outcomes of the subexperiments and of the entire experiment, the tree diagram displays the probabilities of different outcomes of the subexperiments. The numbers on the branches are the weights associated with outcomes of the stage of the experiment. Thus the weights on the branches at the first fork are $\Pr[I]$ and $\Pr[II]$ for outcomes I and II, respectively. We introduced the term branch weights in the last section to refer to these weights. The fork at the end of the branch associated with outcome I of the first experiment has branches whose weights are conditional probabilities, the condition being that outcome I has occurred. These branch weights are $\Pr[A|I]$, $\Pr[B|I]$, and $\Pr[C|I]$ for outcomes A, B, and C, respectively. We note that $\Pr[A|I] + \Pr[B|I] + \Pr[C|I] = 1$, since A, B, and C are all the elements in the sample space for the experiment which is performed after outcome I occurs. Likewise, the branch weights for the branches associated

FIGURE 3-5

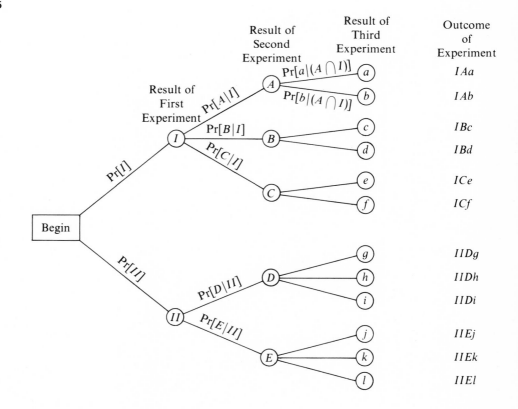

TABLE 3-4

Path	Path Weight
$I A a$	$\Pr[I]\Pr[A\,|\,I]\Pr[a\,|\,(A \text{ and } I)]$
$I A b$	$\Pr[I]\Pr[A\,|\,I]\Pr[b\,|\,(A \text{ and } I)]$
$I B c$	$\Pr[I]\Pr[B\,|\,I]\Pr[c\,|\,(B \text{ and } I)]$
$I B d$	$\Pr[I]\Pr[B\,|\,I]\Pr[d\,|\,(B \text{ and } I)]$

with outcome II of the first experiment are $\Pr[D\,|\,II]$ and $\Pr[E\,|\,II]$ for outcomes D and E, respectively. As before, $\Pr[D\,|\,II] + \Pr[E\,|\,II] = 1$. Also, we note that in a specific example it may happen that either outcome D or outcome E or both are the same as one or more of the outcomes A, B, C. Finally, the branch weights for the branches associated with outcome A after outcome I are $\Pr[a\,|\,(A \text{ and } I)]$ and $\Pr[b\,|\,(A \text{ and } I)]$ for outcomes a and b, respectively. Thus the *path weights* (defined as the product of the probabilities on the branches of the path) associated with the paths with initial outcome I and second outcomes A and B are given in Table 3-4. The path weights associated with the remaining paths can be determined similarly. Recall that we are viewing these three experiments as subexperiments of a larger experiment whose outcomes are determined by the results of each of the three subexperiments. Using the fact that the sum of the branch weights of the branches originating at each fork is 1, we can show that the path weights are an acceptable set of weights for the experiment. Thus the path weights can be used to define a probability measure (called the *tree measure*) for the experiment.

The concepts of branch weights, path weights, and tree measures are important in using tree diagrams to solve probability problems. We summarize our earlier comments in a definition.

Definition

We say that *branch weights* are assigned in a tree diagram if there is an assignment of weights to the set of branches (outcomes) at each fork of the tree. The *path weights* of a tree diagram are the weights assigned to the paths by assigning to each path the product of the branch weights of all branches in that path. The *tree measure* of a multistage experiment is the probability measure which assigns to each outcome the weight of the path associated with that outcome.

EXAMPLE 3-10

A specific instance of an experiment whose tree diagram is that of Fig. 3-5 can be described with the boxes, urns, and balls shown in Fig. 3-6 as follows: Select a box,

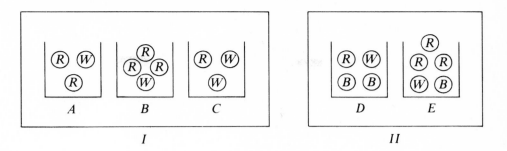

FIGURE 3-6

then select an urn, then select a ball from the urn. If random selections are made in each case, we have, for example,

$$\Pr[I] = \tfrac{1}{2} \qquad \Pr[A\,|\,I] = \tfrac{1}{3} \qquad \Pr[R\,|\,(A \text{ and } I)] = \tfrac{2}{3} \qquad \Pr[W\,|\,(A \text{ and } I)] = \tfrac{1}{3}$$
$$\Pr[B\,|\,I] = \tfrac{1}{3} \qquad \Pr[R\,|\,(B \text{ and } I)] = \tfrac{3}{4} \qquad \Pr[W\,|\,(B \text{ and } I)] = \tfrac{1}{4}$$

Therefore, if we identify outcomes a and c with selecting a red ball (R) and outcomes b and d with selecting a white ball (W), we have the path weights of Table 3-4 equal to the values shown in Table 3-5.

TABLE 3-5

Path	Path Weight
IAR	$\dfrac{1}{2} \cdot \dfrac{1}{3} \cdot \dfrac{2}{3} = \dfrac{1}{9}$
IAW	$\dfrac{1}{2} \cdot \dfrac{1}{3} \cdot \dfrac{1}{3} = \dfrac{1}{18}$
IBR	$\dfrac{1}{2} \cdot \dfrac{1}{3} \cdot \dfrac{3}{4} = \dfrac{1}{8}$
IBW	$\dfrac{1}{2} \cdot \dfrac{1}{3} \cdot \dfrac{1}{4} = \dfrac{1}{24}$

□

In Example 3-10 the probabilities of the two outcomes of the third experiment, the probability of selecting a red ball and the probability of selecting a white ball, are conditional probabilities which are different for different outcomes of the first two experiments. The dependence of probabilities of various outcomes at one stage upon outcomes at earlier stages is generally to be expected. However, some important stochastic processes do not exhibit this dependence (Sec. 3-5).

EXAMPLE 3-11

In the experiment described in Example 3-10, find the probabilities of the following events:

E: a red ball is drawn.
F: a white ball is drawn.
G: a blue ball is drawn.

Solution: The complete tree for this experiment is shown in Fig. 3-7. The path weights shown at the right of this figure can be used to compute the probabilities of events E, F, and G.

The event E is the set of outcomes

$$E = \{IAR, IBR, ICR, IIDR, IIER\}$$

FIGURE 3-7

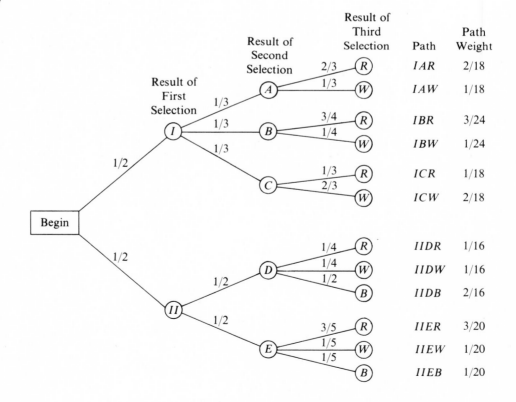

Using the definition of the probability assigned to an event by the tree measure, we have

$$\Pr[E] = \tfrac{2}{18} + \tfrac{3}{24} + \tfrac{1}{18} + \tfrac{1}{16} + \tfrac{3}{20} = \tfrac{121}{240}$$

Similarly,

$$\Pr[F] = \tfrac{1}{18} + \tfrac{1}{24} + \tfrac{2}{18} + \tfrac{1}{16} + \tfrac{1}{20} = \tfrac{77}{240}$$

and

$$\Pr[G] = \tfrac{2}{16} + \tfrac{1}{20} = \tfrac{42}{240}$$

Note that since the events E, F, and G form a partition of the sample space for this experiment, we have

$$\Pr[E] + \Pr[F] + \Pr[G] = \tfrac{121}{240} + \tfrac{77}{240} + \tfrac{42}{240} = 1 \qquad \square$$

EXAMPLE 3-12

Sales representatives for the Downjim Drug Co. are assigned company cars each time they make a trip. There are 2 motor-pool locations, uptown and downtown. The downtown motor pool has Fords and Plymouths in equal numbers. The uptown motor pool has twice as many Fords as Plymouths. One-half the Fords are air-conditioned, and three-quarters of the Plymouths are air-conditioned. If a sales representative is equally likely to be sent to either motor pool and equally likely to be assigned any car in that pool, what is the probability of being assigned an air-conditioned car?

Solution: The tree diagram, complete with branch weights, and the paths and path weights are shown in Fig. 3-8. The following abbreviations are used:

U: Uptown motor pool
D: Downtown motor pool
F: Ford
P: Plymouth
A: Air-conditioned
N: Not air-conditioned

Notice that the probabilities of the outcomes of the third part of the experiment, that is, $\Pr[A]$ and $\Pr[N]$, depend only on the outcome of the second part of the experiment, whether a Ford or a Plymouth is assigned, and not at all on the outcome of the first part of the experiment, which motor pool is used.

Using the path weights from Fig. 3-8, we can compute the probability of the event that an air-conditioned car is assigned. We have

$$\Pr[A] = \tfrac{1}{6} + \tfrac{1}{8} + \tfrac{1}{8} + \tfrac{3}{16} = \tfrac{29}{48}$$

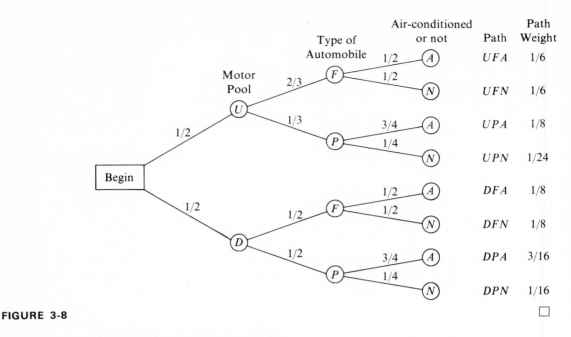

FIGURE 3-8

Exercises for Sec. 3-4

1. In the experiment described in Example 3-10 find the following probabilities:
   a. $\Pr[A]$ **b.** $\Pr[B]$ **c.** $\Pr[A \text{ and } R]$
   d. $\Pr[B \text{ and } W]$ **e.** $\Pr[R \mid A]$ **f.** $\Pr[W \mid B]$

2. Draw a tree diagram for the experiment of flipping a coin which is weighted so heads are twice as likely as tails and then rolling a fair die. Show the branch weights on the tree and compute the path weights. Assume that results of tossing the coin are independent of results of rolling the die.

3. Using the data of Exercise 2, compute the probability of obtaining a tail on the coin and an even number on the die.

4. Using the data of Example 3-12, find the probability that a sales representative is assigned either an air-conditioned Ford or a non-air-conditioned Plymouth.

5. An experiment consists of flipping a coin 3 times. The coin is weighted so that heads are 3 times as likely as tails. Draw a tree diagram to represent the outcomes of this experiment and show the branch weights and path weights (assume the results of the individual tosses of the coin are independent). What is the probability of the event "either 2 heads and 1 tail appear or 2 tails and 1 head appear"?

6. A farmer has 2 bags of corn seed, hybrid *I* and hybrid *II*. Seed of type *I* will germinate with probability .95, and after germination it will become a mature corn plant with probability .8. Seed of type *II* will germinate with probability .9, and after germination it will grow into a mature corn plant with probability .85. Draw a tree diagram to represent the experiment of randomly selecting a type of seed, planting it, and then observing whether it germinates and grows into a mature corn plant. Show all branch weights and path weights for your tree.

7. Using the data of Exercise 6, determine the probability that a seed selected at random will germinate but *not* grow into a mature plant.

8. Consider 2 urns each containing 3 colored balls. Urn *I* contains 2 red and 1 white ball, and urn *II* contains 1 red, 1 white, and 1 blue ball. An experiment consists of randomly drawing a ball from urn *I*, noting its color, placing the ball in urn *II*, and then randomly drawing a ball from urn *II* and noting its color. Draw a tree diagram to represent the outcomes of this experiment, i.e., the ordered pair of colors obtained, and attach the branch weights and path weights to the tree. What is the probability of obtaining 2 red balls?

9. In the experiment described in Exercise 8 determine the conditional probabilities Pr[2 red balls|red on first draw] and Pr[2 white balls|white on first draw].

3-5 INDEPENDENT TRIALS WITH TWO OUTCOMES

In applications of probability theory one frequently considers experiments which consist of a number of independent trials of another experiment, each trial having the same two possible outcomes and the same probabilities for these outcomes. These experiments are known as *binomial* or *Bernoulli* processes* and the individual trials as binomial or Bernoulli trials. These processes are special cases of the stochastic processes considered in Sec. 3-4. In our discussion of binomial processes we shall refer to the two possible outcomes of the individual trials of the basic experiment as *success* and *failure*. As we shall see in the examples and exercises, there are many situations to which this model is applicable.

EXAMPLE 3-13

An experiment consists of 2 repetitions of a binomial trial with the probability of success equal $\frac{1}{3}$ and the probability of failure equal $\frac{2}{3}$. What is the probability that exactly 1 success is obtained in the 2 trials?

Solution: We solve this problem with the aid of the tree diagram for the experiment (Fig. 3-9). The assumption of independence is used in assigning the branch weights. The probabilities of the two outcomes on the second trial are independent of the outcome of the first trial, and they are the same as the probabilities of the two outcomes on the first trial.

The outcomes of the experiment which are of interest to us are *sf* and *fs*. Using the path weights, we conclude that

$$\Pr[\text{exactly 1 success}] = \Pr[\{sf, fs\}] = \tfrac{2}{9} + \tfrac{2}{9} = \tfrac{4}{9} \qquad \square$$

---

* After Jakob Bernoulli (1654–1705), a Swiss mathematician and one of the founders of probability theory.

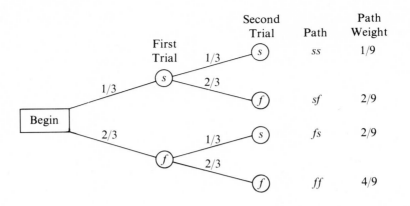

FIGURE 3-9

An example of a specific experiment which can be modeled by Example 3-13 is the following. A fair die is rolled twice and the number of dots on the side which lands uppermost is noted each time. On a single roll of the die, the result is said to be a success if 1 or 6 dots land uppermost and a failure otherwise. Using a symmetry argument, we conclude that on a single roll of the die $\Pr[s] = \frac{1}{3}$ and $\Pr[f] = \frac{2}{3}$. Two rolls of the die can be modeled as two repetitions of a binomial trial, and this is the setting of Example 3-13.

Throughout this section we shall suppose that the probability of success in a single binomial trial is p and the probability of failure is $q = 1 - p$. Let us begin by computing the probability that we have exactly r successes in n trials. We can write each of the elements of the sample space S of the experiment consisting of n trials as a list of n letters, each letter an s or an f. For example, three elements of S are

$$\underbrace{ff \ldots f}_{n \text{ times}} \underbrace{ss \ldots s}_{n \text{ times}} \quad \text{and} \quad \underbrace{ss \ldots s}_{k \text{ times}} \underbrace{ff \ldots f}_{n - k \text{ times}}$$

It may be helpful to consider a specific experiment.

EXAMPLE 3-14

Draw the tree diagram and determine the path weights for a Bernoulli experiment consisting of 4 trials.

Solution: The tree diagram for the experiment consisting of 4 Bernoulli trials is shown in Fig. 3-10. The branch weights result from the fundamental assumption that the probability of s is always p and the probability of f is always q. In

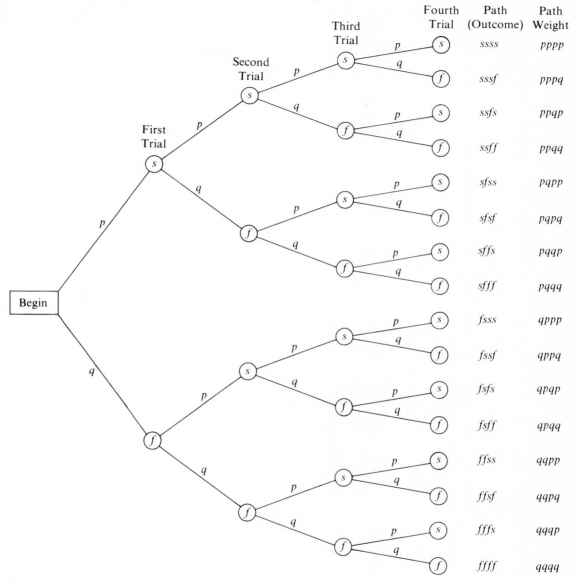

FIGURE 3-10

particular, these probabilities do not depend on which trial in the sequence is being considered.

It is clear from Fig. 3-10 that for the special case of 4 trials the path weight for any outcome which has r successes and $4 - r$ failures is $p^r q^{4-r}$. This follows because the corresponding path has r branches with weight p and $4 - r$ branches with weight q. In order to determine the probability that the experiment will result in an outcome with r successes and $4 - r$ failures, we must count the number of paths in the tree which result in this outcome. For example, in Fig. 3-10 there are 6 paths which result in 2 successes and 2 failures.   □

The argument in the general case of an experiment consisting of n Bernoulli trials is similar to that of Example 3-14. The path weight for any outcome which has r successes and $n - r$ failures is $p^r q^{n-r}$. Again, this results from the observation that the path corresponding to this outcome must have r branches with weight p and $n - r$ branches with weight q. To compute the probability that the experiment has an outcome consisting of r successes and $n - r$ failures it remains only to determine the number of such paths in the tree.

If we view the results of n Bernoulli trials as a sequence of symbols s and f in n boxes, our problem is that of determining the number of different ways that r symbols s can be placed in n boxes. From Chap. 2 we know that we can select r boxes from n boxes in $\binom{n}{r}$ ways. Thus we have derived the following result.

The probability of obtaining exactly r successes in an experiment consisting of n independent trials with success probability p and failure probability $q = 1 - p$ is

$$\binom{n}{r} p^r q^{n-r} \tag{3-5}$$

EXAMPLE 3-15

Solve the problem posed in Example 3-13 by using Eq. (3-5).

Solution: Example 3-13 was concerned with an experiment consisting of two independent trials with success probability $p = \frac{1}{3}$. Therefore $n = 2$, $p = \frac{1}{3}$, and $q = \frac{2}{3}$. The problem asks for the probability of exactly 1 success. Therefore $r = 1$ in Eq. (3-5). We have

$$\Pr[\text{exactly 1 success}] = \binom{2}{1}\left(\frac{1}{3}\right)^1 \left(\frac{2}{3}\right)^{2-1}$$

$$= 2 \cdot \frac{1}{3} \cdot \frac{2}{3} = \frac{4}{9} \qquad \square$$

EXAMPLE 3-16

The Mul T. Decibel Music Company received a shipment of audio tapes which the manufacturer had guaranteed to contain no more than 10 percent defective tapes. A sample of 10 tapes was tested, and 4 were found to be defective. If, in fact, 10 percent of the entire shipment were defective, what is the probability that the test would lead to these results?

Solution: We model the test using a binomial experiment. A success will be defined as the selection of an acceptable tape and a failure as the selection of a defective tape. The data give $p = .9$ and $q = .1$. Since 10 tapes are tested, $n = 10$, and since there are 6 successes, $r = 6$. Using Eq. (3-5), we conclude that the probability of testing 10 tapes and finding 4 defective ones is

$$\binom{10}{6}(.9)^6(.1)^4 = 210 \times .531441 \times .0001$$

or approximately .01. One might reasonably conclude that the manufacturer's claim is suspect. The matter of interpreting this result will be considered in more detail in Chap. 4. □

EXAMPLE 3-17

A sales representative estimates that sales will result from one-fourth of the calls made on a certain company. If the outcomes of repeated calls are independent, what is the probability that at least 2 sales will result from 5 calls?

Solution: The desired probability is the sum of the probabilities that the sales representative will make exactly 2 sales, exactly 3 sales, exactly 4 sales and exactly 5 sales. That is,

$$\text{Pr[at least 2 sales]} = \text{Pr[2 sales]} + \text{Pr[3 sales]} + \text{Pr[4 sales]} + \text{Pr[5 sales]}$$

This follows from property 2 of a probability measure (Sec. 3-2) since the events of 2, 3, 4, and 5 sales are obviously disjoint. Therefore, to answer the question we need to evaluate each of the expressions $\text{Pr}[k \text{ sales}]$ for $k = 2, 3, 4,$ and 5. We model this as an independent trials process and define a success to be making a sale. The data of the problem give $p = \frac{1}{4}$. Since the sales representative makes 5 calls, $n = 5$. Using Eq. (3-5), we have

$$\Pr[\text{2 sales}] = \binom{5}{2}\left(\frac{1}{4}\right)^2\left(\frac{3}{4}\right)^3 = \frac{10(27)}{4^5} = \frac{270}{1024}$$

$$\Pr[\text{3 sales}] = \binom{5}{3}\left(\frac{1}{4}\right)^3\left(\frac{3}{4}\right)^2 = \frac{10(9)}{4^5} = \frac{90}{1024}$$

$$\Pr[\text{4 sales}] = \binom{5}{4}\left(\frac{1}{4}\right)^4\left(\frac{3}{4}\right)^1 = \frac{5(3)}{4^5} = \frac{15}{1024}$$

$$\Pr[\text{5 sales}] = \binom{5}{5}\left(\frac{1}{4}\right)^5\left(\frac{3}{4}\right)^0 = \frac{1}{4^5} = \frac{1}{1024}$$

and consequently

$$\Pr[\text{at least 2 sales}] = \frac{270}{1024} + \frac{90}{1024} + \frac{15}{1024} + \frac{1}{1024} = \frac{376}{1024} \qquad \square$$

In this example it would have been simpler to compute the probability of the complementary event "fewer than 2 sales" and subtract the result from 1. That is,

$$\Pr[\text{at least 2 sales}] = 1 - \Pr[\text{fewer than 2 sales}]$$

The reader is asked to solve the problem in this way in Exercise 3.

EXAMPLE 3-18

In Seaside City approximately seven-tenths of those employed have jobs associated with the fishing industry. A state legislator wishes to conduct an opinion poll among people whose employment is related to fishing. The legislator would like to have opinions from at least 3 such people. If the legislator's technique is to make random telephone calls, how many calls must be made for the probability of reaching 3 people with jobs in the fishing industry to be at least .9?

Solution: We solve the problem by assuming that the telephone solicitation is a Bernoulli process. In particular we assume that seven-tenths of the people called have jobs in the fishing industry. We define a success to be a call placed to such a person. The precise mathematical question is the following. Find n such that

$$\Pr[\text{3 or more successes in } n \text{ trials}] \geq .9$$

We proceed by using a "guess and test" strategy. Clearly there must be at least 3 calls.

If $n = 3$, then

$$\Pr[3 \text{ successes in 3 trials}] = \binom{3}{3}(.7)^3(.3)^0 = .343$$

This is much too small, and consequently more than 3 calls are required.
If $n = 4$, then

$$\Pr[3 \text{ or more successes in 4 trials}] =$$
$$\binom{4}{3}(.7)^3(.3)^1 + \binom{4}{4}(.7)^4(.3)^0 = .412 + .240 = .652$$

If $n = 5$, then

$$\Pr[3 \text{ or more successes in 5 trials}] =$$
$$\binom{5}{3}(.7)^3(.3)^2 + \binom{5}{4}(.7)^4(.3)^1 + \binom{5}{5}(.7)^5(.3)^0 = .309 + .360 + .168 = .837$$

If $n = 6$, then

$$\Pr[3 \text{ or more successes in 6 trials}] = \binom{6}{3}(.7)^3(.3)^3 + \binom{6}{4}(.7)^4(.3)^2$$
$$+ \binom{6}{5}(.7)^5(.3) + \binom{6}{6}(.7)^6(.3)^0$$
$$= .185 + .324 + .303 + .118 = .930$$

We conclude that the legislator should plan on making at least 6 calls in order to be sure that the probability of reaching 3 or more individuals with jobs in the fishing industry is at least .9. $\qquad\square$

Exercises for Sec. 3-5

1. In independent-trials experiments compute the following probabilities:
   a. 3 successes in 6 trials with $p = \frac{1}{4}$
   b. 2 successes in 5 trials with $p = .3$
   c. 4 successes in 7 trials with $q = .2$

2. In independent-trials experiments compute the following probabilities:
   a. At least 5 successes in 7 trials with $p = \frac{1}{2}$
   b. At most 2 successes in 5 trials with $p = .3$
   c. No more than 3 successes in 4 trials with $q = .4$

3. Solve the problem in Example 3-17 by computing the probability of the complementary event and subtracting this number from 1.

4. An unfair coin lands heads uppermost with probability $\frac{2}{3}$. The coin is tossed 5 times.
   a. What is the probability of exactly 3 heads?
   b. What is the probability of at most 3 heads?

5. In Example 3-16 suppose that the manufacturer sent a shipment which had 20 percent defective tapes. What is the probability of obtaining 4 defectives in a test of 10?

6. A fair coin is flipped 6 times. Which is greater, the probability that heads and tails divide 3 and 3 or the probability that they divide 4 and 2?

7. A sales representative estimates that sales result on one-fourth the calls made. How many calls must be made for the probability of at least 1 sale to exceed .5?

8. In a 10-question true-false test, what is the probability of answering at least 7 questions correctly just by guessing?

9. A deepsea fisherman estimates that if he uses 4 lines, the probability of making a catch on any line is .7; if he uses 5 lines, the probability of making a catch on any line is .6, and if he uses 6 lines, the probability of making a catch on any line is .5. If his objective is to catch at least 4 fish, then how many lines should he use to maximize the probability of achieving his goal?

10. A box contains 3 defective widgets and 7 acceptable widgets. Three inspectors sample widgets from the box in sequence as follows: inspector 1 selects a widget from the box at random, tests it, *and returns it to the box*; inspector 2 does the same, followed by inspector 3 using the same method. What is the probability that at least 1 inspector will select a defective widget?

11. In a large city 50 percent of the voters consider themselves to be Democrats, 25 percent consider themselves to be independents, and 25 percent consider themselves to be Republicans. A poll taker asks 10 people at random for their party affiliation. What is the probability that at least 7 are Democrats?

3-6 BAYES PROBABILITIES

One is often confronted with probability problems which at first glance appear to be quite different from those considered in the earlier sections of this chapter. A particular problem of this sort is one in which the information provided makes it easy to determine $\Pr[A|B]$, but $\Pr[B|A]$ is required. As before, we approach these problems both through tree diagrams and formulas. We illustrate the ideas with examples.

EXAMPLE 3-19

The Mul T. Decibel Music Company purchases audio tapes from 2 suppliers; 80 percent of their stock comes from supplier *I*, and 20 percent comes from supplier *II*. It is estimated that 5 percent of the tapes furnished by supplier *I* are defective and that 10 percent of the tapes provided by supplier *II* are defective. If a tape is selected at random from the stock at the Music Company, tested, and found to be defective, what is the probability that it was provided by supplier *I*?

Solution 1: One way to approach this question is to construct a tree diagram and tree measure for the experiment of selecting at random an audio tape from the stock of Mul T. Decibel Music Company. Let us label the first branch with

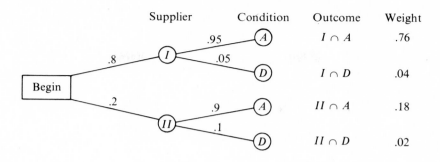

FIGURE 3-11

the supplier, *I* or *II*, and the second branch with *A* or *D* according as the tape is acceptable or defective. The data provide the weights shown in Fig. 3-11. As before, each path weight shown at the right corresponds to the path through the tree terminating immediately adjacent to that weight. The notation used to denote the paths in this tree (outcomes) is slightly different from that used previously. Thus, we use *I* ∩ *A* instead of *IA* to denote the path through *I* and *A*. We make this modification to emphasize that this is the outcome corresponding to a tape which is from supplier *I* and which is acceptable. We see from the tree diagram, for example, that .76 is the weight assigned to path *I* ∩ *A*. Since .04 corresponds to the path *I* ∩ *D* and .02 corresponds to the path *II* ∩ *D*, the total weight of paths associated with defective tapes is .06. Likewise, the total weight of paths associated with acceptable tapes is .94. Thus, the probabilities of selecting at random a defective tape and an acceptable tape are .06 and .94, respectively.

We can now answer the question posed above. We have

$$\Pr[I\,|\,D] = \frac{\Pr[I \cap D]}{\Pr[D]} = \frac{.04}{.06} = \frac{2}{3}$$

To help understand the process used to solve this problem, it is useful to construct a new tree diagram in which the first branch is labeled *A* or *D* and each second branch is labeled *I* or *II*. By what we have just said, the branch weights on the first branches are .06 on *D* and .94 on *A*. Since the total path weights must remain the same (after all the experiment does not change simply because we are drawing the tree diagram differently), we have the information shown in Fig. 3-12. We can now compute the remaining branch weights, conditional probabilities, by using the fact that the product of the weights of the branches in a path must equal the path weight. Therefore, we have

$$.76 = .94 \times \Pr[I\,|\,A] \qquad .18 = .94 \times \Pr[II\,|\,A]$$
$$.04 = .06 \times \Pr[I\,|\,D] \qquad .02 = .06 \times \Pr[II\,|\,D]$$

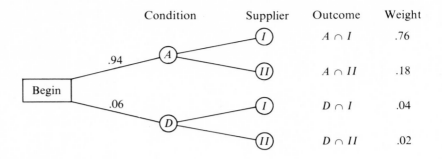

FIGURE 3-12

from which we determine the branch weights to be

$$\Pr[I|A] = \tfrac{38}{47} \qquad \Pr[II|A] = \tfrac{9}{47} \qquad \Pr[I|D] = \tfrac{2}{3} \qquad \Pr[II|D] = \tfrac{1}{3}$$

The complete tree diagram for this analysis of the experiment is shown in Fig. 3-13. As noted above, the answer to our initial question is that the probability that a defective tape was provided by supplier I is $\Pr[I|D] = \tfrac{2}{3}$.

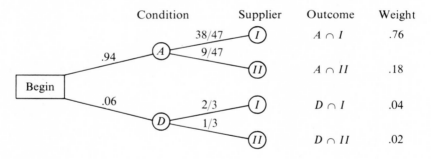

FIGURE 3-13

Solution 2:   There is another way to approach the problem which rests on a formula derived from the definition of conditional probability. The conditional probability we wish to determine is $\Pr[I|D]$. By definition this is equal to $\Pr[I \cap D]/\Pr[D]$ provided $\Pr[D] \neq 0$. To evaluate this ratio we need $\Pr[I \cap D]$ and $\Pr[D]$. Another expression for $\Pr[I \cap D]$, namely $\Pr[D|I]\Pr[I]$, involves only known quantities. It remains to determine $\Pr[D]$.

Since each defective tape originates either with supplier I (in which case it is denoted $D \cap I$) or with supplier II (in which case it is denoted $D \cap II$), and since no tapes come from both suppliers, we have

$$\Pr[D] = \Pr[D \cap I] + \Pr[D \cap II]$$

Again using the definition of conditional probability, we have

$$\Pr[D \cap I] = \Pr[D|I]\Pr[I] \qquad \Pr[D \cap II] = \Pr[D|II]\Pr[II]$$

Therefore, we can write $\Pr[D]$ in terms of known quantities:

$$\Pr[D] = \Pr[D|I]\Pr[I] + \Pr[D|II]\Pr[II]$$

Putting the two steps together, we have finally

$$\Pr[I|D] = \frac{\Pr[D|I]\Pr[I]}{\Pr[D|I]\Pr[I] + \Pr[D|II]\Pr[II]}$$

Using the data given in the problem, we find

$$\Pr[I|D] = \frac{.05 \times .80}{.05 \times .80 + .10 \times .20} = \frac{.04}{.06} = \frac{2}{3}$$

Conditional probabilities computed in either of the ways illustrated in Example 3-19 are known as *Bayes* probabilities*. The formula derived in solution 2 of that example is a general one, and it is an example of Bayes' formula.

Bayes' Formula

Let S be the sample space of an experiment, and suppose that S is partitioned into subsets $S_1, S_2, \ldots, S_k$, such that $\Pr[S_i] > 0$ for $i = 1, 2, \ldots, k$. If A is any event such that $\Pr[A] > 0$, then

$$\Pr[S_i|A] = \frac{\Pr[A|S_i]\Pr[S_i]}{\Pr[A|S_1]\Pr[S_1] + \cdots + \Pr[A|S_k]\Pr[S_k]} \qquad (3\text{-}6)$$

In solution 2 to Example 3-19 the sample space (set of all tapes) was partitioned into those purchased from supplier II and those purchased from supplier II.

The method of computing Bayes probabilities with two trees is most useful when several such probabilities are to be computed. If a single conditional probability is needed, it is usually simpler to solve directly for the one probability needed and to avoid drawing the second tree.

---

* Thomas Bayes (1702–1761) was one of the early investigators of probability theory.

EXAMPLE 3-20

A faculty member at Big State University regularly teaches an evening course in a neighboring city. Each time he makes the trip he is randomly assigned an automobile from the university motor pool. The motor pool consists of 50 percent Chevrolets, 30 percent Fords, and 20 percent Plymouths. Some of the cars are air-conditioned and some are not: 60 percent of the Chevrolets, 50 percent of the Fords, and 70 percent of the Plymouths are air-conditioned. One afternoon the faculty member is assigned an air-conditioned car. What is the probability that it is a Chevrolet?

Again we shall illustrate two methods of solving the problem. Of course the methods are not actually different; the formula simply expresses in a concise way the operations which are being carried out on the tree diagram. The tree diagram is somewhat more cumbersome, but also it provides a convenient means of organizing information.

Solution 1: We begin by constructing a tree diagram with the given data. The first branches in Fig. 3-14 separate the types of automobiles, and the second branches distinguish between air-conditioned, A, and non-air-conditioned, N, vehicles. The path weights are shown at the right-hand side of the tree. The event A that the faculty member receives an air-conditioned car can be de-

FIGURE 3-14

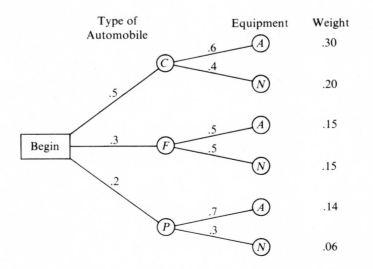

composed into the disjoint events "air-conditioned Chevrolet," "air-conditioned Ford," and "air-conditioned Plymouth." Therefore, using the data on the tree diagram, we have

$$\Pr[A] = \Pr[C \cap A] + \Pr[F \cap A] + \Pr[P \cap A] = .30 + .15 + .14 = .59$$

We can now answer the question posed above. We have

$$\Pr[C|A] = \frac{\Pr[C \cap A]}{\Pr[A]} = \frac{.30}{.59} = \frac{30}{59}$$

The meaning of $\Pr[C|A]$ in a tree diagram can easily be determined. Arguing as in the previous example, we say that the conditional probability $\Pr[C|A]$ is a branch weight in the tree diagram which lists the stages with the order reversed from that of Fig. 3-14. This diagram is shown in Fig. 3-15, and the branch weight $\Pr[C|A]$ is noted.

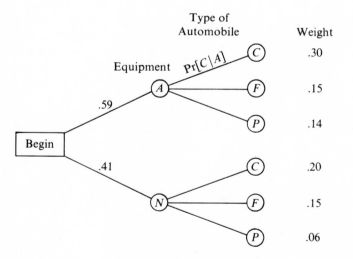

FIGURE 3-15

Solution 2: Since in this example we are interested in a single Bayes probability, it is especially convenient to use formula (3-6). The sample space can be partitioned into the disjoint events C, F, and P (the faculty member is assigned a Chevrolet, Ford, and Plymouth, respectively). Letting these events play the role of

S_1, S_2, and S_3 in formula (3-6), and letting A (the faculty member is assigned an air-conditioned car) be the A in (3-6), we have

$$\Pr[C|A] = \frac{\Pr[A|C]\Pr[C]}{\Pr[A|C]\Pr[C] + \Pr[A|F]\Pr[F] + \Pr[A|P]\Pr[P]}$$

$$= \frac{.6 \times .5}{.6 \times .5 + .5 \times .3 + .7 \times .2}$$

$$= \frac{.30}{.30 + .15 + .14} = \frac{30}{59} \qquad \square$$

EXAMPLE 3-21

There is an epidemic of influenza in Mulberry, and the physicians at the Halenhearty Medical Clinic have examined hundreds of patients with respiratory ailments. There are two types of influenza, types A (Asian) and E (European), in addition to the usual respiratory infections. The initial examination of each patient consists of checking for fever and for inflammation of the upper bronchial tract. If it seems desirable, laboratory tests are performed which will confirm the presence or absence of influenza viruses and whether they are of type A or type E. However, the evaluation of these laboratory tests takes at least 2 days to complete, and in most cases the physician attending a patient will wish to begin treatment much sooner. The data given in Table 3-6 summarize the results of many observations on patients for whom laboratory examinations have confirmed the source of their symptoms. The table gives the probabilities that individuals who were identified as having a specific ailment showed the various possible combinations of symptoms of fever and inflammation. Of the patients who have been examined for respiratory problems, 40 percent have type A influenza, 50 percent have type E influenza, and 10 percent have other respiratory infections.

TABLE 3-6

Symptoms	Type A Influenza	Type E Influenza	Other Respiratory Infections
Fever & inflammation	.5	.8	.5
Fever & no inflammation	.3	.1	.1
No fever & inflammation	.15	.1	.3
No fever & no inflammation	.05	.0	.1

As patients enter the clinic, they are observed to have various combinations of the symptoms of fever and inflammation. For each disease there is a preferred method of treatment, and the physicians would like to prescribe a treatment for each patient which is most likely to be the preferred one for that patient. For example, if a patient enters the clinic and is observed to have inflammation of the upper respiratory tract but no fever, the doctor prescribing a treatment for this patient would like to know which disease is the most likely one to be causing these symptoms. For each of the symptoms identified in Table 3-6, find the disease which is most likely to cause these symptoms.

Solution: In order to solve the problem posed in this example it will be necessary to determine all the Bayes probabilities. Therefore we proceed by using a tree diagram. You are asked to use formula (3-6) to compute one of these Bayes probabilities in Exercise 7.

We suppose that patients who have entered the clinic form a sufficiently large sample to ensure that the observed probabilities of type A and type E influenza and other respiratory infections noted for the sample can be assumed to hold for the entire population of people with respiratory infections. If we assume that the patient in question is a randomly selected individual from this population, we can construct the tree diagram shown in Fig. 3-16 for this situation. In Fig. 3-16 the letter O denotes other respiratory infections, F and NF denote fever and no fever, respectively, and I and NI denote inflammation and no inflammation, respectively. In order to provide the information asked in the problem we also construct the tree diagram as shown in Fig. 3-17.

FIGURE 3-16

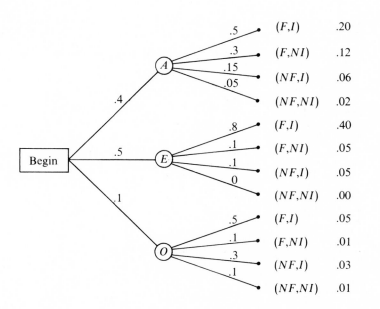

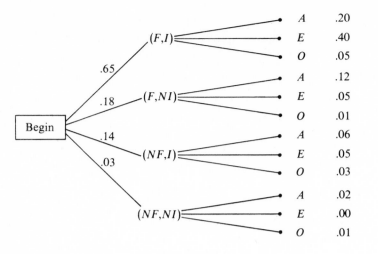

FIGURE 3-17

For example, from Fig. 3-16 we find that $\Pr[(F, I)] = .20 + .40 + .05 = .65$ and $\Pr[(F,NI)]$, $\Pr[(NF,I)]$, and $\Pr[(NF,NI)]$ can be determined similarly. These probabilities give the weights for the first branches in Fig. 3-17. The weights associated with the second branches in Fig. 3-17 can be determined as in solution 1 to Example 3-19. We have

$$\Pr[A\,|\,(NF,I)] = \tfrac{6}{14} \qquad \Pr[E\,|\,(NF,I)] = \tfrac{5}{14} \qquad \Pr[O\,|\,(NF,I)] = \tfrac{3}{14}$$

Thus a patient with inflammation but no fever is most likely to have type A influenza. Similarly, the other probabilities in the tree are easily determined, and the physician's choice of treatment can be made accordingly.   □

It should be pointed out that although our approach to this problem was simplified by using the category "other diseases" and by considering only two symptoms, the method is exactly the same for more symptoms and more diseases. Naturally, the trees are much larger in more detailed studies.

Exercises for Sec. 3-6

1. Determine the remaining branch weights in Fig. 3-15. If the faculty member is assigned a non-air-conditioned car, what is the probability that it is a Ford?

2. The tree diagram of an experiment is shown in Fig. 3-18. Compute $\Pr[I\,|\,A]$.

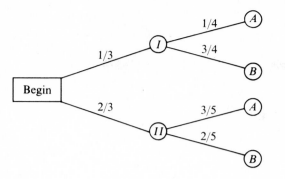

FIGURE 3-18

3. The tree diagram of an experiment is shown in Fig. 3-19. Compute the probability of *I* given that *A* and *a* occur.

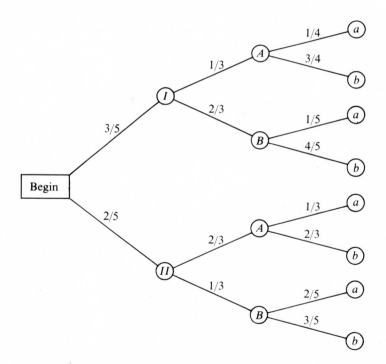

FIGURE 3-19

4. An experiment consists of selecting at random 1 of the 3 urns shown in Fig. 3-20 and then drawing a single ball and noting its color. If the ball selected is red, what is the probability that urn *II* was selected?

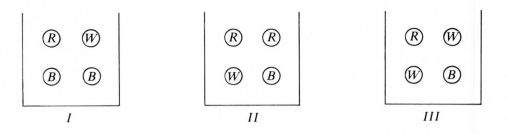

FIGURE 3-20

5. A random selection is made between 2 coins, and the selected coin is flipped twice. One of the coins is fair, and the other is unfair with $\Pr[H] = \frac{2}{3}$. If the result of the experiment is 2 heads, what is the probability that the unfair coin was selected?

6. Using the data of Example 3-21, decide whether the symptoms of fever and inflammation are independent.

7. In Example 3-21, if a patient enters the clinic with the symptoms of fever and inflammation, which disease is the most likely cause of these symptoms? Use formula (3-6) to compute the necessary Bayes probabilities.

8. Two coins are selected at random from a collection consisting of 3 nickels, 3 dimes, and 1 quarter. The sum of their values is an even multiple of 10¢. What is the probability that at least 1 is a nickel?

9. The Fastback Cycle Company produces motorcycles at 3 plants, one each in Germany, Japan, and Brazil. Half their production comes from Germany, and the other half is split equally between Japan and Brazil. Each plant produces the same 3 different sizes of cycles, called the Mini, the Continental, and the Brute. Production of each size is allocated as follows to the plants:

Plant	% Mini	% Continental	% Brute
Germany	30	40	30
Brazil	20	70	10
Japan	30	50	20

Suppose the cycles are distributed randomly to retail stores and someone buys a Continental cycle. What is the probability that the cycle was made in Germany?

10. In Exercise 9, what size cycle should a person buy to have the highest probability of obtaining one made in Japan?

11. Pine seeds of types *A*, *B*, and *C* are randomly scattered in a field. The seeds are 60 percent type *A*, 30 percent type *B*, and 10 percent type *C*. It is known that 30 percent of the type *A* seeds will germinate, 40 percent of the type *B* seeds will germinate, and 70 percent of the type *C* seeds will germinate. If a randomly selected seed has germinated what is the probability that it is of type *A*?

12. There are three candidates in an election, Lefty, Midel, and Rito. Of those people voting 30 percent are leftists, 40 percent are moderates, and 30 percent are rightist. Each of these groups split their vote as follows:

Group	% Lefty	% Midel	% Rito
Leftists	50	40	10
Moderates	30	40	30
Rightists	10	40	50

a. Which candidate received the highest percentage of votes?

b. If a randomly selected person voted for Midel, what is the probability the person is a moderate?

3-7 RANDOM VARIABLES AND EXPECTED VALUE

The manager of the Tin Plate Steel Company's rolling mill knows that production varies from month to month. Quite often monthly production is 10,000 tons. The manager estimates that this is the production in about $\frac{1}{2}$ of the months. If the mill operates at peak production, then 12,000 tons are produced, and this happens about $\frac{1}{10}$ of the time. Finally, when there are material shortages and labor problems, then production falls to 5000 tons, and this is the case in $\frac{2}{5}$ of the months. How much production should the manager expect in a randomly chosen month? Alternatively, what is the average monthly production?

The manager might argue as follows. The plant will produce 10,000 tons $\frac{1}{2}$ of the time; i.e., in any specific month it will produce 10,000 tons with probability $\frac{1}{2}$, 12,000 tons $\frac{1}{10}$ of the time, and 5000 tons $\frac{2}{5}$ of the time. Therefore, the average production, and the amount it is reasonable to expect in any randomly chosen month, is

$$(10{,}000 \times \tfrac{1}{2}) + (12{,}000 \times \tfrac{1}{10}) + (5000 \times \tfrac{2}{5}) = 8200 \text{ tons}$$

The computation above can be viewed as being related to an experiment—selecting a month and noting the production of rolled steel during that month—with sample space {5000, 10,000, 12,000}. The weights associated with these outcomes are $\frac{2}{5}$, $\frac{1}{2}$, and $\frac{1}{10}$, respectively. The problem is to associate one number, an average or expected value, with the experiment.

It is useful to view this problem in a somewhat more general setting. Suppose we have an experiment with sample space $S = \{\mathcal{O}_1, \mathcal{O}_2, \ldots, \mathcal{O}_n\}$, where $\mathcal{O}_1, \mathcal{O}_2, \ldots, \mathcal{O}_n$ are the outcomes. Let $w_1, w_2, \ldots, w_n$ be the weights associated with $\mathcal{O}_1, \mathcal{O}_2, \ldots, \mathcal{O}_n$, respectively.

Definition

A *random variable* is a function defined on a sample space which assigns a real number to each outcome in that space.

For example, if an experiment consists of flipping a coin 4 times and noting each time whether it lands heads or tails, the function which assigns to each outcome the number of heads is a random variable. Also, if we consider the experiment which consists of selecting a sample of 10 audio tapes from a shipment of 1000 tapes and testing each tape in the sample, the function which assigns to each outcome the number of defective tapes defines a random variable.

Several of the important ideas associated with random variables can be illustrated easily with games of chance. Suppose that the following game is proposed to you by a friend. A fair die is to be rolled, and if any of the outcomes 2, 3, 4, or 5 occur, then you receive $1.50 from your friend. If either a 1 or a 6 occurs, you pay your friend an amount equal to the outcome. Thus, if a 1 comes up, you pay your friend $1.00, and if a 6 comes up, you pay $6.00. Should you play this game? Is the game to your advantage or to your friend's advantage? Also, how much of an advantage do you (or your friend) have? If you play the game many times, what sort of gain (or loss) should you expect?

We begin our analysis of this game by forming a table which contains the outcomes of the experiment (a play of the game), the gain (or loss) to you which is associated with each outcome, and the probability of each outcome's occurring. This information is contained in Table 3-7.

It is impossible to predict the outcome of any specific play of this game. Sometimes you will lose $1.00, sometimes you will lose $6.00, and sometimes (actually most of the time) you will win $1.50. Although it is impossible to predict a single outcome of the game, it is possible to make predictions about the results of many plays of the game. If there are many plays of the game (rolls of the die), about $\frac{1}{6}$ of these plays (rolls) should result in a loss of $1.00, another $\frac{1}{6}$ should result in a loss of $6.00, and the remaining $\frac{4}{6}$ of the outcomes should each result in a win of $1.50. These values reflect the probabilities of the various outcomes given in Table 3-7. Thus, suppose that you play this game

TABLE 3-7

Outcome	Gain $(+)$ or Loss $(-)$	Probability
1	-1.0	1/6
2	$+1.5$	1/6
3	$+1.5$	1/6
4	$+1.5$	1/6
5	$+1.5$	1/6
6	-6.0	1/6

600 times and each outcome of the die occurs exactly 100 times. Then 100 times you lose $1.00, 100 times you lose $6.00, and on each of the other 400 times you win $1.50. Since your losses add up to $700 and your gains to $600, your net return is a loss of $100, or $-$100. Since you played 600 times, your average return per play is

$$-\frac{\$100}{600} = -\frac{\$1.00}{6}$$

Summarizing these computations, we find the average return to be

$$V = \frac{-\$100 + \$150 + \$150 + \$150 + \$150 - \$600}{600} = -\frac{\$1.00}{6}$$

In 600 actual plays of this game it is unlikely that each outcome on the die will occur exactly 100 times. However, it is likely that each outcome will occur about 100 times, and thus this result is about what you would expect for a fair die. In the general situation it is to be expected that each outcome on the die will occur about $\frac{1}{6}$ of the time and hence the average net return (in dollars) is

$$V = (-1.00)(\tfrac{1}{6}) + (1.50)(\tfrac{1}{6}) + (1.50)(\tfrac{1}{6}) + (1.50)(\tfrac{1}{6}) + (1.50)(\tfrac{1}{6}) + (-6.00)(\tfrac{1}{6})$$

$$= \frac{-7.00 + 6.00}{6} = -\frac{1}{6}$$

Therefore in any large number of plays of this game you should expect to lose about one-sixth of a dollar per play of the game. Your friend has a definite advantage!

The analysis for this simple game can be described in a more general setting by using random variables. Indeed let a random variable X be defined by assigning the value -1.00 to outcome 1, the value 1.50 to each of the outcomes 2, 3, 4, and 5, and the value -6.00 to outcome 6. The information in Table 3-7 provides information about the random variable X, and it can be represented as shown in Table 3-8. The fourth column of Table 3-8, for which there is no analog in Table 3-7, contains the weighted gain (or loss) associated with each value of the random variable. The average gain (V above) is the sum of the entries in the fourth column of Table 3-8.

The technique of grouping together all outcomes for which the random variable assumes the same value is very useful. We now apply the idea in a more general situation. We continue to assume that the sample space contains only a finite number of elements and consequently that a random variable assumes only a finite number of values. Suppose that the values assumed by a random variable X are $x_1, x_2, \ldots, x_k$. For $j = 1, 2, \ldots, k$, let E_j be the event consisting of all outcomes to which the random variable X assigns the number x_j. That is, E_j is the

TABLE 3-8

Event	Value of Random Variable X	Probability	Product
$\{1\}$	-1.00	$1/6$	$-\dfrac{1.00}{6}$
$\{2, 3, 4, 5\}$	$+1.50$	$4/6$	$+\dfrac{6.00}{6}$
$\{6\}$	-6.00	$1/6$	$-\dfrac{6.00}{6}$

event such that X takes the value x_j for each outcome in E_j and it takes the value x_j for no other outcomes. In the game example discussed just above the random variable takes three values $(-1, 1.5, -6)$, and $k = 3$. If we define

$$x_1 = -1 \qquad x_2 = 1.5 \qquad x_3 = -6 \tag{3-7}$$

then $$E_1 = \{1\} \qquad E_2 = \{2, 3, 4, 5\} \qquad E_3 = \{6\}$$

Notice that x_j is a value of the random variable and E_j is an event, i.e., a set of outcomes.

Definition

The *probability function* of a random variable X is the function which assigns to each value x_j taken by the random variable X the probability of the event E_j on which the function X takes the value x_j:

$$\Pr[X = x_j] = \Pr[E_j]$$

We shall denote the probability associated with x_j by p_j

$$\Pr[X = x_j] = p_j \qquad \text{for} \qquad j = 1, 2, \ldots, k$$

Note that each p_j is nonnegative and $p_1 + p_2 + \cdots + p_k = 1$.

The probability function of a random variable X is also known as the *density function* of X.

In the game example we have

$$p_1 = \Pr[E_1] = \tfrac{1}{6} \qquad p_2 = \Pr[E_2] = \tfrac{4}{6} \qquad p_3 = \Pr[E_3] = \tfrac{1}{6} \tag{3-8}$$

In what follows we consistently use uppercase letters, X, Y, Z, W, etc., to denote random variables and the associated lowercase letters x, y, z, w to denote values taken by those random variables.

EXAMPLE 3-22

A certain production process is known to produce defective audio tapes 10 percent of the time. An experiment consists of selecting a sample of 3 tapes, testing the tapes, and noting the number of defective ones. Let the random variable X be defined by assigning to each outcome the number of defective tapes. Assume that this is an independent trials process, and find the probability function of X.

Solution: The probability of the event for which X takes the value 1, that is, the set of outcomes with 1 defective tape, can be obtained by using (3-5) with $n = 3$, $r = 1$ and $p = .1$. We have

$$\Pr[X = 1] = p_1 = \binom{3}{1}(.1)^1(.9)^2 = .243$$

Likewise,

$$\Pr[X = 0] = p_0 = \binom{3}{0}(.1)^0(.9)^3 = .729$$

$$\Pr[X = 2] = p_2 = \binom{3}{2}(.1)^2(.9) = .027$$

and

$$\Pr[X = 3] = p_3 = \binom{3}{3}(.1)^3(.9)^0 = .001 \qquad \square$$

We usually display the probability function for a random variable X as shown in Table 3-9. The values of the random variable are listed in the left-hand column and the corresponding probabilities in the right-hand column.

TABLE 3-9

x	$\Pr[X = x]$
0	.729
1	.243
2	.027
3	.001

Definition

If X is a random variable with values x_1, x_2, ..., x_k and with a probability function with values p_1, p_2, ..., p_k, then the *expected value* of X is defined to be

$$E[X] = x_1 p_1 + x_2 p_2 + \cdots + x_k p_k = \sum_{j=1}^{k} x_j p_j \qquad (3\text{-}9)$$

We note that the expected value of the random variable in the game example is just the quantity we referred to there as the average return. Indeed, using (3-7) and (3-8) and the definition (3-9), we have

$$E[X] = (-1)(\tfrac{1}{6}) + (1.5)(\tfrac{4}{6}) + (-6)(\tfrac{1}{6}) = -\frac{1}{6}$$

which is exactly the average return of the game example.

Notice that in this case the value of $E[X]$ is not one of the values assumed by the random variable X. Indeed, since $E[X]$ is a weighted average of values taken by X, there is no reason to expect that $E[X]$ will be one of these values, and in general it will not be.

EXAMPLE 3-23

Find the expected value of the random variable X of Example 3-22.

Solution: From (3-9) we see that the expected value of X is a sum each term of which is the product of a value of X and the probability associated with this value. But these are the quantities which occur in the two columns of Table 3-9. Therefore each term in the sum is the product of the two entries in the same line of Table 3-9. It is convenient to add another column to the table which contains these products. This was done in Table 3-8 for the game example. In this case the expanded table is shown in Table 3-10.

TABLE 3-10

x	$\Pr[X = x]$	**Product**
0	.729	0.0
1	.243	.243
2	.027	.054
3	.001	.003

The expected value of X is the sum of the entries in the third column of Table 3-10

$$E[X] = 0 + .243 + .054 + .003 = .300 \qquad \square$$

EXAMPLE 3-24

A friend invites you to play the following game: 2 coins are to be selected at random from 6 coins: 2 nickels, 3 dimes, and 1 quarter. If the sum of the values of the coins is an even multiple of 10¢, your friend will pay you 25¢, otherwise you must pay your friend the value of the coins. What is your expected gain per play of the game?

Solution: If we let N, D, and Q denote nickel, dime, and quarter, respectively, then (since the order in which the coins are drawn is unimportant) the sample space of the experiment can be represented $\{NN, DD, NQ, ND, DQ\}$. The gain to you is 25¢ for each of the first three outcomes and -15¢ and -35¢ for the fourth and fifth outcomes, respectively. If X denotes the random variable which assigns to each outcome its value to you in cents, the probability function of X is given in Table 3-11. The expected value $E[X]$ of X is (in cents)

$$E[X] = \tfrac{150}{15} - \tfrac{90}{15} - \tfrac{105}{15} = -\tfrac{45}{15} = -3$$

Therefore, you should expect to average a loss of 3¢ per play of the game.

TABLE 3-11

x	$\Pr[X = x]$	Product
25	$\dfrac{\binom{2}{2}}{\binom{6}{2}} + \dfrac{\binom{3}{2}}{\binom{6}{2}} + \dfrac{\binom{2}{1}\binom{1}{1}}{\binom{6}{2}} = \dfrac{6}{15}$	$\dfrac{150}{15}$
-15	$\dfrac{\binom{2}{1}\binom{3}{1}}{\binom{6}{2}} = \dfrac{6}{15}$	$-\dfrac{90}{15}$
-35	$\dfrac{\binom{1}{1}\binom{3}{1}}{\binom{6}{2}} = \dfrac{3}{15}$	$-\dfrac{105}{15}$

□

EXAMPLE 3-25

Suppose there are 2 defective items in a lot of 10 items. If a sample of 3 items is selected, what is the expected number of defectives?

Solution: Let X denote the random variable which assigns to each outcome (set of 3 items) the number of defective items. The probability function for X is given in Table 3-12.
Therefore, the expected value of X is

$$E[X] = 0 + \tfrac{56}{120} + \tfrac{16}{120} = \tfrac{72}{120}$$

TABLE 3-12

x	$\Pr[X = x]$	Product
0	$\dfrac{\binom{8}{3}}{\binom{10}{3}} = \dfrac{56}{120}$	0
1	$\dfrac{\binom{8}{2}\binom{2}{1}}{\binom{10}{3}} = \dfrac{56}{120}$	$\dfrac{56}{120}$
2	$\dfrac{\binom{8}{1}\binom{2}{2}}{\binom{10}{3}} = \dfrac{8}{120}$	$\dfrac{16}{120}$

Exercises for Sec. 3-7

1. An experiment consists of flipping a fair coin 3 times and noting the number of heads and the number of tails. A random variable is defined which assigns to each outcome $2 \times$ (number of heads) $+ 3 \times$ (number of tails).
   a. What are the values assumed by this random variable?
   b. What is its probability function?

2. An experiment consists of selecting 2 coins at random from a collection consisting of 2 nickels and 3 dimes. A random variable is defined which assigns to each outcome the value of the coins in cents.
   a. What are the values assumed by this random variable?
   b. What is its probability function?

3. An inspector estimates that 90 percent of the widgets produced by a machine are acceptable. An experiment consists of selecting a sample of 3 widgets at random and checking them to determine how many of the sample are acceptable. Define a random variable by assigning to each outcome the number of acceptable widgets. Consider the sampling to be 3 binomial trials.
   a. What are the values assumed by this random variable?
   b. What is its probability function?

4. An experiment consists of rolling 2 dice and noting the sum of the uppermost numbers. A random variable is defined which assigns to an outcome
   The number 1 if the sum is odd and the sum if the sum is even
   a. What are the values assumed by this random variable?
   b. What is its probability function?

5. A box contains 9 acceptable and 3 defective widgets. A sample of 5 widgets is selected at random from the box. A random variable is defined which assigns to each outcome the number of defective widgets.
   a. What are the values assumed by this random variable?
   b. What is its probability function?

6. An experiment consists of drawing a card at random from a completely shuffled bridge deck. A random variable is defined which assigns to each outcome the number on the card if it is a numbered card and 10 if it is a face card.
   a. What are the values assumed by this random variable?
   b. What is its probability function?

Find the expected value of the random variable defined in:

7. Exercise 1   **8.** Exercise 2   **9.** Exercise 3

10. Exercise 4   **11.** Exercise 5   **12.** Exercise 6

13. An unfair coin is weighted so that when it is flipped the probability of a head landing uppermost is $\frac{2}{3}$. If the coin is flipped until a head or 4 consecutive tails appear, what is the expected number of flips?

14. A motor pool contains 10 automobiles, 2 of which have defective fuel gauges. A salesman is assigned an automobile at random on 4 occasions. What is the expected number of automobiles with defective fuel gauges assigned to him?

15. Consider two urns each of which contains 3 colored balls. Urn *I* contains 2 red and 1 white ball, and urn *II* contains 1 red, 1 white, and 1 blue ball. An experiment consists of randomly drawing a ball from urn *I*, noting its color, placing that ball in urn *II*, and then randomly drawing a ball from urn *II* and noting its color. What is the expected number of red balls drawn from the 2 urns?

16. A sales representative estimates that on each call there is a probability $\frac{1}{4}$ of making a sale. A bonus of \$100 is paid for the first and second sale each day, and a bonus of \$200 is paid for each additional sale. If it is possible to make 3 calls each day, what is the expected income from bonuses?

IMPORTANT TERMS

You should be able to describe, define, or give examples of each of the following:

Probability measure
Assignment of weights
Mutually exclusive events
Conditional probability of A given B
Branch weight
Path weight
Stochastic process
Tree measure

Disjoint events
Independent trials
Bernoulli or binomial process
Bayes probabilities
Random variable
Density function
Expected value

REVIEW EXERCISES

1. An experiment has the sample space $S = \{\mathcal{O}_1, \mathcal{O}_2, \mathcal{O}_3, \mathcal{O}_4, \mathcal{O}_5\}$ and weights $w_1 = .09$, $w_2 = .23$, $w_3 = .41$, $w_4 = .14$, and $w_5 = .13$. Define $E_1 = \{\mathcal{O}_1, \mathcal{O}_2, \mathcal{O}_4\}$, $E_2 = \{\mathcal{O}_3, \mathcal{O}_4, \mathcal{O}_5\}$, and $E_3 = \{\mathcal{O}_2, \mathcal{O}_3\}$. Find:
a. $\Pr[\tilde{E}_1]$
b. $\Pr[E_1 \cup E_2]$
c. $\Pr[E_1 \cap E_3]$
d. $\Pr[E_2 \cap \tilde{E}_3]$

2. Suppose E and F are events with $\Pr[E] = .6$ and $\Pr[E \cup F] = .9$. Find $\Pr[F]$ if:
a. E and F are mutually exclusive.
b. E and F are independent.

3. A slot machine which costs 25¢ to play has a payoff of X silver dollars when you win and nothing when you lose. If it is 15 times more likely that you lose than that you win, what is the smallest value X can have so that in the long run you do not lose money?

4. Two fair dice are rolled and the number of dots on the uppermost face of each die is noted. Find the probability that:
   a. The sum is 8 given that at least one of the numbers is even.
   b. At least one of the numbers is even given that the sum is 8.
   c. At least one of the numbers is even given that the sum is 7.

5. Referring to Exercise 4, define the events E, "the sum is 8," and F, "at least one of the numbers is even." Are E and F independent?

6. An unfair die is weighted so that an even number of dots is twice as likely to land on top as an odd number of dots. The die is rolled 5 times.
   a. Find the probability of exactly 3 odd numbers landing on top.
   b. Find the probability of at most 3 odd numbers landing on top.
   c. Find the expected number of odd numbers.

7. An experiment consists of selecting 3 coins at random from a collection of 3 dimes and 5 quarters.
   a. Find the probability that the value of the coins is at least 40¢.
   b. Find the expected number of dimes selected.
   c. Find the expected total value of the 3 coins selected.

8. A fair coin is flipped 10 times. Which is more likely:
   a. Exactly 5 heads?
   b. Either 4 or 6 heads?

9. Of 20 bridge players at a party, 10 are good players, 6 are average, and 4 are poor. If 2 players are randomly assigned as partners find the probability that:
   a. Both are good players.
   b. Neither is a poor player.
   c. At least 1 is an average player.
   d. At least 1 is not an average player.
   e. Both are good given that at least 1 is a good player.

10. The 20 bridge players described in Exercise 9 are randomly assigned into 10 pairs. Find the probability that:
    a. All poor players are assigned poor partners.
    b. No poor player is assigned a poor partner.
    c. At least 2 good players are assigned to play together.

11. A student has 8 blue socks, 6 brown socks, and 2 gray socks mixed together in a drawer. On a dark morning she selects two socks at random from the drawer. They form a pair of the same color. What is the probability that they are brown?

12. An unfair coin with $\Pr[H] = p > 0$ is flipped 10 times. If the probability of exactly 4 heads is equal to the probability of exactly 5 heads, what is p?

13. A 3-digit number is formed from the digits 1, 2, 4, 6, 9 by using 3 different digits. If the digits are selected and ordered at random, what is the probability that:
    a. The number is even?
    b. The number is less than 500?
    c. The number is even given that it is less than 500?
    d. The number is less than 500 given that it is even?

14. A distributor purchases widgets from factories in Chicago and Detroit: $\frac{1}{3}$ of them from Chicago and the remainder from Detroit. Of the widgets produced in Chicago, $\frac{1}{2}$ are of normal size, $\frac{1}{4}$ are oversize, and $\frac{1}{4}$ are undersize; $\frac{1}{3}$ of the widgets produced in Detroit are normal, and the remainder are oversize. What is the probability that a widget selected at random from the distributor's stock is of normal size?

15. Of the vehicles checked at an inspection station, 80 percent are passed, 18 percent are found to be defective but repairable, and 2 percent are found to be defective and irrepairable. Of those passed, $\frac{3}{4}$ are less than 5 years old; of those found to be defective but repairable $\frac{1}{2}$ are less than 5 years old; and of those found to be defective and irrepairable only 1 in 20 is less than 5 years old. What is the probability that your friend's 4-year-old car (about which you know only its age) will be passed by this inspection station?

STATISTICS

FOUR

4-1 INTRODUCTION

Statistics is one of the most valuable and widely applied of the mathematical tools used in business and the life and social sciences. Our understanding of many standard problems in these areas is improved by using statistical methods and ideas, and there are some problems whose study depends wholly or in an essential way on the use of statistics. Of particular importance is the use of statistics in decision theory, i.e., the evaluation of, and selection between, various alternatives.

The statistical topics considered here were selected both because of their connection with earlier chapters and because of their use in applications. The selection is intentionally restricted to a few aspects of the use of statistics in decision making. For our purposes it is convenient to view statistics as the study of problems which are in a sense the reverse of those studied in probability. In probability theory one is given a sample space of outcomes and a probability measure on this space. With this information one seeks to determine the probabilities to be assigned to certain events. In statistics, on the other hand, one begins with estimates of the probabilities of certain events. These estimates are frequently obtained from real-world data. Using this information, one seeks to identify a sample space and a probability measure defined on that space which, according to some criterion, is consistent with the known probabilities.

Even with our restriction to topics which relate to decision making, we cannot hope to provide a comprehensive survey in this chapter. We must further limit our study in both scope and depth. For this reason we concentrate on two types of experiments with their associated sample spaces and probability measures. Further, we shall not try to choose between the two different measures but instead seek to determine the likelihood that a proposed measure is an appropriate one. For example, in one case we determine the likelihood that given data are consistent with an assumption that the situation can be modeled as a binomial experiment with a certain success probability. We usually work in the setting of decision making in business, but we also consider examples from biology and the social sciences.

We conclude this introduction with an illustration of the type of problem to be considered in this chapter. In Sec. 4-2 we give the definitions of the binomial and normal distributions, and in Sec. 4-3 we discuss some of the basic facts about these distributions. The relationship between the normal and binomial distribution, a topic of considerable importance in statistical applications, is discussed in Sec. 4-4. Section 4-5 contains applications, including a solution to the problem posed below.

EXAMPLE 4-1

An electronics firm buys transistors in lots of 50. Each transistor is either acceptable or defective. The firm does not wish to accept a lot with more than 10 defective items or reject a lot with fewer than 5 defective items (both these events result in unacceptable financial losses to the firm, while the alternatives do not). Since it is expensive to test all the transistors in a lot, the firm decides to test a subset of the transistors in each lot and on the basis of the results of this test either to accept or reject the lot. The problem for the firm is to decide how many items to test and how to interpret the results. For example, if 10 items are tested and 1 is defective, should the shipment be accepted or rejected?   □

4-2 THE BINOMIAL AND NORMAL DISTRIBUTIONS

A binomial experiment (see Sec. 3-5) consists of a number of independent trials of another experiment, each trial having the same two possible outcomes and the same probabilities for these outcomes. The two outcomes are called success and failure, and the probability of success on a single trial is denoted by p. It was shown in Sec. 3-5 that if a binomial experiment consists of n trials, the probability of exactly r successes is

$$\binom{n}{r} p^r (1 - p)^{n-r} \tag{4-1}$$

This formula defines an assignment of weights to the elements of a sample space consisting of all possible outcomes of a binomial experiment with n trials. Remember that an outcome is defined by the number of successes and not the order in which the successes and failures occur in the sequence of trials. Example 3-22 introduced another way of viewing this situation. The result of repeating an independent trial n times is a certain number, say r, of successes and another number, $n - r$, of failures. If we associate with this outcome the number of successes r, we have defined a random variable on the sample space. The probability function (or density function) of this random variable (see Sec. 3-7) is given by formula (4-1). More precisely, the sample space consists of a set of outcomes, each specified by a number of successes. It is therefore reasonable to identify the sample space with the set

$$S = \{0, 1, 2, \ldots, n\}$$

and to use the formula (4-1), with $r = 0, 1, \ldots, n$, to obtain the probabilities that the random variable takes the values $0, 1, \ldots, n$, respectively.

EXAMPLE 4-2

An experiment consists of 5 repetitions of a binomial trial with $p = \frac{1}{4}$. Determine the probability density function for the random variable which associates with each outcome the number of successes.

Solution: When the random variable is denoted by X, the values taken by X are given in the first column of Table 4-1 and the respective probabilities [determined by (4-1)] are given in the second column of that table.

TABLE 4-1

x	$\Pr[X = x]$	Cumulative Sum of Probabilities
0	$\left(\frac{3}{4}\right)^5 = \dfrac{243}{1024}$	$\dfrac{243}{1024}$
1	$5\left(\frac{1}{4}\right)\left(\frac{3}{4}\right)^4 = \dfrac{405}{1024}$	$\dfrac{648}{1024}$
2	$10\left(\frac{1}{4}\right)^2\left(\frac{3}{4}\right)^3 = \dfrac{270}{1024}$	$\dfrac{918}{1024}$
3	$10\left(\frac{1}{4}\right)^3\left(\frac{3}{4}\right)^2 = \dfrac{90}{1024}$	$\dfrac{1008}{1024}$
4	$5\left(\frac{1}{4}\right)^4\left(\frac{3}{4}\right)^1 = \dfrac{15}{1024}$	$\dfrac{1023}{1024}$
5	$\left(\frac{1}{4}\right)^5 = \dfrac{1}{1024}$	$\dfrac{1024}{1024} = 1$

The third column (which we discuss more fully below) verifies that the probabilities of column 2 do in fact sum to 1. □

Definition

If X is a random variable which assigns to each outcome of a binomial experiment the number of successes, then X is called a *binomial random variable*. If the experiment consists of n independent trials, each with success probability p, then the values of X are 0, 1, ..., n and the associated probabilities are

$$\Pr[X = r] = \binom{n}{r} p^r (1 - p)^{n-r} \qquad r = 0, 1, \ldots, n \tag{4-2}$$

The density function defined by (4-2) is the *binomial density function*. Note that the binomial density function depends on two parameters, n and p.

Examples of the graphs of binomial density functions are given in Fig. 4-1.

Associated with each density function there is a cumulative sum function which is known as the *distribution function*. We first describe this function in the context of the special case of Example 4-1. The distribution function is defined on the set of values taken by a random variable; i.e., the distribution function is defined on the same set as the density function. In Example 4-1 the distribution function is defined by the values in the third column of Table 4-1. If the distribution function is denoted by d, we have, for example,

$$d(0) = \tfrac{243}{1024} \qquad d(1) = \tfrac{648}{1024} \qquad d(2) = \tfrac{918}{1024}$$
$$d(3) = \tfrac{1008}{1024} \qquad d(4) = \tfrac{1023}{1024} \qquad d(5) = \tfrac{1024}{1024} = 1$$

The general case is covered by the following definition.

Definition

Let X be a random variable with values $x_1, x_2, \ldots, x_n$, where $x_1 < x_2 < \cdots < x_n$, and let p_i be the probability of the event on which X takes the value x_i, $i = 1, 2, \ldots, n$. The *distribution function* d of X is defined on the set $\{x_1, x_2, \ldots, x_n\}$ by the formula

$$d(x_i) = p_1 + p_2 + \cdots + p_i \qquad i = 1, 2, \ldots, n \tag{4-3}$$

Thus $d(x_1) = p_1, d(x_2) = p_1 + p_2, \ldots, d(x_n) = p_1 + p_2 + \cdots + p_n = 1$. Notice that d takes values between 0 and 1 and that $d(x_n) = 1$.

Note that if one knows either the density function or the distribution function for a random variable, one can determine the other. The determination of the distribution function from the density function is obvious from the definition

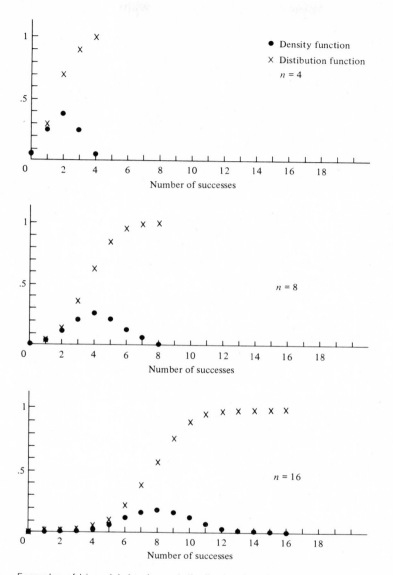

FIGURE 4-1   Examples of binomial density and distribution functions for $p = \frac{1}{2}$.

given above. On the other hand, if X is as described in that definition and its distribution function d is known, the density function can be determined as follows:

$$p_1 = \Pr[X = x_1] = d(x_1)$$
$$p_i = \Pr[X = x_i] = d(x_i) - d(x_{i-1}) \qquad i = 2, \ldots, n$$

Also notice that

$$d(x_i) = \Pr[X \le x_i]$$

Thus the distribution function gives the probabilities that the random variable takes values less than or equal to specified ones. As we shall see later in this chapter, such probabilities arise naturally in many situations.

EXAMPLE 4-3

Find the distribution function for the random variable X whose density function is given in Table 4-2.

TABLE 4-2

x	$\Pr[X = x]$
1	.3
5	.1
10	.4
20	.2

Solution: The distribution function d is defined on the set $\{1, 5, 10, 20\}$, and

$$d(1) = .3$$
$$d(5) = .3 + .1 = .4$$
$$d(10) = .3 + .1 + .4 = .8$$
$$d(20) = .3 + .1 + .4 + .2 = 1.0$$

□

EXAMPLE 4-4

Let X be a binomial random variable with parameters n (number of trials) and p (success probability). Find a formula for the distribution function of X. This is known as the *binomial distribution* with parameters n and p.

Solution: The set of values taken by X is $\{0, 1, 2, \ldots, n\}$. Using this and the definition of a distribution function, we have

$$d(r) = \Pr[X = 0] + \Pr[X = 1] + \cdots + \Pr[X = r]$$

for $r = 0, 1, 2, \ldots, n$. This can be made more explicit by using formula (4-1). We have

$$d(r) = \binom{n}{0}(1 - p)^n + \binom{n}{1}p(1 - p)^{n-1} + \cdots + \binom{n}{r}p^r(1 - p)^{n-r}$$

for $r = 0, 1, 2, \ldots, n$. It is easy to check that this formula gives the numbers which appear in the third column of Table 4-1 ($n = 5$ and $p = \frac{1}{4}$). ☐

Binomial distribution functions for $p = \frac{1}{2}$ and $n = 4$, 8, and 16 are graphed in Fig. 4-1.

The Normal Random Variable

Binomial random variables are especially useful, and there are many problems in which they arise. However, there are computational difficulties in evaluating the density and distribution functions when the parameter n (the number of trials) is large. Fortunately, when n is large, these computational difficulties can often be avoided by working with another random variable whose density function is a good approximation to the binomial density function. This random variable, the *normal random variable*, is of great importance, and the remainder of this section is devoted to it. In Sec. 4-4 we describe the relationship between the normal and binomial random variables.

The density functions graphed in Fig. 4-1 are representative of the graphs of many density functions which arise in nature and in industrial settings. The common features of these graphs, symmetry about a central value and steadily decreasing values away from the central value, are summarized by saying that these graphs are *bell-shaped*. The domain of each of these density functions is a finite set of points. The domain of the density function of the normal random variable, on the other hand, is the entire real axis. For a binomial random variable the value taken by the density function at a point x in the domain is the probability that the random variable takes the value x on a specific performance of the experiment. On the other hand, the probability that a normal random variable assumes any particular value is zero, and we consider instead the probability that a normal random variable takes a value in an interval. This probability is defined as a portion of the area under the standard normal curve.

Definition   The graph of the function

$$f(x) = \frac{1}{\sqrt{2\pi}}\, e^{-x^2/2} \tag{4-4}*$$

for $-\infty < x < \infty$ (shown in Fig. 4-2) is known as the *standard normal curve*.

---

* The letter e in formula (4-4) represents the number $e \approx 2.71828$ (where $\approx$ is read "is approximately equal to"), which is the base of the natural logarithms. It is a truly ubiquitous number, occurring (sometimes quite unexpectedly) in many mathematical settings and applications.

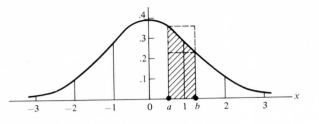

FIGURE 4-2

Definition

The *standard normal random variable* can take any number as a value. The probability that this random variable takes a value between the numbers a and b, $a < b$, is the area under the standard normal curve between a and b (see the shaded area in Fig. 4-2).

In general one needs to use techniques from calculus to compute the area under a portion of a curve like that of Fig. 4-2. However, since the standard normal curve is so important in statistics, areas under it have been computed with great accuracy and are readily available in tables. Such a table of values is given in Appendix A at the end of this book. We illustrate the use of the table to compute the probability that the standard normal random variable takes values in an interval in the next example. However, before proceeding to that example, another observation is necessary. Consider Fig. 4-2, and note that if the point a is held fixed and b is moved toward a, the area under the curve between a and b decreases. Indeed, this area is less than the area of the larger rectangle shown with the line segment with the endpoints a and b as base, and it is more than the area of the smaller rectangle shown with the same base. As the point b moves closer and closer to a, the areas of both of these rectangles become smaller and smaller. In fact the areas of both rectangles approach zero. With this motivation we define the area under a point on the standard normal curve to be zero.

Definition

If X is the standard normal random variable, and if x is any number, then

$$\Pr[X = x] = 0 \qquad (4\text{-}5)$$

This allows us to conclude that for any number x we have

$$\Pr[X \leq x] = \Pr[X < x]$$

a result which we use frequently and without further comment in the remainder of this chapter.

Also, the area under the entire curve is 1, and by symmetry the area under the curve on each side of $x = 0$ is $\frac{1}{2}$. In order to verify symmetry, replace x by $-x$ in the formula defining the standard normal curve and note that $f(-x) = f(x)$.

If X is the standard normal random variable, then

$$\Pr[X \leq 0] = \tfrac{1}{2} \qquad \text{and} \qquad \Pr[X \geq 0] = \tfrac{1}{2} \qquad\qquad (4\text{-}6)$$

Notice that since $\Pr[X = 0] = 0$, we can write inclusive inequalities, i.e., including the case of equality, in both cases.

EXAMPLE 4-5

Let X be the standard normal random variable. Find the following probabilities:

(a) $\Pr[X \text{ less than } .43] = \Pr[X \leq .43]$
(b) $\Pr[X \text{ at least } 1.7] = \Pr[X \geq 1.7]$
(c) $\Pr[X \text{ less than } -1.2] = \Pr[X \leq -1.2]$
(d) $\Pr[X \text{ between } .2 \text{ and } 1.2] = \Pr[.2 \leq X \leq 1.2]$
(e) $\Pr[X \text{ between } -1.2 \text{ and } .2] = \Pr[-1.2 \leq X \leq .2]$

Solution: We use Appendix A and Eqs. (4-5) and (4-6) to find these probabilities. The entries in the table of Appendix A are the areas under the standard normal curve from 0 to t, where the units and tenths digit of t are given to the left of the table and the hundredths digit is given at the top of the table. For example, if $t = 1.57$, we look in the row labeled 1.5 and the column labeled .07:

	.06	↓ .07	.08
1.4	.4279	.4292	.4306
→ 1.5	.4406	.4418	.4429
1.6	.4608	.4525	.4535

We find that the area under the standard normal curve from 0 to 1.57 is .4418.

(a) We must determine the area under the standard normal curve to the left of .43. We think of this area as consisting of two pieces, that to the left of 0 and that from 0 to .43. We use Eq. (4-6) to determine the former and the table to determine the latter. By (4-6) we have $\Pr[X \leq 0] = .5$. To determine $\Pr[0 \leq X \leq .43]$ we use the row labeled .4 and the column labeled .03 in the table. The area under the standard normal curve from 0 to .43 is .1664. That is, $\Pr[0 \leq X \leq .43] = .1664$. Finally

$$\Pr[X \leq .43] = \Pr[X \leq 0] + \Pr[0 \leq X \leq .43] = .5000 + .1664 = .6664$$

(b) We begin with $\Pr[X \geq 0] = .5$. We then divide the area to the right of $x = 0$ into two pieces at $x = 1.7$, and we obtain

$$\Pr[0 \leq X \leq 1.7] + \Pr[1.7 \leq X] = .5$$

By the table the term $\Pr[0 \leq X \leq 1.7]$ is equal to .4554. Therefore

$$.4554 + \Pr[1.7 \leq X] = .5000$$

or
$$\Pr[1.7 \leq X] = .0446$$

(c) We use the fact that the standard normal curve is symmetric about the vertical line through 0, and consequently the area under the curve to the left of -1.2 is equal to the area under the curve to the right of $+1.2$. These areas are shaded in Fig. 4-3. The area to the right of 1.2, $\Pr[X \geq 1.2]$, can be determined just as in part (b) We have

$$\Pr[0 \leq X \leq 1.2] + \Pr[1.2 \leq X] = .5$$
$$.3849 + \Pr[1.2 \leq X] = .5000$$

and
$$\Pr[1.2 \leq X] = .1151$$

We conclude that $\Pr[X \leq -1.2] = .1151$.

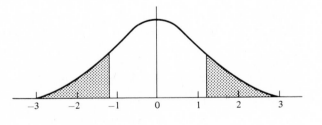

FIGURE 4-3

(d) To find the area under the standard normal curve from .2 to 1.2 we use the table to find the areas from 0 to 1.2 and from 0 to .2 and then subtract the latter from the former. That is,

$$\Pr[.2 \leq X \leq 1.2] = \Pr[0 \leq X \leq 1.2] - \Pr[0 \leq X \leq .2]$$
$$= .3849 - .0793 = .3056$$

(e) To find the area under the curve between -1.2 and .2 we compute the area between -1.2 and 0 and add to this the area from 0 to .2. Using the symmetry of the curve, we have

$$\Pr[-1.2 \leq X \leq .2] = \Pr[-1.2 \leq X \leq 0] + \Pr[0 \leq X \leq .2]$$
$$= \Pr[0 \leq X \leq 1.2] + \Pr[0 \leq X \leq .2]$$
$$= .3849 + .0793 = .4642 \qquad \square$$

Many applications involve random variables which are "normal" in the sense that their density functions are related in a simple way to the density function of the standard normal random variable. To amplify this idea, we first describe how one random variable can be used to define another with a formula connecting the values of the two random variables. The setting we consider is an elementary one, and the technique can be applied in much more complex situations. Suppose we have a random variable X. It may be helpful to think of a random variable with a density function which has a graph similar to that of Fig. 4-2, that is, a bell-shaped curve. If we are given two constants μ and σ, $\sigma > 0$, then we can define a new random variable Z by the formula $Z = (X - \mu)/\sigma$.* That is, associated with each value of x of X there is a value $(x - \mu)/\sigma$ of Z. It follows that if X takes values in an interval with endpoints a and b, $a < b$, then Z takes values in an interval related to the original one by a change of variables. In particular, the inequality $a \le X \le b$ is equivalent to the two inequalities $a \le X$ and $X \le b$. Subtracting μ from both sides of both inequalities, we have $a - \mu \le X - \mu$ and $X - \mu \le b - \mu$. Also, dividing both sides of both inequalities by σ, we have (since $\sigma > 0$)

$$\frac{a - \mu}{\sigma} \le \frac{X - \mu}{\sigma} \quad \text{and} \quad \frac{X - \mu}{\sigma} \le \frac{b - \mu}{\sigma}$$

These two inequalities can be combined into

$$\frac{a - \mu}{\sigma} \le \frac{X - \mu}{\sigma} \le \frac{b - \mu}{\sigma}$$

We conclude that if X takes values in the interval with endpoints a and b, then $(X - \mu)/\sigma$ takes values in the interval with endpoints $(a - \mu)/\sigma$ and $(b - \mu)/\sigma$. Moreover, since the set of values of X described by $a \le X \le b$ is exactly the same as the set of values of $(X - \mu)/\sigma$ described by

$$\frac{a - \mu}{\sigma} \le \frac{X - \mu}{\sigma} \le \frac{b - \mu}{\sigma}$$

we have

$$\Pr[a \le X \le b] = \Pr\left[\frac{a - \mu}{\sigma} \le \frac{X - \mu}{\sigma} \le \frac{b - \mu}{\sigma}\right]$$

With this background we can give a definition of a normal random variable.

---

* The symbols μ and σ are the Greek letters mu and sigma, respectively. These symbols are the customary ones to use in this situation, and we shall continue the custom.

Definition

A random variable X is said to be *normal* if there are numbers μ and σ, $\sigma > 0$, such that the random variable Z defined by $Z = (X - \mu)/\sigma$ is the standard normal random variable.

The significance of the constants μ and σ of this definition in terms of the original random variable X will be considered in Sec. 4-3. With this definition, the preceding discussion can be summarized as a theorem.

Theorem

If X is a normal random variable, and if Z is the standard normal random variable, then for every a and b with $a < b$ we have

$$\Pr[a \leq X \leq b] = \Pr\left[\frac{a - \mu}{\sigma} \leq Z \leq \frac{b - \mu}{\sigma}\right] \tag{4-7}$$

where μ and σ are the numbers for which $(X - \mu)/\sigma$ is the standard normal random variable.

EXAMPLE 4-6

The density function graphed in Fig. 4-4 is the density function of a normal random variable X since the random variable Z defined by $Z = (X - 110)/10$ is the standard normal random variable. In this example μ and σ are 110 and 10, respectively.   □

The equality (4-7) relates the probability that X takes values in an interval to the probability that the standard normal random variable takes values in a different interval. This relationship and the probabilities in the table in Appendix A can be used to compute the probability that a normal random variable takes values in any interval. The technique is illustrated in the next example.

FIGURE 4-4

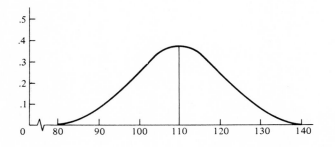

EXAMPLE 4-7

Let X be a normal random variable, and suppose that $Z = (X - 110)/10$ is the standard normal random variable. Find the following probabilities for the random variable X:

(a) $\Pr[110 \leq X \leq 120]$
(b) $\Pr[100 \leq X \leq 120]$
(c) $\Pr[90 \leq X \leq 130]$
(d) $\Pr[90 \leq X \leq 140]$

Solution: Since the random variable Z defined by $Z = (X - 110)/10$ is the standard normal random variable, it has the probabilities given in the table of Appendix A.

In part (a) we are interested in values of X in the interval $110 \leq X \leq 120$. If X has values in this interval, $X - 110$ has values in the interval $0 \leq X - 110 \leq 10$, and $(X - 110)/10$ has values in the interval $0 \leq (X - 110)/10 \leq 1$. But $(X - 110)/10$ is the standard normal random variable. We conclude that

$$\Pr[110 \leq X \leq 120] = \Pr[0 \leq Z \leq 1]$$

where Z is the standard normal random variable. From the table in Appendix A we have $\Pr[0 \leq Z \leq 1] = .3413$, and consequently

$$\Pr[110 \leq X \leq 120] = .3413$$

To find the probability sought in part (b) we note that the inequality $100 \leq X \leq 120$ can be used to deduce the inequality

$$\frac{100 - 110}{10} \leq \frac{X - 110}{10} \leq \frac{120 - 110}{10}$$

which is the same as

$$-1 \leq Z \leq 1$$

Since the standard normal curve is symmetric about the origin,

$$
\begin{aligned}
\Pr[100 \leq X \leq 120] &= \Pr[-1 \leq Z \leq 1] \\
&= \Pr[-1 \leq Z \leq 0] + \Pr[0 \leq Z \leq 1] \\
&= .3413 + .3413 = .6826
\end{aligned}
$$

The answers to parts (c) and (d) are obtained from Appendix A in exactly the same way as those of parts (a) and (b). We have

$$Pr[90 \leq X \leq 130] = Pr[-2 \leq Z \leq 2] = .4772 + .4772 = .9544$$
$$Pr[90 \leq X \leq 140] = Pr[-2 \leq Z \leq 3] = .4772 + .4987 = .9759 \qquad \square$$

Exercises for Sec. 4-2

1. Two fair dice are rolled. Let X be the number of sixes obtained. What are the values of the density function for X and the distribution function for X?

2. A box contains 10 pairs of jeans, 3 of which have the wrong size marked on them. A sample of 3 pairs of jeans is selected at random. What is the density function for the random variable which is the number of missized pairs in the sample? What is the distribution function?

3. An independent third party runs a candidate for governor in each of 5 states. Suppose each candidate has 1 chance in 5 of being elected, and suppose the elections are independent. Let X be the random variable which is the number of governors elected by this party. Find the density function for X and the distribution function.

4. A fair die is rolled, and a random variable X is defined to be the square of the number on the top of the die. Let $p(x)$ and $d(x)$ be the density function and distribution function, respectively, for this random variable. What are $p(1)$, $p(4)$, $d(4)$, and $d(16)$?

5. Let X be a random variable, and suppose the distribution for X is the standard normal distribution. Find the following probabilities:
   a. $Pr[0 \leq X \leq .5]$
   b. $Pr[.5 \leq X \leq 1.5]$
   c. $Pr[1.5 \leq X]$
   d. $Pr[-1.5 \leq X \leq .5]$

6. Let X be a random variable which is normally distributed, and suppose the distribution of the random variable $Z = (X - 10)/3$ is the standard normal distribution. Find the following probabilities:
   a. $Pr[X \leq 10]$
   b. $Pr[7 \leq X \leq 13]$
   c. $Pr[X \geq 16]$
   d. $Pr[X \leq 4 \text{ or } X \geq 16]$

7. An investment club buys 100 shares of stock in each of 5 different companies. Assume that the stocks behave independently and the probability of making a profit on any given stock is .6.
   a. What is the probability that the club will make a profit on all 5 stocks?
   b. What is the probability that the club will make a profit on at least 1 stock?

4-3 MEAN AND STANDARD DEVIATION

The expected value of a random variable (introduced in Sec. 3-7) is also known as the *mean* of that random variable. The latter name is frequently used in applications, especially applications with a statistical flavor. The mean of a random variable is, by definition, a number determined both by the values of the random variable and by its density function. Equivalently, since the density function of a random variable is determined by the probability measure assigned to the sample space, the mean of a random variable depends on this probability measure. In many problems the probability measure and density function are unknown, and the task is to find or estimate these functions. One method of deducing information about these functions is to obtain (from real-world data) estimates on the mean of the random variable and then to compare these estimates with the values resulting from computing the mean of the random variable with various assumed density functions. The goal is to find, if possible, a density function for which the computed value of the mean of the random variable is close to the value obtained from real data. It may happen that there are no simple assumed density functions which give good agreement, or it may happen that good agreement is obtained with several different assumed density functions. To use this process we must know the means of random variables with various density functions. Also, we must be able to obtain estimates of the mean of a random variable, and we must know something about the accuracy of these estimates. We begin this section with a discussion of the means of binomial and normal random variables. Later we also consider measures of the distribution of the values of a random variable about its mean.

The Mean of a Binomial Random Variable

Let X be the binomial random variable for an experiment consisting of n repetitions of independent trials each of which has success probability p. The random variable X takes the values $0, 1, 2, \ldots, n$ with the probabilities given by formula (4-2); that is,

$$\Pr[X = r] = \binom{n}{r} p^r (1 - p)^{n-r}$$

for $r = 0, 1, \ldots, n$. According to the definition of expected value (Sec. 3-7) we have

$$E[X] = \sum_{k=0}^{n} k \binom{n}{k} p^k (1 - p)^{n-k} \tag{4-8}$$

Although (4-8) appears to be a complicated sum, it can be greatly simplified. We illustrate the simplification for $n = 1$ and $n = 2$. If $n = 1$, then

$$E[X] = 0(p^0)(1 - p)^1 + 1(p^1)(1 - p)^0 = p$$

and if $n = 2$, then

$$E[X] = 0(p^0)(1 - p)^2 + 1(2p^1)(1 - p)^1 + 2(p^2)(1 - p)^0 = 2p$$

The simplification of (4-8) for all values of n can be established by mathematical induction or by using the fact that if X_1 and X_2 are random variables, then $E[X_1 + X_2] = E[X_1] + E[X_2]$.

Theorem   If X is a binomial random variable for an experiment consisting of n repetitions of independent trials with success probability p, then

$$E[X] = np \qquad (4\text{-}9)$$

EXAMPLE 4-8

A binomial experiment consists of 100 repetitions of independent trials each with success probability $\frac{1}{4}$. What is the expected number of successes?

Solution: Using Eq. (4-9), we have the mean number of successes equal to $100(\frac{1}{4}) = 25$. □

The theorem and the illustration of Example 4-8 have conclusions which correspond to our intuitive notion of expected value. If an experiment results in a certain outcome which we call success one-fourth of the time, and if the experiment is repeated 100 times, we expect that outcome to occur about 25 times. Naturally, in any specific sequence of 100 trials the number of successes may differ from 25. In some cases 20 successes may occur, and in other cases 30 may occur. A reasonable question to ask is: How often will the number of successes be close to 25; say between 20 and 30? Questions like this will be considered in Sec. 4-4.

The computation of the mean of a normal random variable requires techniques which are not available to us (they are based on calculus). However, it is possible to analyze the situation partially using an elementary argument based on symmetry. The symmetry of the graph of the density function of a standard normal random variable about a vertical line through zero can be interpreted as meaning that the probability that the random variable assumes positive values in any interval is exactly equal to the probability that the random variable assumes negative values of equal magnitudes. Thus the expected value, which in this case is a sort of generalized "sum" of values multiplied by probabilities, consists of exactly balancing positive and negative terms and is therefore equal to zero.

Theorem   The mean of the standard normal random variable is zero.

A normal random variable X is defined to be one for which there are numbers μ and σ such that the random variable $(X - \mu)/\sigma$ is the standard normal random variable. The formula $(X - \mu)/\sigma$ simply shifts the values of X about μ to values about zero. That is, μ plays the same role for X as zero does for the standard normal random variable.

Theorem   If X is a normal random variable, and if $(X - \mu)/\sigma$ is the standard normal random variable, then μ is the mean of X.

Variation about The Mean

It is helpful to begin our discussion of the distribution of the values of a random variable about the mean with an example.

EXAMPLE 4-9

Suppose we have three experiments, labelled I, II, and III, and three random variables, X_I, X_{II}, X_{III}. Moreover suppose that the density functions of these random variables are as shown in Table 4-3. Find the mean of each random variable.

TABLE 4-3

x	$\Pr[X_I = x]$	x	$\Pr[X_{II} = x]$	x	$\Pr[X_{III} = x]$
$+5$	$\dfrac{1}{2}$	$+1$	$\dfrac{1}{4}$	$+5$	$\dfrac{1}{10}$
-5	$\dfrac{1}{2}$	0	$\dfrac{1}{2}$	0	$\dfrac{4}{5}$
		-1	$\dfrac{1}{4}$	-5	$\dfrac{1}{10}$

Solution: The expected value of a random variable, i.e., the mean, is the sum of the products obtained by multiplying each value of the random variable by the probability that the random variable assumes that value. Thus,

$$E[X_I] \ = \tfrac{1}{2}(5) + \tfrac{1}{2}(-5) = 0$$

$$E[X_{II}] \ = \tfrac{1}{4}(1) + \tfrac{1}{2}(0) + \tfrac{1}{4}(-1) = 0$$

$$E[X_{III}] = \tfrac{1}{10}(5) + \tfrac{4}{5}(0) + \tfrac{1}{10}(-5) = 0 \qquad \qquad \square$$

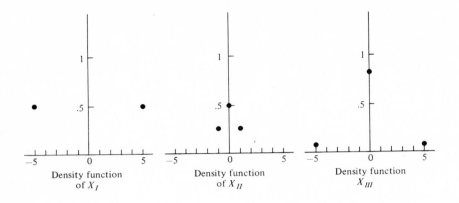

FIGURE 4-5

In Example 4-9 the random variables, X_I, X_{II}, and X_{III}, each have mean zero. However, it is clear by inspecting the density functions of Table 4-3 or the graphs of Fig. 4-5 that each of these random variables is quite different from the others. In particular, the values of the random variables and their respective probabilities differ from experiment to experiment. One measure of this variability is simply the difference between the largest value and the smallest value assumed by the random variable. If we denote this quantity by $s(X)$ for a random variable X and call it the *spread* of the random variable, then

$$s(X_I) \ = 5 - (-5) = 10$$
$$s(X_{II}) \ = 1 - (-1) = 2$$
$$s(X_{III}) = 5 - (-5) = 10$$

Thus, the spreads of X_I and X_{III} are equal. However, from Table 4-3 and Fig. 4-5 it is clear that the random variables X_I and X_{III} are not actually similar with respect to their variability, and indeed in a sense X_{II} is more like X_{III} than is X_I, even though the spread of X_{II} is 2 and the spreads of X_I and X_{III} are 10. This follows because in most trials the random variables X_{II} and X_{III} take values close to 0, while the random variable X_I never takes the value 0 and in fact always takes a value 5 units from 0. This points out the difficulty in using the spread as a measure of the variability of a random variable; namely, it does not take into account the probabilities with which the random variable takes on various values.

A second method of measuring the variability of a random variable is to consider the amounts by which the values of the random variable differ from the mean and the frequency with which these differences occur. When this type of measure is used, X_I, for example, would show more variation than X_{III}. Indeed X_{III} usually (with probability $\frac{4}{5}$) takes the same value as the mean, while X_I always takes a value which is 5 units from the mean of X_I. There are various methods and formulas which can be used to describe this sort of a variation from the mean. For instance, one can simply add up the differences between the mean and the values assumed by the random variable with each such difference multiplied by the probability that this difference occurs. However, such a sum may be very small or even zero in cases with large variations because the differences have different signs (some positive and some negative) and they cancel out. One way to avoid this problem is to ignore the sign of the differences between the values of the random variable and the mean, i.e., to sum the products of the probabilities and the absolute values of the differences. Although such a sum will certainly give a useful measure of the variability of a random variable, in general sums of absolute values are difficult to work with. Thus, instead of using the sum of the absolute value of the differences one usually uses the sum of the squares of the differences. The precise definition is as follows:

Definition   Let X be a random variable which assumes the values x_1, x_2, ..., x_n with probabilities p_1, p_2, ..., p_n, respectively. Let $\mu = E[X]$ be the mean of X. The *variance* of X is the number

$$v[X] = p_1(x_1 - \mu)^2 + p_2(x_2 - \mu)^2 + \cdots + p_n(x_n - \mu)^2 \qquad (4\text{-}10)$$

The *standard deviation* of X is the number

$$\sigma[X] = \sqrt{v[X]} \qquad (4\text{-}11)$$

EXAMPLE 4-10

Compute the variance and standard deviation of the random variables X_I, X_{II}, and X_{III} of Example 4-9.

Solution: Using the definitions (4-10) and (4-11), we have

$$v[X_I] = \tfrac{1}{2}(5 - 0)^2 + \tfrac{1}{2}(-5 - 0)^2 = 25$$
$$v[X_{II}] = \tfrac{1}{4}(1 - 0)^2 + \tfrac{1}{2}(0 - 0)^2 + \tfrac{1}{4}(-1 - 0)^2 = \tfrac{1}{2}$$
$$v[X_{III}] = \tfrac{1}{10}(5 - 0)^2 + \tfrac{4}{5}(0 - 0)^3 + \tfrac{1}{10}(-5 - 0)^2 = 5$$

and $$\sigma[X_I] = 5 \qquad \sigma[X_{II}] = \frac{1}{\sqrt{2}} \qquad \sigma[X_{III}] = \sqrt{5} \qquad \square$$

The importance of the concepts of the variance and the standard deviation of a random variable will be discussed in Sec. 4-4. For the moment we simply note that the standard deviation of X_I is greater than that of X_{III} and both are greater than that of X_{II}. At least in this sense, the standard deviation reflects the dispersion of the values of a random variable.

Binomial and normal random variables will be used in the examples of the remainder of this chapter, and we turn next to a discussion of the standard deviations of these random variables.

Theorem

If X is a binomial random variable for an experiment consisting of n repetitions of independent trials with success probability p, then the variance $v[X]$ of X is

$$v[X] = np(1 - p) \tag{4-12}$$

and the standard deviation $\sigma[X]$ is

$$\sigma[X] = \sqrt{np(1 - p)} \tag{4-13}$$

The theorem can be easily verified for small values of n. In the experiment described, the mean of the random variable X is np. If we write q for $1 - p$, it follows from the definition of the variance that

$$v(X) = (0 - np)^2 \binom{n}{0} p^0 q^n + (1 - np)^2 \binom{n}{1} pq^{n-1}$$

$$+ (2 - np)^2 \binom{n}{2} p^2 q^{n-2} + \cdots + (n - np)^2 \binom{n}{n} p^n q^0$$

To confirm this, recall that the values of X are 0, 1, 2, ..., n and the probabilities of these values are given by Eq. (4-2). The verification of the theorem in the case $n = 1$ is straightforward. We have (since $p + q = 1$)

$$v[X] = (0 - p)^2 \binom{1}{0} p^0 q^1 + (1 - p)^2 \binom{1}{1} p^1 q^0$$

$$= p^2 q + q^2 p = pq(p + q) = pq$$

Mathematical induction can be used to complete a proof of the theorem.

EXAMPLE 4-11

Literature prepared by O. A. Row Seed Company asserts that the germination rate for its best-selling Robust Red Tomato seeds is 98 percent. Using this, determine the mean and standard deviation of the random variable X, which is defined as the number of seeds which germinate into seedlings in a planting of 2000 seeds. Assume germinations of seeds are independent events.

Solution: The random variable X is a binomial random variable with $n = 2000$ and $p = .98$. Using formulas (4-9) and (4-12), we have

$$E[X] = (.98)(2000) = 1960$$

$$\sigma[X] = \sqrt{(2000)(.98)(.02)} = \sqrt{39.2} = 6.26$$

where the square root is computed to two decimal places.   □

The variance and standard deviation of a normal random variable can be found by applying the following theorem.

Theorem   If X is the standard normal random variable, then $v[X] = 1$ and $\sigma[X] = 1$. If X is a normal random variable, and if $(X - \mu)/\sigma$ is the standard normal random variable, then σ is the standard deviation of X.

In Example 4-6 of Sec. 4-2 the random variable X is such that $(X - 110)/10$ is the standard normal random variable. We conclude from this theorem that the standard deviation of X is 10.

We have shown that the standard deviation of a random variable is a measure of how its values are distributed about the mean. We now wish to be more specific about the information which can be obtained from the mean and standard deviation of a random variable. We concentrate here on normal random variables, and in the next section we consider binomial random variables. We begin with two examples to indicate the nature of the results which follow.

EXAMPLE 4-12

Consider an experiment and a random variable X with the values and probabilities given in Table 4-4.

The experiment is performed, and X is measured. What is the probability that X takes a value within 1 standard deviation of its mean value? Here and in the future we interpret the term *within* as including values of the random variable which are exactly 1 standard deviation from the mean.

TABLE 4-4

x	$\Pr[X = x]$
1	.4
2	.3
3	.2
4	.1

Solution: To answer the question we must first compute $\mu = E[X]$ and $v[X]$. We have, using the definition of expected value,

$$\mu = E[X] = 1(.4) + 2(.3) + 3(.2) + 4(.1) = 2$$

and, using Eq. (4-10),

$$v[X] = \sigma^2[X] = (1 - 2)^2(.4) + (2 - 2)^2(.3) + (3 - 2)^2(.2) + (4 - 2)^2(.1) = 1$$

or
$$\sigma(X) = 1$$

The set of values to be considered is the interval with the endpoints $\mu - \sigma = 2 - 1 = 1$ and $\mu + \sigma = 2 + 1 = 3$. There are three values of the random variable in this interval (values 1, 2, 3). The probability that the random variable assumes one of these values is $Pr[X = 1] + Pr[X = 2] + Pr[X = 3] = .4 + .3 + .2 = .9$. Therefore, the probability that the random variable takes a value in the interval from 1 to 3 is .9. We note also that in this example all values of X lie within 2 standard deviations of the mean; i.e., they lie in the interval from $\mu - 2\sigma = 0$ to $\mu + 2\sigma = 4$.   □

EXAMPLE 4-13

Consider an experiment and a random variable X with values and probabilities as given in Table 4-5.

The experiment is performed, and X is measured. What is the probability that X takes a value within 1 standard deviation of its mean? Within 2 standard deviations?

TABLE 4-5

x	$Pr[X = x]$
1	.05
2	.05
3	.20
4	.10
5	.05
6	.40
8	.15

Solution: We have

$$\mu = E[X] = 1(.05) + 2(.05) + 3(.2) + 4(.1) + 5(.05) + 6(.4) + 8(.15)$$
$$= 5$$

and

$$\sigma^2[X] = (4)^2(.05) + (3)^2(.05) + (2)^2(.2) + 1(.1) + 1(.4) + (3)^2(.15)$$
$$= 3.9$$

so $\sigma[X] = \sqrt{3.9}$. Since $1.9 < \sqrt{3.9} < 2.0$, the interval with endpoints $\mu - \sigma$ and $\mu + \sigma$, that is, the interval with endpoints $5 - \sqrt{3.9}$ and $5 + \sqrt{3.9}$, contains the values 4, 5, and 6 (but not the value $3 = 5 - 2$). The probability that the random variable takes one of these values is .55. The interval with endpoints $\mu - 2\sigma$ and $\mu + 2\sigma$ contains the values 2, 3, 4, 5, 6, and 8. The probability that the random variable takes one of these values is .95. $\square$

We now consider a general version of the question raised in Examples 4-12 and 4-13. Namely, given the mean and the standard deviation of a random variable, what can be said about the values of the random variable? In particular what is the probability that the random variable will assume a value within 1 standard deviation of the mean? 2 standard deviations of the mean? 3? We shall answer the question for normal random variables and indicate how to estimate the answer for many binomial random variables.

The following result for a normal random variable can be obtained by using the fact that $(X - \mu)/\sigma$ is a standard normal random variable together with the values in Appendix A.

If X is a normal random variable with mean μ and standard deviation σ, then

$$\Pr[\mu - \sigma \ \le X \le \mu + \sigma] \ = .6826$$
$$\Pr[\mu - 2\sigma \le X \le \mu + 2\sigma] = .9554 \qquad (4\text{-}14)$$
$$\Pr[\mu - 3\sigma \le X \le \mu + 3\sigma] = .9974$$

From (4-14) we see that about 68 percent of the values of a normal random variable are within 1 standard deviation of the mean, 95 percent are within 2 standard deviations, and over 99 percent are within 3 standard deviations.

The information provided by (4-14) is information about normal random variables, and in general it cannot be expected to hold for other random variables. However, there are special classes of random variables (in particular many binomial random variables) which have density functions similar to those of normal random variables. In such cases (4-14) and the probabilities of Appendix A can be used. This is a very significant advantage since the probabilities in (4-14) are very difficult to compute directly for most random variables. For example, consider the computations involved in finding $\Pr[\mu - \sigma \le X \le \mu + \sigma]$, where X is a binomial random variable with $n = 1000$ and $p = .4$. For this random variable $\mu = np = 400$ and $\sigma = \sqrt{np(1-p)} = \sqrt{(1000)(.4)(.6)} = 15.49$ (with accuracy to two decimal places). Thus

$$
\begin{aligned}
\Pr[\mu - \sigma \le X \le \mu + \sigma] &= \Pr[384.51 \le X \le 415.49] \\
&= \Pr[X = 385] + \Pr[X = 386] \\
&\quad + \cdots + \Pr[X = 414] + \Pr[X = 415]
\end{aligned}
$$

Each entry in this sum involves a very large number of multiplications, and thus a simpler method of computing probabilities such as $\Pr[\mu - \sigma \le X \le \mu + \sigma]$ can be very helpful. Such a method is the topic of the next section.

Exercises for Sec. 4-3

1. Compute the mean, spread, variance, and standard deviation for each of the following random variables.

a. x	Pr[X = x]	b. x	Pr[X = x]	c. x	Pr[X = x]
1	1/5	1	2/5	1	.1
2	1/5	2	1/15	2	.2
3	1/5	3	1/15	3	.3
4	1/5	4	1/15	4	.4
5	1/5	5	2/5		

2. In a class with 10 students the scores on a mathematics exam are 85, 65, 40, 60, 90, 95, 80, 70, 85, 65. Find the mean and standard deviation of the random variable which assigns to each student that student's score on the exam.

3. Give an example of two different random variables with the same mean and the same standard deviation.

4. Find the mean and standard deviation for the random variable of Exercise 1, Sec. 4-2.

5. Find the mean and standard deviation for the random variable of Exercise 2, Sec. 4-2.

6. Suppose that shooting free throws in basketball is considered to be a binomial experiment, and consider a player whose probability of success on any specific attempt is .73. If the player shoots 150 free throws each season, what is the expected number of free throws made per year, and what is the standard deviation for the number of successes per year?

7. In Exercise 6 how many free throws should the player shoot to have an expected number of successes which is at least 100?

8. Consider a binomial random variable X with $n = 10$ and $p = .25$. What is the probability that X will take a value within 1 standard deviation of the mean? That is, find $\Pr[\mu - \sigma \le X \le \mu + \sigma]$.

9. For each of the random variables in Exercise 1 compute the probability $\Pr[\mu - \sigma \le X \le \mu + \sigma]$.

10. Let the random variable X have the following values and associated probabilities:

x	$\Pr[X = x]$
-1	.1
2	.5
4	.3
10	.1

Find the mean and standard deviation for this random variable and compute the probability that the random variable takes a value in the interval from $\mu - \sigma$ to $\mu + \sigma$. Also compute the probability that it takes a value in the interval from $\mu - 2\sigma$ to $\mu + 2\sigma$.

11. Let the random variable X have the following values and associated probabilities:

x	$\Pr[X = x]$
-10	.1
-1	.2
0	.4
$+1$	.2
$+10$	.1

Graph the density function for this random variable. Also compute the mean and standard deviation. What percent of the values of the random variable lie within 3 standard deviations of the mean?

4-4 THE NORMAL APPROXIMATION TO A BINOMIAL RANDOM VARIABLE

In many problems involving a binomial random variable it is of interest to know how the values of that random variable are distributed. In particular, we frequently need to determine the probability that a binomial random variable takes values in a given set. If the binomial random variable has a large n (number of trials), and if the set of values of interest is large, it is a tedious task to compute the values of the density function. Consequently, it is an involved computation to determine the desired probability directly. In this section we show that for many binomial random variables there is a convenient approximation which enables us to use Appendix A to estimate such probabilities. A more general result concerning the distribution of the values of an arbitrary random variable about its mean is given in the exercises. The result is known as *Chebyschev's theorem,* and it is the topic of Exercises 9 to 11.

To begin our comparison of the density functions of binomial random variables with those of normal random variables we consider an example.

EXAMPLE 4-14

The density function for the binomial random variable associated with 10 repetitions of independent trials with success probability .4 is given in Table 4-6 and graphed in Fig. 4-6. We compare this density function with an appropriate normal density function.

The most obvious difference between the graph of the density function in Fig. 4-6 and the graph of a normal density function (Fig. 4-2, for example) is that the former consists of discrete points while the latter consists of a curve. Also, the probabilities of binomial density functions are associated with numbers (points on the x axis) while the probabilities of normal density functions are associated with intervals. To help emphasize the similarities of the two situations, let us represent the probabilities in the discrete (binomial) case somewhat differently. Since probabilities are represented by areas in the case of a normal random variable, for purposes of comparison we do the same in the case of a binomial random variable. Instead of assigning a probability, say p, to a point, say k, we assign probability p to an interval $(k - .5, k + .5)$. In this example we assign probability .0060 to the interval $(-.5, .5)$ instead of to the point (value) 0; we assign probability .0403 to the interval $(.5, 1.5)$; . . . ; we assign probability .0001 to the interval $(9.5, 10.5)$. Since each of these intervals is of length 1, we can associate the appropriate probability with an area by constructing a rectangle with the interval as base and with height equal to the desired probability. The density function for this binomial random variable graphed in the way just described is shown in Fig. 4-7, where the area of the shaded rectangle with base of unit length centered at 4 is .2508. The height of that rectangle is also .2508.

This method of representing the probabilities in the discrete case clearly increases the similarities of the two situations. However, there are still some dissimilarities: (1) the graph of the binomial density function is a stairstep curve, while the graph of the normal density function is smooth (it has no corners); and

FIGURE 4-6

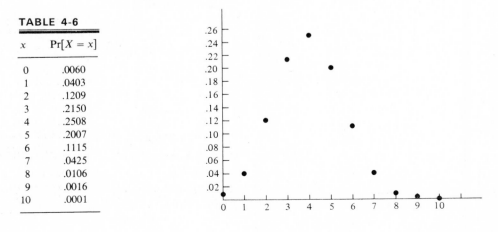

TABLE 4-6

x	$Pr[X = x]$
0	.0060
1	.0403
2	.1209
3	.2150
4	.2508
5	.2007
6	.1115
7	.0425
8	.0106
9	.0016
10	.0001

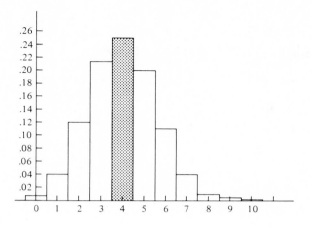

FIGURE 4-7

(2) there is an obvious lack of symmetry in the graph of the binomial density function. Since the mean and standard deviations of this binomial random variable are 4 and $\sqrt{2.4}$, respectively, we compare the graph of Fig. 4-7 with the graph of a normal random variable with this mean and standard deviation. The graph of such a normal random variable is shown in Fig. 4-8, where it is superimposed on the graph of Fig. 4-7. The area under the graphs of both density functions is 1. There are points at which the graph of the normal density function is above that of the binomial density function, and vice versa. In this example it may not appear that the normal density function is a very good approximation to the binomial density (the curve consisting of the tops of the rectangles) because the two curves are not always close together. However, it is not the height of the curves at any single point which is of interest to us but the area under the

FIGURE 4-8

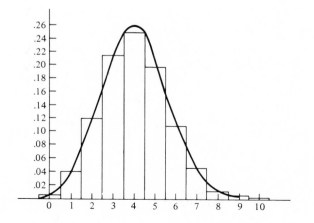

curves and between the points such as 1.5 and 2.5. These areas, which represent the values of the binomial density function, are often quite close to the areas under the normal density curve between the same points. For example, using Appendix A, we can obtain the areas under the normal density curve with mean 4 and standard deviation $\sqrt{2.4}$. These computations require interpolation in the table of Appendix A. The process of interpolation is explained in Appendix A. The areas under the curve and between values of the form $k - .5$ to $k + .5$ are shown in Table 4-7, where they are compared with the areas under the binomial density function.

TABLE 4-7

Interval	Area under Binomial Density	Area under Normal Density
$(-.5,.5)$	.0060	.0101
$(.5,1.5)$	.0403	.0390
$(1.5,2.5)$	.1209	.1156
$(2.5,3.5)$	.2150	.2032
$(3.5,4.5)$	.2508	.2606
$(4.5,5.5)$	.2007	.2032
$(5.5,6.5)$	.1115	.1156
$(6.5,7.5)$	.0425	.0390
$(7.5,8.5)$	.0106	.0101
$(8.5,9.5)$	.0016	.0016
$(9.5,10.5)$	.0001	.0002

The values in Table 4-7 indicate that some of the areas under the normal density are very close to the corresponding areas under the binomial density curve while others (interval 2.5 to 3.5, for example) are not quite so close. We are interested in identifying those cases in which all the corresponding areas have values which are very close together, and to pursue this idea we shall have to consider other values of n and p (here $n = 10$ and $p = .4$).   □

The accuracy of the approximation described in the preceding example varies with the parameters n and p and is illustrated in Fig. 4-9. Notice that for a fixed value of p ($p = \frac{1}{2}$ in Fig. 4-9) the graphs of the density functions become broader and flatter as n increases. Although the density function of the binomial random variable is always a stairstep function and that of the normal is always smooth, the values of the two functions do not differ as much for large n as for small n. Also, for any k and for large values of n the area under the normal curve from $k - .5$ to $k + .5$ (the probability that the normal random variable takes a value in the interval with endpoints $k - .5$ and $k + .5$) is very nearly equal to the area under the graph of the binomial density function from $k - .5$ to $k + .5$ (the probability that the binomial random variable takes the value k). We pursue the area-probability relation when we state a method of computing binomial probabilities using the normal density function. However, for the moment we

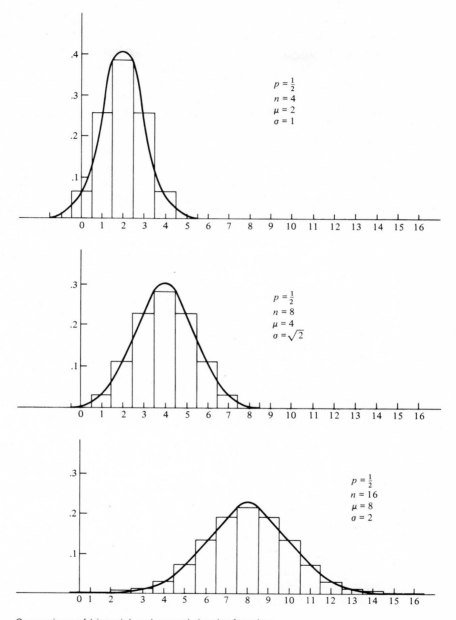

FIGURE 4-9   Comparison of binomial and normal density functions.

content ourselves with the observation, based on the previous argument regarding Fig. 4-9, that for suitable values of the parameters n and p the probabilities computed using a normal density function are acceptable approximations for the binomial probabilities. If this is the case, then we say that the *normal approximation to a binomial random variable is valid*. Such an approximation is not always valid, and it is important to exercise care in the use of this technique. Various rules of thumb have been proposed as a guide to the use of the normal density in place of a binomial. We use the following rule.

The normal approximation to the binomial may be used to compute probabilities for a binomial experiment consisting of n repetitions of independent trials with success probability p whenever

$$np(1 - p) \geq 12 \qquad (4\text{-}15)$$

EXAMPLE 4-15

For which of the following binomial experiments is the normal approximation valid?

(a) $n = 100$ $p = .2$
(b) $n = 100$ $p = .1$
(c) $n = 200$ $p = .1$

Solution: Computing $np(1 - p)$ in each case, we have

(a) $100(.2)(.8) = 16$
(b) $100(.1)(.9) = 9$
(c) $200(.1)(.9) = 18$

We conclude from (4-15) that, using our rule, the normal approximation is valid in cases (a) and (c) but not in case (b).   □

We proceed to a discussion of the method by which one uses the normal approximation to a binomial random variable. The basis of the method is a comparison of the areas under the graphs of the density functions of a binomial random variable and an appropriate normal random variable. If the approximation is to be acceptable, we expect the two random variables to have the same (or nearly the same) mean and standard deviation. Therefore, given a binomial random variable X with parameters n and p [so that X has mean np and standard deviation $\sqrt{np(1 - p)}$], we always choose as a normal approximation the normal random variable with mean np and standard deviation $\sqrt{np(1 - p)}$.

In what follows it is useful to have notation for that part of the area under

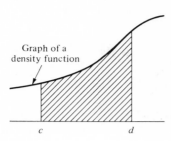

Graph of a
density function

c   *d*

FIGURE 4-10

the graph of a density function which is bounded on the left and right by vertical lines and below by the horizontal axis. Such an area is shown in Fig. 4-10, where the intercepts of the vertical lines are denoted by c and d.

Definition   Let X be a binomial random variable whose density function is graphed as in Fig. 4-7, and let Y be the normal random variable with the same mean and standard deviation as X.

$A_B(c,d)$ denotes the area under the graph of the density function of X from c to d.

$A_N(c,d)$ denotes the area under the graph of the density function of Y from c to d.

It follows immediately from the way the density function of X was graphed in Fig. 4-7 that

$$\Pr[X = k] = A_B(k - .5, k + .5)$$

and if $k < m$, then

$$\Pr[k \le X \le m] = A_B(k - .5, m + .5) \tag{4-16}$$

The latter equality follows from

$$\Pr[X = k] = A_B(k - .5, k + .5)$$
$$\Pr[X = k + 1] = A_B(k + .5, k + 1.5)$$
$$\cdots\cdots\cdots\cdots\cdots\cdots\cdots\cdots\cdots\cdots\cdots$$
$$\Pr[X = m] = A_B(m - .5, m + .5)$$

and the fact that the sum of the terms on the left is equal to $\Pr[k \le X \le m]$ and the sum of the terms on the right is $A_B(k - .5, m + .5)$.

Also, from the definition of a normal random variable,

$$\Pr[c \le Y \le d] = A_N(c,d)$$

The Method of Approximation

The problem is to determine $\Pr[k \leq X \leq m]$ for a given binomial random variable X. The method is as follows:

1. **Use Eq. (4-16) to represent the desired probability as the area under the graph of the density function of X.**
2. **If condition (4-15) is satisfied, use $A_N(k - .5, m + .5)$ as an approximation for $A_B(k - .5, m + .5)$. Here $A_N(k - .5, m + .5)$ is an area under the graph of the density function of the normal random variable Y with the same mean and standard deviation as X.**
3. **Use the methods introduced in Sec. 4-2 to evaluate $A_N(k - .5, m + .5)$.**

We illustrate the method with an example.

EXAMPLE 4-16

Consider a binomial experiment consisting of 100 repetitions of independent trials with success probability $p = .2$. What is the probability that the binomial random variable takes a value between 25 and 29, inclusive?

Solution: Let X denote the binomial random variable with $n = 100$, $p = .2$. The mean and standard deviation of X are $np = 20$ and $\sqrt{np(1 - p)} = \sqrt{16} = 4$, respectively. Let Y be the normal random variable with mean 20 and standard deviation 4.

The problem is to find $\Pr[25 \leq X \leq 29]$. By Eq. (4-16) this is equal to $A_B(24.5, 29.5)$. Since $np(1 - p) = 16 > 12$, condition (4-15) is satisfied and we can use $A_N(24.5, 29.5)$ as an approximation to $A_B(24.5, 29.5)$.

From the definition of a normal random variable we have

$$A_N(24.5, 29.5) = \Pr[24.5 \leq Y \leq 29.5]$$

Also, since the normal random variable Y has mean 20 and standard deviation 4, we know that the random variable Z defined by

$$Z = \frac{Y - 20}{4}$$

is the standard normal random variable. The inequality

$$24.5 \leq Y \leq 29.5$$

for Y can be transformed into an inequality on Z by using the techniques introduced in Sec. 4-2:

$$4.5 = 24.5 - 20 \leq Y - 20 \leq 29.5 - 20 = 9.5$$

$$\frac{4.5}{4} \leq \frac{Y - 20}{4} \leq \frac{9.5}{4}$$

In terms of Z this inequality is

$$1.12 \leq Z \leq 2.38$$

to two-decimal-place accuracy. We have, as in Sec. 4-2,

$$\Pr[24.5 \leq Y \leq 30.5] = \Pr[1.12 \leq Z \leq 2.38]$$

The probability on the right-hand side of this equation can be evaluated using Appendix A and the techniques of Sec. 4-2. Since

$$\Pr[0 \leq Z \leq 1.12] = .3686 \qquad \Pr[0 \leq Z \leq 2.38] = .4913$$

it follows that

$$\Pr[1.12 \leq Z \leq 2.38] = \Pr[0 \leq Z \leq 2.38] - \Pr[0 \leq Z \leq 1.12]$$
$$= .4913 - .3686 = .1227$$

The answer to our original problem, namely, the probability that X takes a value from 25 to 29, inclusive, is approximately .1227. Increased accuracy can be obtained by retaining three decimal places in the values of the normal random variables and using interpolation in the table of Appendix A. $\square$

EXAMPLE 4-17

Let X be the random variable defined in Example 4-16. What is the probability that X takes a value within 1.5 standard deviations of its mean?

Solution: Let Y and Z be defined as in Example 4-16. Since $\mu = 20$ and $\sigma = \sqrt{np(1 - p)} = 4$, the problem is to find the probability that X takes a value within 6 units of 20, that is, $\Pr[14 \leq X \leq 26]$. By Eq. (4-16) this is equal to $A_B(13.5, 26.5)$.

As before, $np(1 - p) = 16 > 12$, and the normal approximation is valid. We use $A_N(13.5,26.5)$ as an approximation to $A_B(13.5,26.5)$. Since

$$A_N(13.5,26.5) = \Pr[13.5 \leq Y \leq 26.5]$$

our problem reduces to a computation of the latter probability. Again, the inequality $13.5 \leq Y \leq 26.5$ for Y translates into an inequality for Z, where $Z = (Y - 20)/4$. The inequality for Z is

$$-1.62 \leq Z \leq 1.62$$

Using the techniques of Sec. 4-2, we have

$$\begin{aligned}
\Pr[13.5 \leq Y \leq 26.5] &= \Pr[-1.62 \leq Y \leq 1.62] \\
&= 2\Pr[0 \leq Z \leq 1.62] \\
&= 2(.4474) = .8948
\end{aligned}$$

We conclude that the probability that X takes a value within 1.5 standard deviations of its mean is approximately .8948.   □

In Examples 4-16 and 4-17 we included all the steps in illustrating the method of applying the normal approximation to compute binomial probabilities. We did so to help explain the basis of the method. In practice, however, it is customary to use an abbreviated version of the method, which can be summarized as follows:

Let X be a binomial random variable with parameters n (number of trials) and p (success probability), and suppose $np(1 - p) \geq 12$, so that the normal approximation is valid. Then

$$\Pr[k \leq X \leq m] \approx \Pr\left[\frac{k - .5 - np}{\sqrt{np(1 - p)}} \leq Z \leq \frac{m + .5 - np}{\sqrt{np(1 - p)}}\right] \qquad (4\text{-}17)$$

where $\approx$ means "is approximately equal to" and Z is the standard normal random variable.

Two comments are in order: (1) probabilities involving the standard normal random variable can be computed using Appendix A; and (2) it may happen that instead of the probability of a value's being between k and m we are interested in either the probability of its being less than m or greater than k. In such a case (4-17) becomes

$$\Pr[X \leq m] \approx \Pr\left[Z \leq \frac{m + .5 - np}{\sqrt{np(1 - p)}}\right] \tag{4-18}$$

or

$$\Pr[k \leq X] \approx \Pr\left[\frac{k - .5 - np}{\sqrt{np(1 - p)}} \leq Z\right] \tag{4-19}$$

respectively. We employ the abbreviated method, in this case formula (4-19), in the last example of this section.

EXAMPLE 4-18

Shortshocks Electronics, Inc., manufactures hand calculators, and company records indicate that when the production process is operating normally, defective circuit boards occur at random with a frequency of about 1 in 5. The company selects 100 circuit boards per day and tests them, and if 30 or more defectives are found, the production process is halted for adjustment. What is the probability that the company will shut down the process even though it is operating normally?

Solution: Suppose that the process is operating normally. We consider each day's tests to be a binomial experiment with $n = 100$ and $p = .2$ (a *defective* circuit board is a success). Let X denote the associated binomial random variable. The mean number of successes is 20, and the standard deviation is 4. Since $np(1 - p) = 16 > 12$, the normal approximation is valid.

We wish to compute the probability that X takes a value of 30 or more, that is, $\Pr[X \geq 30]$. Using (4-19), we determine that this probability is approximately

$$\Pr\left[\frac{30 - .5 - 20}{4} \leq Z\right] = \Pr[2.38 \leq Z]$$

Using the techniques of Sec. 4-2 and Appendix A, we find

$$\Pr[2.38 \leq Z] = .5 - \Pr[Z \leq 2.38]$$
$$= .5 - .4913 = .0087$$

To summarize, if the production process is operating normally, the probability that it will be halted as a consequence of obtaining 30 or more defectives in a test is approximately .0087. $\square$

Exercises for Sec. 4-4

1. Suppose $n = 100$ and $p = .2$ for a binomial experiment. Find μ and σ and convert each of the following inequalities for X into an inequality for $Z = (X - \mu)/\sigma$.
   a. $X \leq 30$ b. $15 \leq X \leq 25$
   c. $X \geq 25$ d. $\mu - \sigma \leq X \leq \mu + \sigma$

2. Consider a binomial experiment with 100 trials and probability $p = .5$ of success. Find the mean and standard deviation for the random variable X (the number of successes). Also find $\Pr[47 \leq X \leq 49]$ using the normal approximation to the binomial. How would you find this probability without using the normal approximation?

3. Construct a table for $n = 10$ and $p = .2$ analogous to Table 4-7. Then compare the areas under the binomial density (using rectangles) between values $k - .5$ to $k + .5$ with the corresponding areas under a normal density curve with mean $np = 2$ and standard deviation $\sqrt{npq} = \sqrt{1.6}$. Is the normal approximation in this case better or worse than in Table 4-7?

4. A bag with 1600 fair pennies is dumped onto a table. Using the normal approximation to the binomial density function, estimate the probabilities of the following events:
   a. The number of heads is between 760 and 840.
   b. The number of heads is at least 780.

5. Suppose that with a certain flu vaccine about 10 percent of all those inoculated receive undesirable side effects. If 500 people are inoculated on a certain day, what is the probability that at least 60 of them will receive undesirable side effects? (*Hint:* Consider each inoculation to be a binomial trial.)

6. Suppose that the average height of 1000 freshmen male students is 5 feet 10 inches and the standard deviation is 3 inches. Suppose also that the students' heights closely follow a normal distribution. How many of these students would you expect to be 6 feet 2 inches or taller? 5 feet 4 inches or shorter?

7. On an exam in an economics class the mean score was 70, and the standard deviation was 10. Suppose that the scores were normally distributed and grades were assigned as follows: over $88 = A$, from 75 to $88 = B$, from 60 to $74 = C$, from 48 to $59 = D$, below $48 = F$. What percent of the students received each grade?

8. In Example 4-18 suppose only 80 units are tested, but in normal operation 1 in 4 is defective. How many defective items should be found before the process is halted if the company wants to be sure that the process will be halted when it is running normally with probability less than .025?

9. The results of this section apply to only the binomial distribution and the normal distribution. A more general result concerning how the values of an arbitrary random variable are distributed about the mean is the following:

Chebyshev's Theorem

If X is a random variable with mean μ and standard deviation σ, then the probability that X takes a value in the interval from $\mu - h\sigma$ to $\mu + h\sigma$ is at least $1 - 1/h^2$. That is,

$$\Pr[|X - \mu| < h\sigma] \geq 1 - \frac{1}{h^2}$$

For $h = \frac{3}{2}$, this theorem states that the random variable takes a value in the interval from $\mu - \frac{3}{2}\sigma$ to $\mu + \frac{3}{2}\sigma$ with probability at least

$$1 - \frac{1}{(\frac{3}{2})^2} = \frac{5}{9}$$

Using the random variable and probabilities shown below, verify Chebyshev's theorem for $h = 2$. That is, show that the random variable takes a value within 2 standard deviations of the mean with probability at least $1 - 1/2^2 = .75$.

x	$\Pr[X = x]$
-12	1/3
-6	1/6
0	1/12
6	1/6
12	1/4

10. Use Chebyshev's theorem to answer the question: What value of h has the property that at least 80 percent of the values of the random variable with mean μ and standard deviation σ lie in the interval from $\mu - h\sigma$ to $\mu + h\sigma$?

11. Consider a random variable X which takes the values $-k$, 0, and k with probabilities p, $1 - 2p$, and p, respectively.
   a. Find the mean μ and standard deviation σ of X in terms of k and p.
   b. What is σ when $p = \frac{1}{2}$?
   c. For which value of p is $\sigma = k/2$? For which value of p is $\sigma = k/3$?
   d. Show that for any positive number h there is a random variable X such that the probability that X takes a value at least $h\sigma$ units from the mean of X is $1/h^2$. Thus show that in some cases Chebyshev's theorem gives the best possible result.

12. Consider a binomial experiment with probability p of success and probability q of failure and with n trials. Show that if $npq \geq 12$, then $np \geq 5$ and $nq \geq 5$. The two conditions $np \geq 5$ and $nq \geq 5$ are also used as a criterion for deciding when the normal approximation to the binomial is valid. As this problem indicates, this criterion is weaker than that of $npq \geq 12$.

4-5 DECISION PROBLEMS

In many situations of interest in business and the social and life sciences statistical arguments are used to obtain estimates on unknown probability distributions. For example, if we are concerned with the effectiveness of a new vaccine, initially (before any tests are conducted) the probability that the vaccine will prove effective is unknown. This probability, as well as probabilities of side effects, must be determined through experimentation. In this generality the problem is very complex. We consider in this section a very simple version of a related question. If we assume that a situation can be modeled as a binomial experiment, how should we evaluate the reasonableness of various assumed values of the success probability p?

EXAMPLE 4-19

The Pop-a-Pill Pharmaceutical Company has developed a vaccine which it claims reduces the risk of catching mononucleosis. The company is asked by FEDREG (the government agency which evaluates such claims) to substantiate the claim, and the company proposes an experiment involving 10,000 college students. The company's claim is that the vaccine reduces the normal risk of infection by a factor of at least $\frac{3}{4}$. Thus, if the probability of catching mononucleosis is p without the vaccine, then it is not more than $p/4$ with the vaccine. FEDREG personnel are reluctant to accept the claim on the basis of a single test, but they agree to permit the company to use the claim in its advertising if very few students are infected; specifically, if the results of the experiment would occur with probability less than .01 for any infection probability larger than the one claimed. In addition, they will prohibit the company from using the claim if many students are infected; specifically if the results of the experiment would occur with probability less than .05 for any infection probability less than that claimed. Otherwise more testing will be required. From health records it is determined that the probability that a randomly selected student at Big State University will contract mononucleosis during December is .02. The company proposes to vaccinate 10,000 randomly selected students and to view the subsequent testing as a binomial experiment. How should the results of the test be evaluated?

Solution: First, we determine the criterion (number of infected students) under which FEDREG will accept the claim. Since the normal infection probability (for unvaccinated students) is .02, the company claims that its vaccine reduces the infection probability to .02/4 = .005 or less. The claim will be accepted if the observed number of infected students would occur with probability $\leq .01$ for any infection rate greater than or equal to the claimed value, .005. Since the strictest acceptance criterion is associated with the smallest infection probability (subject to being $\geq .005$), we consider the infection probability to be equal to .005. The expected number of infected students is then $(.005)(10,000) = 50$, the variance ($= npq$) is 49.75, and the standard deviation is approximately 7.05. Since $npq > 12$, we are justified in using the normal approximation to the binomial random variable. Adopting the notation of Sec. 4-4, we let X, Y, and Z denote the binomial random variable, the normal random variable with the same mean and standard deviation as the binomial, and the standard normal random variable, respectively. Our goal is to determine the largest integer K such that

$$\Pr[X \leq K] \leq .01$$

Using the techniques of Sec. 4-4, this is equivalent to determining the largest integer K such that

$$\Pr[Y \le K + .5] \le .01$$

or

$$\Pr\left[Z \le \frac{K + .5 - 50}{7.05}\right] \le .01$$

From the table in Appendix A we find that (to two decimal-place accuracy)

$$\frac{K - 49.5}{7.05} \le -2.33$$

or

$$K \le 49.5 - 16.43 = 33.07$$

We conclude that the event of 33 or fewer students being infected would have probability $\le .01$ for any infection probability greater than or equal to .005. If the experiment results in 33 or fewer infected students, then according to the agreement, FEDREG should permit the company to use its claim in advertising.

Next we determine the criterion (number of infected students) under which FEDREG will reject the claim. The claim will be rejected if the observed number of infected students would occur with probability $\le .05$ for infection rates $\le .005$. In this case the strictest criterion is associated with the largest infection probability (subject to being $\le .005$), and again we find infection probability .005 the crucial one to consider. Retaining the notation introduced above, we seek the smallest integer K such that

$$\Pr[X \ge K] \le .05$$

This is equivalent to determining the smallest integer K such that

$$\Pr[Y \ge K - .5] \le .05$$

or

$$\Pr\left[Z \ge \frac{K - 50.5}{7.05}\right] \le .05$$

Using Appendix A, we have

$$\frac{K - 50.5}{7.05} \ge 1.65$$

or

$$K \ge 50.5 + 11.63 = 62.13$$

We conclude that the event of 63 or more infected students would have probability $\leq .05$ for any infection probability less than or equal to .005. If the experiment results in 63 or more infected students, then according to the agreement, FEDREG should reject the claim and the company should not be permitted to use it in its advertising.

It is interesting to note that if the vaccine had no effect and the infection probability remained at .02, then 152 or fewer infected students occur with probability less than .001. Consequently, even if there are, say, 65 infected students and the company's claim is rejected, it is very unlikely that the vaccine had no beneficial effect.   □

EXAMPLE 4-20   (CONTINUATION OF THE EXAMPLE OF SEC. 4-1)

In the situation described in Example 4-1 the firm receives shipments of transistors in lots of 50, and it is decided that a lot should be rejected if it contains more than 5 defective transistors. The company plans to test a subset of the 50 transistors in each lot and to accept or reject the shipment on the basis of this test. If the company is willing to assume no more than a 5 percent risk of accepting a lot with more than 5 defective transistors, how many (out of 50) should be tested, and how should the results be evaluated?

Solution: It is reasonable to assume that the company incurs a cost in testing transistors and that this cost increases as the number of transistors tested increases. Otherwise the company should simply test the transistors in each lot until either more than 5 defective ones are found or until it is clear that the lot contains no more than 5 defective ones. Therefore, we assume that the company seeks to test as few transistors as possible consistent with the intention of rejecting shipments which contain more than 5 defective ones. In the same spirit, it is also assumed that the company is willing to tolerate the rejection of some (preferably few) lots which actually contain 5 or fewer defective transistors but which, on the basis of the test, appear likely to contain more than 5.

We make these assumptions precise in the following form. The company is willing to assume no more than a 5 percent risk of accepting a lot with more than 5 defective transistors. That is, a test is sought with the property that if a lot contains more than 5 defectives, then with probability .95 it will be rejected. In this problem both (1) the number of items to be tested and (2) the criterion applied to the results of the test are unknown. Of course, these two aspects of the problem are not independent. The criterion may well vary with the number tested (the sample size).

Since the company is interested in testing the smallest possible number of transistors, we approach this problem by considering various choices for the number of items to be tested, and then we compute the probability of accepting a bad lot for different acceptance criteria. Suppose first that we test only 3 items, and suppose also that the lot being tested is bad, i.e., has more than 5 defective items. The larger the number of defective items, the greater the probability of finding them in a test. Thus in order to develop a criterion which is applicable to *all* bad lots we suppose that there are only 6 defective transistors in the lot of 50 being tested. In this way we obtain the smallest probabilities of finding the defective items in the test and thus rejecting the lot. Table 4-8 contains the probabilities of finding 0, 1, 2, or 3 defective items in a test of 3 items selected from a lot with 6 defective items and 44 acceptable items. These probabilities are obtained by using the formulas developed in Chap. 3; for example,

$$\Pr[2 \text{ defectives}] = \frac{\binom{6}{2}\binom{44}{1}}{\binom{50}{3}} \approx .0337$$

We note from Table 4-8 that the probability of finding no defectives in the sample of three transistors is very high. We conclude from this that such a small sample cannot be used since even the criterion "reject the lot unless all items are acceptable" would result in a probability of .6757 of accepting a lot with 6 defective items.

TABLE 4-8

Number of Defective Transistors	Probability
0	.6757
1	.2896
2	.0337
3	.0010

Next suppose we test 6 items. If we again consider a lot with 6 defective items, then the probability that no defective items will be found is

$$\Pr[0 \text{ defectives in test of 6}] = \frac{\binom{44}{6}}{\binom{50}{6}} = .4442$$

Similarly, if we consider testing 10, 15, or 20 items, we have

$$\Pr[0 \text{ defectives in test of 10}] = \frac{\binom{44}{10}}{\binom{50}{10}} = .2415$$

$$\Pr[0 \text{ defectives in test of 15}] = \frac{\binom{44}{15}}{\binom{50}{15}} = .1021$$

$$\Pr[0 \text{ defectives in test of 20}] = \frac{\binom{44}{20}}{\binom{50}{20}} = .0374$$

We see that to meet the company's restriction of a 5 percent risk, it is inadequate to test 15 items but overcautious to test 20. The results of testing 18 and 19 are

$$\Pr[0 \text{ defectives in test of } 18] = \frac{\binom{44}{18}}{\binom{50}{18}} = .0570$$

$$\Pr[0 \text{ defectives in test of } 19] = \frac{\binom{44}{19}}{\binom{50}{19}} = .0463$$

From this we conclude that in order to meet the company's restriction of a 5 percent risk, the smallest number to be tested is 19 and the evaluation criterion is to reject the shipment unless all tested transistors are acceptable.

Since this choice involves testing nearly 40 percent of the items in a lot, it may be that the company's decision to assume at most 5 percent risk is an excessive restriction. For example, a test of only 10 items and an evaluation criterion of "reject the lot unless all are acceptable" will identify a bad lot with probability .75. Thus the company must weigh the costs of testing items against the cost of accepting bad lots (and rejecting good ones) in deciding what risk to assume. Probabilities of rejecting good lots are considered in Exercises 4 and 5 below.   □

EXAMPLE 4-21

In this example we consider the situation described in Chap. 1 concerning the Downjim Drug Company. To summarize (see Example 1-2 for details) the company produces batches of vaccine, and on the average 1 batch in 20 contains live viruses. All batches with live viruses must be identified by testing, and the company is considering a testing plan based on pooled testing. In pooled testing samples taken from a number of batches are combined and tested together. As many as 10 batches can be pooled and tested together. If none of the batches which are tested together contain live viruses, the test will be negative and these batches can be cleared for distribution. On the other hand, if the test of the pooled samples is positive, 1 or more of the individual batches contains live viruses and all these batches must then be tested individually to determine which ones are contaminated. The costs are $8 for each test conducted plus $2 per batch tested. Thus to test 1 batch alone it coses $10 and to test 3 batches pooled it costs $14. The question asked by the company is: How many batches should be pooled and tested together if total testing costs are to be a minimum?

Solution: In this problem the basic statistical work has been done by the company in determining that, on the average, 1 batch in 20 is contaminated (contains live viruses). Thus the testing of a batch of vaccine can be considered to be a binomial trial with $p = \frac{1}{20}$. The answer to the company's question can be obtained by figuring the average cost to test each batch for each method of pooling the batches. To do this we let n represent the number of batches to be pooled together, and we let $C(n)$ be the average cost to test a single batch when testing is initially done by pooling samples from n batches. We know that $1 \le n \le 10$ from data given by the company. Also, if only one batch is tested at a time, the cost is $\$8 + \$2 = \$10$, so $C(1) = \$10$. We now compute $C(2)$. If 2 batches are pooled, either they contain live viruses or they do not. If they do not contain live viruses, the test of the pooled samples is negative and the total cost to test the 2 batches is $\$8 + \$2 + \$2 = \$12$. This happens with probability $(\frac{19}{20})^2$. On the other hand, if either (or both) of the batches contain live viruses, the pooled test has a positive result and both batches must be tested individually. The cost in this case is $(\$8 + \$2 + \$2) + 2(\$8 + \$2) = \$32$. This event occurs with probability $1 - (\frac{19}{20})^2$. Since there are 2 batches being tested in each case, the average cost per batch is given by

$$C(2) = \frac{\text{expected cost with 2 samples pooled}}{2}$$

$$= \frac{(\$12)(\frac{19}{20})^2 + \$32[1 - (\frac{19}{20})^2]}{2}$$

$$= \frac{\$10.83 + \$3.12}{2} = \$6.97$$

Since $C(2)$ is less than $C(1)$, the company is better off to test the batches in pairs than to test them alone. However, it may be that it is even better to pool more batches for the initial test. To decide we must compute $C(n)$ for $n = 3, 4, \ldots,$ 10. The method of computation is the same in each case; one computes the expected cost of testing starting with a pooled sample from n batches and then divides by n to obtain $C(n)$. This gives the formula (in dollars)

$$C(n) = \frac{(8 + 2n)(\frac{19}{20})^n + (8 + 2n + 10n)[1 - (\frac{19}{20})^n]}{n}$$

$$= \frac{8}{n} + 12 - 10\left(\frac{19}{20}\right)^n$$

The values of $C(n)$ for $n = 1, 2, \ldots, n$ are shown in Table 4-9.

From Table 4-9 we see that the company should pool 4 batches at a time for initial testing. In this way the average cost to test a batch is just a bit over half the cost to test the batches individually. $\qquad\square$

TABLE 4-9

n	$C(n)$
1	$10.00
2	6.97
3	6.09
4	5.85
5	5.86
6	5.98
7	6.16
8	6.37
9	6.59
10	6.81

Exercises for Sec. 4-5

1. Suppose that the Wonder Widget Corporation manufactures widgets in such a way that the probability of a defective widget is .1. Find numbers a and b such that the number of defective widgets found in 200 independent tests will be between $a - b$ and $a + b$ with probability .95.

2. The Wonder Widget Corporation uses the following rule to control the operation of its new widget-making machine. Select and test a sample of 400 widgets each week. If the number of defective widgets is 15 or more, stop the machine for adjustments; otherwise continue operation. What is the probability that:
   a. The machine is stopped for adjustment when it produces, on the average, 4 percent defective widgets?
   b. Operation of the machine continues when it produces, on the average, 8 percent defective widgets?

3. The manufacturer of Growfast Grass advertises that its Premium Quality Bluegrass seed contains not more than 2 percent weed seeds. After growing a yard full of weeds using the seed, Sam Homeowner decides to evaluate the claim. A portion of a box of Premium Quality Bluegrass seed is dumped on a table and Sam begins to count weed seeds (Sam is an amateur horticulturist and he can distinguish weed seeds from grass seeds). He rapidly counts 40 weed seeds. If Sam's efforts are viewed as binomial trials in a sample whose size is the number n of seeds dumped on the table,
   a. How small must n be in order for Sam to be 90 percent certain that the company's claim is false?
   b. How large must n be for there to be a 90 percent chance that the company's claim is true?

4. (Continuation of Example 4-20) If the company tests 10 items, what is the probability that a good lot will be rejected under the criterion "reject the lot if any defective items are found?

5. (Continuation of Example 4-20) Referring to Exercise 4, what if 20 items are tested and the same criterion is used?

6. (Continuation of Example 4-21) Solve the problem of Example 4-21 if the probability of a randomly selected batch of vaccine containing live viruses is .1. All other data of the problem remain the same.
7. (Continuation of Example 4-21) Solve the problem of Example 4-21 if the costs of conducting tests are $10 for each test plus $1.50 per batch tested. All other data of the problem remain the same.

IMPORTANT TERMS

You should be able to describe, define, or give examples of each of the following:

Binomial random variable
Binomial density function
Distribution function
Binomial distribution function with
  parameters n and p
Standard normal random variable
Normal random variable

Standard normal curve
Mean
Variance
Standard deviation
Normal approximation to a binomial
  random variable
Chebyshev's theorem

REVIEW EXERCISES

1. Two fair dice are rolled, and the sum of the numbers of dots on the tops of the dice is noted. A random variable X is defined by associating with each outcome this sum.
   a. Make a table which lists each outcome, the value of the random variable for that outcome, and the probability of that outcome's occurring.
   b. What values are assumed by X, and what are the events on which these values are assumed?
   c. What is the density function of X?
   d. What is the distribution function of X?

2. Two fair dice are rolled, and the numbers of dots on the tops of the dice are noted. A random variable Y is defined by associating with each outcome the number of times an even number of dots occurs.
   a. What values are assumed by Y, and what are the events on which these values are assumed?
   b. What is the density function of Y?
   c. What is the distribution function of Y?

3. A box contains 8 widgets, of which 3 are oversize, 1 is undersize, and the remainder are normal. A sample of 4 widgets is selected at random, and the sizes of the widgets are noted. A random

variable X is defined by associating with each sample the number of oversize widgets.
   a. What are the values assumed by X, and what are the events on which these values are assumed?
   b. What is the density function of X?
   c. What is the mean of X?
   d. What is the variance of X?

4. A box of 8 widgets contains 3 oversize, 1 undersize, and 4 normal widgets. A sample of 4 widgets is selected at random, and the sizes of the widgets are noted. A random variable Y is defined by associating with each sample the number of oversize widgets minus the number of undersize widgets.
   a. What are the values of Y, and what are the events on which these values are assumed?
   b. What is the density function of Y?
   c. What is the mean of Y?

5. Let X be a random variable whose density function is given in Table 4-10.
   a. Find the mean μ of X.
   b. Find the variance v of X
   c. Find the standard deviation σ of X
   d. Find the probability that X takes a value in the interval from $\mu - \sigma$ to $\mu + \sigma$.

TABLE 4-10

x	$\Pr[X = x]$
-1	.20
0	.35
1	.10
2	.15
4	.20

6. An unfair coin has $\Pr[H] = .4$. An experiment consists of flipping the coin 250 times. A random variable X associates with each outcome of the experiment the number of times that heads lands uppermost.
   a. What is the mean of X?
   b. What is the variance of X?
   c. What is the standard deviation of X?

7. Let X be the standard normal random variable. Find
   a. $\Pr[-.92 \le X \le 1.35]$
   b. $\Pr[X \le -.40]$
   c. $\Pr[X \le 1.05]$

8. Let Y be a normal random variable, and suppose that $Z = (Y - 10)/4$ is the standard normal random variable. Find:
   a. $\Pr[6 \le Y \le 12]$
   b. $\Pr[Y \ge 16]$
   c. $\Pr[E]$ where E is the event on which Y takes values less than 7 or greater than 15

9. Let X be a random variable whose density function is given in Table 4-11.
   a. Find the mean of X.
   b. Find the standard deviation of X.

TABLE 4-11

x	$\Pr[X = x]$
1	1/12
3	6/12
5	3/12
7	2/12

10. If 20 percent of the widgets produced by a new high-speed widget-making machine are defective, how large a sample should be taken to have the mean number of acceptable, i.e., not defective, widgets at least 1000?

11. Let Y be a binomial random variable with $n = 162$ and $p = 1/3$.
    a. Find the mean μ and the standard deviation σ of Y.
    b. Show that the normal approximation to Y is legitimate.
    c. Use the normal approximation to evaluate $\Pr[45 \le Y \le 66]$.

12. A report on Vik cigarette lighters asserts that the mean number of lights from a randomly selected lighter is 500 and the variance is 25. If the report is accurate, then with probability .99 a Vik lighter will give between _____ and _____ lights.

13. A major league baseball player has a batting average (number of hits divided by number of times at bat) of .200. If the player's turns at bat are viewed as binomial trials with $p = .2$, what is the probability that in 100 times at bat he has at least 50 hits?

14. One-fifth of the widgets produced by a machine are oversize. If 400 widgets are selected at random and measured for size, what is the probability that the number of oversize ones is no more than 60?

15. A student takes a 100 question true-false test. A passing grade is 70 or more correct answers.
    a. What is the probability that the student passes the test by guessing? When guessing, the probability of a correct answer on any question is .5.
    b. If the student studies and believes that she has probability .8 of answering any question correctly, what is the probability that she does *not* pass the test?

PART TWO

LINEAR MODELS

LINEAR PROBLEMS IN TWO VARIABLES AND THEIR GRAPHS

CHAPTER

FIVE

5-1   INTRODUCTION

A great many situations which arise in the life, management, and social sciences can be described satisfactorily with linear models. In some cases a linear model leads to predictions which agree quite closely with observations, while in other cases a linear model provides only a good first approximation. One of the major virtues of linear models is their mathematical tractability; i.e., the mathematical systems which arise in linear models can be analyzed, and useful numerical information can be obtained with fairly elementary techniques. It is easy to identify situations in which a linear model provides quite accurate predictions. For example, if 3 rats require 4 ounces of food for a period of time, then we expect that 6 rats require 8 ounces of food and 15 rats require 20 ounces of food for the same period of time. In the same vein, if 1 widget requires material costing 30¢, then we expect n widgets to require material costing $30n$ cents. In both these cases we are adopting a linear model for the situation. Somewhat different linear models were formulated in Example 1-1.

The linear models introduced in this chapter are closely related to the familiar concepts of lines and intersections of lines in the plane. Indeed, one of the reasons for considering linear problems in two variables separately from linear problems in more than two variables is that the former can be given a direct interpretation in terms of plane analytic geometry. For example, in the situation involving rats and food described above, there is a function which associates 3 rats with each 4 ounces of food. A formula for this function is

$$f(x) = \tfrac{3}{4}x$$

where $f(x)$ is the number of rats associated with x ounces of food and x is required to be a positive multiple of 4, that is, $x \in \{4,\ 8,\ 12,\ \ldots\}$. The graph of this function is a set of points which lie on a straight line. This simple geometric fact is the reason for the name linear model.

This chapter begins (Secs. 5-2 and 5-3) with a review of two-dimensional cartesian coordinate systems and graphing of lines. These sections contain background material for the remainder of the chapter, and they may be omitted by those who are familiar with coordinate systems and lines in two dimensions. The main topics of the chapter, systems of equations, inequalities, and linear optimization problems in two variables, begin in Sec. 5-4.

5-2   THE CARTESIAN COORDINATE SYSTEM

The cartesian coordinate system is a particularly convenient means of pictorially representing information and relationships between quantities which can be measured, or at least described, with real numbers. This coordinate system is based on the fact that the real numbers can be put in one-to-one correspondence with the points on a straight line. Such a correspondence can be carried out by identifying two distinct points on a straight line and denoting one by 0 and the other by 1. The distance between 0 and 1 is taken as the basic unit of distance. With each real number x a point p on the line is associated in the following way: If x is positive, then take p to be the point whose distance from 0 is x units and which lies on the half line originating at 0 and containing the point 1. If x is negative, then take p to be the point whose distance from 0 is $-x$ units and which lies on the half line originating at 0 and not containing 1. Associate the real number zero with the point 0.

The usual picture is that shown in Fig. 5-1. In such pictures the point 0 associated with the number zero is called the *origin of the line*. A line with this identification of points with real numbers is referred to as a *real axis*. The half line originating at 0 which contains points associated with the positive (negative) real numbers is referred to as a *positive (negative) axis*.

A cartesian coordinate system is constructed by superimposing two real axes at right angles in such a way that the 0 points on the two axes coincide.

FIGURE 5-1

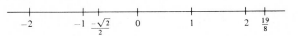

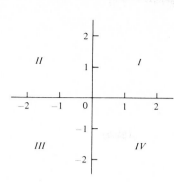

FIGURE 5-2

It is often, but not necessarily, the case that the units of distance are the same on the two axes. In the problems which follow we shall encounter both situations, i.e., axes with the same units and axes with different units of distance. It is common to orient the axes as shown in Fig. 5-2 and to identify the four *quadrants* bounded by parts of the axes as I, II, III, IV, as shown. With respect to a fixed coordinate system, each point P in the plane can be associated with a unique ordered pair of real numbers (x,y), and vice versa. To determine the ordered pair to be associated with a given point P we proceed as follows (see Fig. 5-3*a*). The first member of the ordered pair, x in this case, is by convention the real number obtained when P is projected vertically onto the horizontal axis. Likewise, the second member of the ordered pair is the real number obtained when P is projected horizontally onto the vertical axis. To determine the point P to be associated with an ordered pair (x,y), we define P to be the intersection of the perpendicular lines erected at the points x on the horizontal axis and y on the vertical axis, respectively. Several points are labeled with the appropriate ordered pairs in Fig. 5-3*b*. In diagrams like this it is common to refer to the horizontal axis as the *x axis* and the vertical axis as the *y axis*. The elements of the ordered pair (x,y) are the *coordinates* of the point corre-

FIGURE 5-3

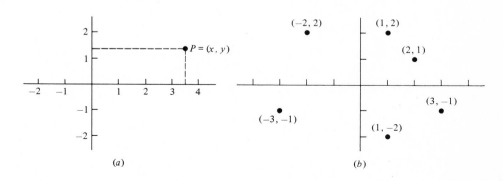

(*a*)   (*b*)

sponding to this ordered pair. The first is the x coordinate and the second is the y coordinate.

As noted earlier, cartesian coordinate systems are especially useful in representing the relationships between quantities measured in terms of real numbers. The following two examples illustrate this use of coordinate systems.

EXAMPLE 5-1

Suppose that the yearly death rate per 1000 individuals for inhabitants of Megapolis is given in Table 5-1 for the period 1970–1977. Construct a coordinate system in which the time in years is measured on the horizontal axis and the number of deaths per thousand is measured on the vertical axis and use it to represent the information in Table 5-1.

TABLE 5-1

Year	Deaths per 1000
1970	12.5
1971	12.1
1972	12.7
1973	11.7
1974	11.5
1975	11.0
1976	10.5
1977	10.8

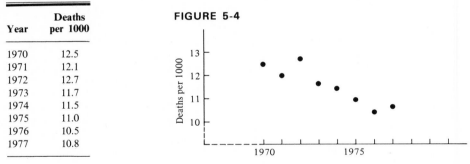

FIGURE 5-4

Solution: See Fig. 5-4. Note that in this example a different unit of distance, i.e., *scale*, is used on each axis. Also, in order to make the representation as convenient as possible, the initial part of each axis is compressed and shown only as a dotted line. This is a common practice when only a small part of each axis is needed to represent the information. ☐

EXAMPLE 5-2

A sociologist conjectures that in Megapolis the number of armed robberies per month is related to the unemployment rate by a function given by the formula

$$\text{Number of armed robberies} = 15 + 600 \times \text{unemployment rate}$$

If the function is denoted by g and an unemployment rate by r, then

$$g(r) = 15 + 600r$$

is the conjectured number of armed robberies in a month with unemployment rate r. Using this function, graph the conjectured number of robberies for unemployment rates of .04, .08, .12, .16, and .20. (Note that an unemployment rate of .04 means that 4 percent of the work force is unemployed.)

Solution: We construct a coordinate system in which the unemployment rate is measured on the horizontal axis and the conjectured number of armed robberies is measured on the vertical axis. Thus, for example, if the unemployment rate is .12, the conjectured number of armed robberies is $g(.12) = 15 + 600(.12) = 87$. The desired data are graphed in Fig. 5-5.

FIGURE 5-5

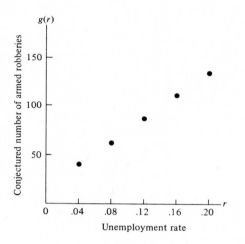

Exercises for Sec. 5-2

1. Graph the data given in Table 1-1 on a cartesian coordinate system.
2. Olympic records in women's freestyle swimming in 1976 are given below. Graph these data on an appropriately scaled cartesian coordinate system.

Distance (meters)	Time (to nearest second)
100	56 seconds
200	1 minute, 59 seconds
400	4 minutes, 10 seconds
800	8 minutes, 37 seconds

3. Compute the average speed (in meters per second) in each of the above events. Using this average speed, graph the points (distance, speed) for the four distances for which data are given on a cartesian coordinate system.

4. The unemployment rates in January and June of the years 1972 through 1976 are shown below. Graph these points on an appropriate cartesian coordinate system.

Date	Rate	Date	Rate
January 1972	5.9	June 1972	5.5
January 1973	5.0	June 1973	4.8
January 1974	5.2	June 1974	5.2
January 1975	8.2	June 1975	8.6
January 1976	7.8	June 1976	7.5

5. On a cartesian coordinate system graph the pairs of points (x,y), where x has the values $-3, -2, -1, 0, 1, 2, 3$, and y is related to x by the formula $y = 2x - 1$.

6. Repeat Exercise 5 with y related to x by the formula $y = 2x^2 - 1$.

7. Repeat Exercise 5 with y related to x by the formula $y = 2x^3 - 1$.

8. A fruitfly colony is observed once a week for 6 weeks, and the following data are obtained for the number of flies in the colony at each of the observations. Graph these data on an appropriate cartesian coordinate system.

Week	Number of Flies
1	10
2	40
3	120
4	250
5	310
6	335

9. Temperatures in Fahrenheit can be changed to temperatures in Celsius by the formula

$$C = \tfrac{5}{9}(F - 32)$$

Use a cartesian coordinate system to represent pairs of values of the form (Fahrenheit, Celsius), where the Fahrenheit values are $-40, 0, 32, 77, 100$, and $220°$.

10. Many savings accounts pay interest which is compounded quarterly (every 3 months). Suppose that $10,000 is invested in such an account and interest is 8 percent per year (2 percent per quarter). Then the total value of the investment after t quarters, i.e., original investment plus interest, is given by the formula

$$V = (10,000)(1 + .02)^t$$

Use a cartesian coordinate system to represent the pairs (number of quarters invested, value of investment) for investments of 4, 8, 12, 20, and 40 quarters.

5-3   THE EQUATION OF A LINE

In our study of linear models in two variables we shall find it very convenient to use both algebraic and geometric representations of lines. The algebraic representation is an equation, and the geometric representation is a graph on a two-dimensional cartesian coordinate system.

Suppose that we are interested in graphing the set of all points (x,y) which satisfy the equation $x - 3y = 6$. We begin by constructing a table of some of the values of x and y for which $x - 3y = 6$ (Table 5-2) and graphing these points (Fig. 5-6). It appears from the figure that all the points lie in a line. Indeed, a ruler or other straightedge can be positioned so that all these points lie along one edge. Moreover, if we were to take any other point which also satisfies the equation, we would find that it also lies along the same line as the points graphed in Fig. 5-6. Finally, if we were to lay a straightedge along the points shown in Fig. 5-6 and take any other point which also lies along the straightedge, we would find that the coordinates x and y of this point satisfy the equation $x - 3y = 6$.

TABLE 5-2

x	y
-3	-3
0	-2
1	$-\dfrac{5}{3}$
2	$-\dfrac{4}{3}$
3	-1
6	0

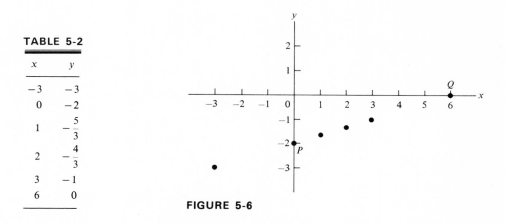

FIGURE 5-6

We have used a ruler or straightedge in this discussion as an approximation of the geometric notion of a line. Although the geometric idea of a line is very intuitive, there are difficulties in making it precise, and we prefer to avoid these problems. Accordingly we define lines algebraically. In this example the line is defined to be the set $\{(x,y):\ x - 3y = 6\}$. We adopt a similar point of view in the general case.

Definition   For any real numbers A, B, and C with A and B not both 0 the set of points $\{(x, y):\ Ax + By = C\}$ is a *line*. If (x_1, y_1) and (x_2, y_2) are two points on this line, i.e., if $Ax_1 + By_1 = C$ and $Ax_2 + By_2 = C$, the line is referred to as the *line through (x_1, y_1) and (x_2, y_2)*.

Remark: It is customary to refer to the equation $Ax + By = C$ as *a line*, and we shall retain this custom where it is convenient. We mean, of course, the set $\{(x,y): Ax + By = C\}$.

The latter part of this definition can also be given intuitive justification. We do so by returning to Fig. 5-6 and noting that a straightedge positioned so that any two specific points lie along the same edge has the property that all other points lie along the same edge. The line determined by the equation $x - 3y = 6$ is shown in Fig. 5-7. The equation $Ax + By = C$ is known as the *general equation of a line* and also as a *linear equation* in two variables. One of the problems that will turn up frequently is that of determining values for the coefficients A, B, and C from given information of various types. It is to this problem that we turn next.

FIGURE 5-7

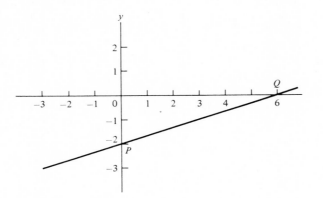

Points P and Q illustrated in Figs. 5-6 and 5-7 have the property that one of the coordinates is zero: $P = (0,-2)$ and $Q = (6,0)$. Points of this form are especially easy to determine from the equation. For example, if we seek a point of the form $(0,y)$ which satisfies $x - 3y = 6$, we set $x = 0$ and solve for y

$$0 - 3y = 6 \quad \text{or} \quad y = \frac{6}{-3} = -2$$

The point $(0,-2)$ is the point which is common to the line and the y axis, and -2 is known as the y intercept of the line. Likewise, 6 is the x intercept of the line. In the general case we have the following definition:

Definition   The line defined by the equation $Ax + By = C$ has an x intercept if $A \neq 0$ and the x *intercept* is the number C/A; the line has a y intercept if $B \neq 0$ and the y intercept is the number C/B.

Note that the x intercept specifies a point on the x axis and the y intercept specifies a point on the y axis. The two intercepts coincide at the point $(0,0)$ if (and only if) both intercepts exist and $C = 0$.

There is a form of the equation of a line in which the y intercept appears explicitly. To derive it we suppose $B \neq 0$. First divide the equation $Ax + By = C$ by B and then subtract $(A/B)x$ from both sides of the result. We have

$$y = -\frac{A}{B}x + \frac{C}{B}$$

This is known as the *slope-intercept* form of the equation, and it is usually written

$$y = mx + b$$

where $m = -A/B$ and $b = C/B$. This constant term, b, is the y intercept introduced above, and m, the coefficient of x, is a quantity which we have not considered before.

Definition

If $B \neq 0$, then $m = -A/B$ is the *slope* of the line $Ax + By = C$. If $B = 0$, the slope of the line is not defined.

EXAMPLE 5-3

Find the slope of the line $x - 3y = 6$.

Solution: This is the equation of a line with $A = 1$, $B = -3$, and $C = 6$. Therefore the slope m is $-A/B = -1/(-3) = \frac{1}{3}$. □

Suppose that (x_1,y_1) and (x_2,y_2) are two distinct points on the line defined by $Ax + By = C$, $B \neq 0$ (Fig. 5-8). That is, $Ax_1 + By_1 = C$ and $Ax_2 + By_2 = C$. If we subtract the first of these equations from the second, we have

$$Ax_2 + By_2 - (Ax_1 + By_1) = 0 \quad \text{or} \quad A(x_2 - x_1) = -B(y_2 - y_1)$$

Since the points (x_1,y_1) and (x_2,y_2) are distinct and $B \neq 0$, we have $x_2 \neq x_1$ (Exercise 13), and we can divide the last equation by $-B$ and by $x_2 - x_1$. The result, recalling the definition of slope m, is

$$m = -\frac{A}{B} = \frac{y_2 - y_1}{x_2 - x_1} \qquad (5\text{-}1)$$

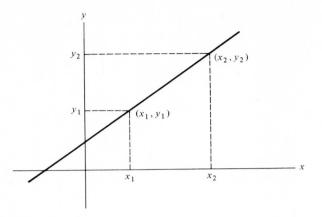

FIGURE 5-8

For lines which have slopes, this provides the following interpretation:

The slope of a line is the ratio of the difference of the y coordinates to the difference of the x coordinates of any two distinct points on the line.

Example 5-4

Compute the slope of the line $x - 3y = 6$ by using the points $(1, -\frac{5}{3})$ and $(2, -\frac{4}{3})$.

Solution: We set $(x_1, y_1) = (1, -\frac{5}{3})$ and $(x_2, y_2) = (2, -\frac{4}{3})$ and use Eq. (5-1):

$$m = \frac{-\frac{4}{3} - (-\frac{5}{3})}{2 - 1} = \frac{\frac{1}{3}}{1} = \frac{1}{3} \qquad \square$$

Another interpretation of the slope can be obtained by considering two points on a line with x coordinates differing by 1 unit. Suppose $B \neq 0$ and two points (x_1, y_1) and $(x_1 + 1, y_2)$ lie on the line. Then, applying Eq. (5-1), we have

$$m = \frac{y_2 - y_1}{(x_1 + 1) - x_1} = y_2 - y_1$$

The interpretation is as follows:

The slope of a line is the change in the y coordinate of a point on the line as the x coordinate of the point increases by 1 unit.

EXAMPLE 5-5

Determine the change in the y coordinate of a point on the line $x - 3y = 6$ as the x coordinate of the point increases by 1 unit.

Solution: Consider the point $(2, -\frac{4}{3})$. The point on the line $x - 3y = 6$ with x coordinate 1 unit larger is $(3, -1)$ (see Table 5-2). Therefore the change in y coordinate is $-1 - (-\frac{4}{3}) = -1 + \frac{4}{3} = \frac{1}{3}$. This is the slope of the line $x - 3y = 6$.

□

The geometric meaning of the slope can be illustrated by graphing several lines with different slopes passing through the same point. Lines $y = mx + 1$ with $m = -1, 0, \frac{1}{2}, 1, 3$ are graphed in Fig. 5-9. For comparison, lines of the form $y = 2x + b$ for $b = -2, 0, 1, \frac{5}{2}$ are graphed in Fig. 5-10.

FIGURE 5-9

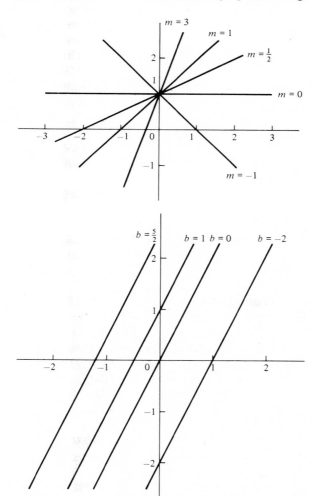

FIGURE 5-10

Notice that the equation of a line with slope 0 is $By = C$, with $B \neq 0$ (otherwise the solution set of the equation is not a line). Every point of the form $(x, C/B)$ satisfies the equation. That is, if the equation is $-3y = 6$, then $(-1, -2)$, $(3, -2)$, and $(4, -2)$ are all solutions of the equation. The graph of the line determined by the equation $-3y = 6$ is a line perpendicular to the y axis which intersects that axis in the point -2 (Fig. 5-11a).

It remains to consider the equation $Ax = C$, where we assume $A \neq 0$. Every point of the form $(C/A, y)$ satisfies the equation. For example, if the equation is $4x = 5$, the points $(\frac{5}{4}, -2)$, $(\frac{5}{4}, 0)$ and $(\frac{5}{4}, 3)$ are solutions of the equation. The graph of the line determined by $4x = 5$ is the line perpendicular to the x axis which intersects that axis in the point $\frac{5}{4}$ (Fig. 5-11b).

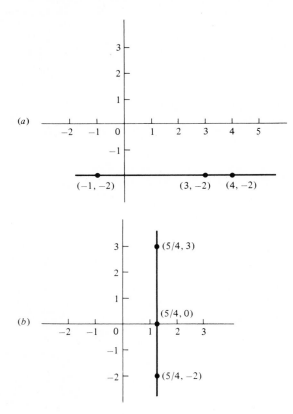

FIGURE 5-11

Definition   Two lines are *parallel* if they have the same slope or if the slope is undefined for both.

This definition can be shown to be equivalent to the usual geometrical notion of parallel lines. If two lines are not parallel, they intersect. The problem of determining the point (or points) of intersection is considered in the next section.

Equations of Lines

In addition to the general equation of a line,

$$Ax + By = C$$

and the slope-intercept equation,

$$y = mx + b$$

there are two other forms of the equation of a line which are useful, and we derive them next. The expression (5-1) for the slope can be used to derive the equation of the line which contains a specified point and has a specified slope. Let (x_1, y_1) and m be the given point and slope, respectively. If (x, y) is any other point on the line, then according to (5-1), the points (x, y) and (x_1, y_1) must satisfy

$$m = \frac{y - y_1}{x - x_1}$$

or, equivalently,
$$y - y_1 = m(x - x_1) \qquad (5\text{-}2)$$

Equation (5-2) is known as the *point-slope* form of the equation of a line.

Next, suppose that we are given two points (x_1, y_1), (x_2, y_2) and we seek the equation of the line through them. Again from (5-1) we have

$$m = \frac{y_2 - y_1}{x_2 - x_1}$$

and using this value of m in Eq. (5-2) we have

$$y - y_1 = \frac{y_2 - y_1}{x_2 - x_1}(x - x_1) \qquad (5\text{-}3)$$

Equation (5-3) is known as the *two-point* form of the equation of a line. The names derive from the given information: one point and the slope in (5-2) and two points in (5-3).

EXAMPLE 5-6

Find the equation of the line through the points $(3,2)$ and $(-1,0)$.

Solution: We examine the given information and determine that Eq. (5-3) is the appropriate one to use (we are given two points). We take $(x_1,y_1) = (3,2)$ and $(x_2,y_2) = (-1,0)$ (the other way would give the same result) and we have

$$y - 2 = \frac{0 - 2}{-1 - 3}(x - 3) \qquad \text{or} \qquad y = \tfrac{1}{2}(x - 3) + 2 = \tfrac{1}{2}(x + 1)$$

The equation of the line through $(3,2)$ and $(-1,0)$ is $y = \tfrac{1}{2}(x + 1)$.   □

EXAMPLE 5-7

Find the equation of the line which contains the point $(1,1)$ and which is parallel to the line determined by the equation $4x - 2y = 3$.

Solution: We solve this problem in two steps. First, we determine the slope of the line whose equation is $4x - 2y = 3$. Second, we find the equation of the line through the point $(1,1)$ with this slope.

Using the definition of slope, we find the slope of the line $4x - 2y = 3$ to be $4/[-(-2)] = 2$.

Using (5-2) with $m = 2$ and $(x_1,y_1) = (1,1)$, we find the equation of the line through $(1,1)$ with slope 2 to be

$$y - 1 = 2(x - 1) \qquad \text{or} \qquad y = 2x - 1$$ □

EXAMPLE 5-8

Suppose that the cost of an airline ticket is related to the distance traveled by a linear equation. If the cost of a 200-mile flight is \$38 and the cost of a 350-mile flight is \$50, find the equation relating cost and distance, and find the cost of a 275-mile flight.

Solution: We identify distance (in miles) with x and cost (in dollars) with y. Then the given points are $(200,38)$ and $(350,50)$. Using Eq. (5-3), we find that the relation between x and y is

$$y - 38 = \frac{50 - 38}{350 - 200}(x - 200)$$

or, after simplification,

$$y = \tfrac{2}{25}x + 22 \tag{5-4}$$

In terms of cost and distance this is

$$\text{Cost} = \tfrac{2}{25}(\text{distance}) + 22 \qquad\qquad (5\text{-}5)$$

Using this equation, we find the cost of a 275-mile flight to be

$$\tfrac{2}{25}(275) + 22 = 44$$

in units of dollars.

The constants $\tfrac{2}{25}$ and 22 in Eqs. (5-4) and (5-5) have a natural interpretation. The coefficient $\tfrac{2}{25}$ is the operating cost per mile (in dollars), and the 22 is the fixed cost of the airline (in dollars) apportioned to each flight. □

Exercises for Sec. 5-3

1. Find the slope of the line through each of the following pairs of points.
   a. (2,3/2), (1,1/2)
   b. (0,1), (1,0)
   c. (1/3,2/3), (0,0)
   d. (3,1/2), (6,1/2)

2. Find the equation of the line through each of the following pairs of points and graph each line.
   a. (1,−1), (2,3)
   b. (0,0), (−1,2)
   c. (1,2), (1,4)
   d. (−3,−1), (1,3)

3. Find the equation of the line with:
   a. Slope −2 and passing through (1,0)
   b. Slope 0 and containing the point (1,−1)
   c. Slope not defined and containing the point (2,2)

4. Find the equation of the line through the point (−2,3) and parallel to the line with equation $-x = 4$. Graph each of these lines.

5. Find the equation of the line with
   a. Slope −2, y intercept = 2
   b. Slope 2, y intercept = −2

6. Find the x and y intercepts of each of the following lines:
   a. $y = 2x + 1$
   b. $y = 5$
   c. $x = 2y + 1$
   d. $x = 5$

7. A production supervisor believes that the number of defective widgets produced each day is related to the total number produced by a linear equation. Suppose this to be the case. If 10 defective widgets are produced in a total of 200 on one day and 15 defective widgets are produced in a total of 225 on another day, how many defective widgets will be expected on a day when the total production is 250?

8. Use the data of Exercise 7 and suppose the production supervisor wants to produce as many widgets as possible without having more than 25 defective widgets produced. What total production should be scheduled?

9. In Example 5-8 what is the length of a flight which costs $262?

10. Find the equation of the line through the points (32,0) and (212,100) and graph this line. What is the slope of this line? Using the information in Exercise 9 of Sec. 5-2, how could the two axes be labeled for this graph?

11. The following data give the year-by-year resale values of a car which cost $5000 when it was new. Decide whether these data are linear, i.e., decide whether the pairs of points (age, resale value) lie on a line.

Age (years)	Resale Value (dollars)
1	4000
2	3200
3	2600
4	2200
5	2000

12. A certain car can be rented on either of two bases:
    a. $20 per day and 20 ¢ per mile driven
    b. $30 per day and 10¢ per mile driven
    Write equations which describe the costs (in dollars) of driving x miles in one day under each of these plans. Which plan is less expensive for someone who plans to drive 75 miles per day?

13. Suppose that a line is defined by the equation $Ax + By = C$ with $B \neq 0$. Show that if (x_1,y_1) and (x_2,y_2) are two distinct points on this line, then $x_1 \neq x_2$.

14. Work Example 5-6 with $(x_1,y_1) = (-1,0)$ and $(x_2,y_2) = (3,2)$.

5-4   SYSTEMS OF LINES

It is common for linear models to contain more than one linear equation. In this section we consider systems of two equations, their solutions, and their graphs.

EXAMPLE 5-9

A manufacturer produces glass of two types, dark green bottle glass and heat-resistant glass. Both types require silica and soda, together with other materials. To produce 100 pounds of bottle glass requires 70 pounds of silica and 15 pounds of soda, and to produce 100 pounds of heat-resistant glass requires 80 pounds of silica and 4 pounds of soda. The manufacturer has 500 pounds of silica and 48 pounds of soda to be used in the production of glass. Find a linear mathematical model which describes the use of silica and soda in glass production.

Solution: We begin by letting

x = amount (in hundreds of pounds) of bottle glass to be produced

y = amount (in hundreds of pounds) of heat-resistant glass to be produced

Consider the facts regarding the use of silica. To produce 100 pounds of bottle glass requires 70 pounds of silica, and therefore to produce x hundred pounds of bottle glass requires $70x$ pounds of silica. Likewise, to produce y hundred pounds of heat resistant glass requires $80y$ pounds of silica. There are 500 pounds of silica to be used, and consequently x and y must be selected so that

$$70x \quad + \quad 80y \quad = \quad 500 \qquad\qquad (5\text{-}6)$$

Silica for   **Silica for heat-**   **Total Silica**
Bottle Glass   **resistant Glass**

Next, we consider the use of soda. It requires 15 pounds of soda to produce 100 pounds of bottle glass, and therefore it requires $15x$ pounds of soda to produce x hundred pounds of bottle glass. Similarly, it requires $4y$ pounds of soda to produce y hundred pounds of heat-resistant glass. Since there are 48 pounds of soda to be used, x and y are to be selected so that

$$15x + 4y = 48 \qquad\qquad (5\text{-}7)$$

The linear mathematical model for the situation is the system of two linear equations

$$70x + 80y = 500$$
$$15x + 4y = 48 \qquad\qquad (5\text{-}8)$$

□

If we were interested in finding a production schedule (x,y) which uses the total amount of raw materials, we need to find x and y such that both equations in the system (5-8) are satisfied. Since these are both linear equations, the set of points satisfying each of them is a line. Therefore, from a geometrical point of view, we are looking for a point $P = (x,y)$ which lies on the line (5-6) and on the line (5-7). Although there is no assurance (in advance of solving the problem) that there is any such point, in this example there is a point P common to the two lines (see Fig. 5-12). Notice also that in this problem for the answer to be meaningful we must have $x \geq 0$ and $y \geq 0$.

We turn now to the problem of finding a point (x,y) for which (5-8) is satisfied. The technique we introduce is most applicable to two equations in two variables, where it is very efficient and easy to use. We begin by multiplying the first equation in (5-8) by 15 (the coefficient of x in the second equation) and

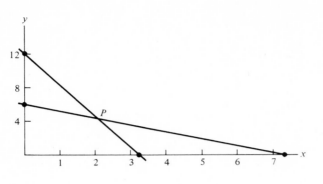

FIGURE 5-12

the second equation in (5-8) by 70 (the coefficient of x in the first equation). The result is the system (5-9):

$$1050x + 1200y = 7500$$
$$1050x + 280y = 3360 \tag{5-9}$$

The system (5-9) is equivalent to system (5-8) in the sense that if (x,y) satisfies either (5-8) or (5-9), it also satisfies the other. Also, if (x,y) satisfies both the equations in (5-9), it also satisfies their difference. That is, it must satisfy the equation which results from subtracting the second equation of (5-9) from the first. The difference is the equation

$$920y = 4140 \tag{5-10}$$

From the last equation we conclude that y must be 4.5. That is, if (x,y) satisfies the system (5-8), then y must be 4.5. Notice that the disappearance of the x term results from the specific choice of multipliers used to obtain (5-9). Substituting $y = 4.5$ into the equation $70x + 80y = 500$, we obtain

$$70x + 80(4.5) = 500$$
$$70x = 140$$
$$x = 2$$

Therefore, the point $(2, 4.5)$ satisfies (5-9) and therefore (5-8). Geometrically, $(2,4.5)$ is the intersection of the two lines determined by the equations in (5-8). The values $x = 2$ and $y = 4.5$ can be interpreted in terms of the original situation; the manufacturer should produce 200 pounds of bottle glass and 450 pounds of heat-resistant glass to completely use the supplies of silica and soda.

In this example the pair of linear equations has exactly one solution. This corresponds geometrically to the two lines intersecting in a single point (Fig. 5-12).

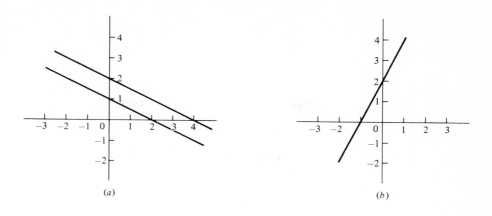

(a)   (b)

FIGURE 5-13

In other examples it may happen that there are no solutions to the equations or that there are infinitely many solutions. In the former case the geometrical picture is that of two parallel lines (Fig. 5-13a), and in the latter case it is of coincident lines, i.e., lines which are identical (Fig. 5-13b). The lines graphed in Fig. 5-13a are determined by the system

$$x + 2y = 2$$
$$x + 2y = 4$$

and those graphed in Fig. 5-13b are determined by the system

$$2x - y = -2$$
$$-6x + 3y = 6$$

There is an easy means of distinguishing the case in which there is a unique point of intersection from the other two cases: *if the two lines have different slopes, then they have exactly one point of intersection; otherwise they are either parallel or coincident.*

The technique used above to find the solution of (5-8) is completely general. In fact, if the equations

$$Ax + By = C$$
$$Dx + Ey = F \tag{5-11}$$

determine lines which are neither parallel nor coincident, there is a unique point (x,y) which satisfies both equations. When one or more of the coefficients A, B, D, E are zero, the system can be analyzed directly. If there is a solution, one of the equations can be solved immediately for x or y, and this value can be substituted in the other equation, which then contains a single unknown.

EXAMPLE 5-10

Find all points (x,y) which satisfy

$$3x - 4y = 12$$
$$5y = 9$$

Solution: Since the slope of the line determined by the first equation is $\frac{3}{4}$ and the slope of the line determined by the second equation is 0, there is a unique point (x,y) which satisfies both equations. From the second equation we have $y = \frac{9}{5}$. Substituting this into the first equation, we have

$$3x - \tfrac{36}{5} = 12$$
$$3x = \tfrac{96}{5}$$

and finally, $x = \frac{32}{5}$. Thus, $\left(\frac{32}{5}, \frac{9}{5}\right)$ is the desired point.   □

In the general case (in which all four coefficients A, B, D, E are nonzero) we proceed as follows:

1. **Form an equivalent system by multiplying the first equation in (5-11) by D and the second equation by A:**

$$DAx + DBy = DC$$
$$ADx + AEy = AF$$

2. **Subtract the second equation from the first, giving**

$$(DB - AE)y = DC - AF \qquad \text{(5-12)}$$

If the term $DB - AE \neq 0$, we can solve for y and obtain

$$y = \frac{DC - AF}{DB - AE}$$

Note: The condition $DB - AE \neq 0$ is equivalent to the condition that the two lines have different slopes (Exercise 10). If $DB - AE = 0$, the two lines have the same slope. In this case if the right-hand side of Eq. (5-12) is not zero, the lines are parallel but not coincident and there is no solution of the system (5-11); if the right-hand side of Eq. (5-12) is zero, the lines are coincident and any point on the common line satisfies the system (5-11).

3. **Substitute the value of y obtained in step 2 in one of the equations obtained in step 1 or in one of the equations of the original system (5-11) and obtain the associated value of x. The pair (x,y) satisfies both equations in the system (5-11).**

In general one does not check to determine whether the two lines in the system (5-11) have the same slope before beginning this solution process. Instead, the process is begun, and if the lines have the same slope, i.e., if $DB - AE = 0$, then the process stops at step 2 as described in the note.

The reader is encouraged to use the *method* outlined in steps 1, 2, and 3 and not the formula derived in step 2.

EXAMPLE 5-11

Find all solutions of the system of equations

$$5x - 3y = 4$$
$$3x + 7y = 8$$

and graph the lines determined by these equations.

Solution: We begin with step 1 of the process outlined above and we form the equivalent system

$$15x - 9y = 12$$
$$15x + 35y = 40$$

Continuing with step 2 and subtracting the second equation from the first, we obtain

$$-44y = -28 \quad \text{or} \quad y = \frac{-28}{-44} = \frac{7}{11}$$

Using step 3, we substitute $y = \frac{7}{11}$ into $5x - 3y = 4$. We find

$$5x - 3\left(\tfrac{7}{11}\right) = 4 \quad \text{or} \quad 5x = 4 + \tfrac{21}{11} = \tfrac{65}{11}$$

from which we obtain $x = \frac{65}{55} = \frac{13}{11}$. The unique solution of the given system is $\left(\frac{13}{11}, \frac{7}{11}\right)$. The lines determined by the system are graphed in Fig. 5-14. □

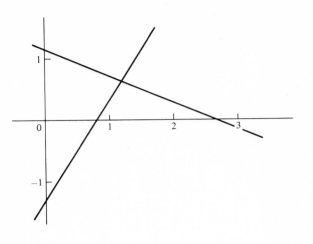

FIGURE 5-14

EXAMPLE 5-12

Find all solutions of the system of equations

$$2x - 5y = 1$$
$$6x - 15y = 3$$

Solution: An equivalent system is

$$12x - 30y = 6$$
$$12x - 30y = 6$$

and subtracting the second from the first, we have $0 = 0$. This is one of the situations described in the note to step 2 of the solution method given above. We have both $DB - AE = 0$ [$6(-5) - 2(-15) = 0$ in this example] and also $DC - AF = 0$ [$6(1) - 2(3) = 0$ in this example], and therefore the lines are coincident. The coincidence is also easily verified by noting that both lines have slope $\frac{2}{5}$ and x intercept $\frac{1}{2}$. We conclude that every point on the common line is a solution of the given system. For any value t of the variable x the value $y = \frac{2}{5}t - \frac{1}{5}$ is such that the point $(x,y) = (t, \frac{2}{5}t - \frac{1}{5})$ is on the common line. Therefore, the set of solutions of the system is the set of points $(t, \frac{2}{5}t - \frac{1}{5})$, where t is any real number. The situation is pictured in Fig. 5-15.   □

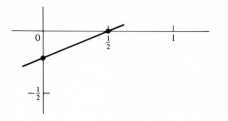

FIGURE 5-15

EXAMPLE 5-13

Find all solutions of the system of equations

$$2x - 5y = 1$$
$$6x - 15y = 5$$

and graph the lines determined by these equations.

Solution: An equivalent system is

$$12x - 30y = 6$$
$$12x - 30y = 10$$

and subtracting the second equation from the first, we have $0 = -4$. The system has no solutions (see note in step 2). The lines are parallel and not coincident and are shown in Fig. 5-16. □

FIGURE 5-16

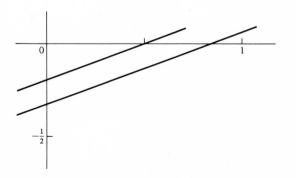

Exercises for Sec. 5-4

1. Decide which of the pairs of lines determined by the following equations are parallel but not coincident, coincident, or intersect in exactly one point. Graph each pair of lines.

   a. $3x - 2y = 6$
       $4x\ \ \ \ \ = 12$

   b. $-2x + \frac{1}{2}y = \frac{3}{2}$
       $x - 3y = 2$

   c. $3x - \ \ y = -1$
       $9x - 3y = -3$

   d. $4x + y = 12$
       $4x - y = 12$

2. Find all solutions of each of the following pairs of equations:

   a. $5x - 2y = 0$
       $3x + 4y = 2$

   b. $-3x - 3y = 3$
       $6x + 6y = -6$

   c. $x + y = -1$
       $x - y = 2$

   d. $.1x + .3y = .5$
       $.4x - .1y = .7$

3. Suppose that for tax purposes a $100,000 building is depreciated at a rate of 5 percent per year of its original value and a $150,000 building is depreciated at a rate of 8 percent per year of its original value. Thus, after 1 year the first building is valued at $95,000 and after 2 years at $90,000. Let t represent the number of years a building has been depreciated and v be the value of the building after t years. Write linear equations relating v and t for the two buildings. What are the slopes of these lines? When do the two buildings have the same value?

4. The formula for converting Fahrenheit temperatures to Celsius is $C = \frac{5}{9}(F - 32)$. At what temperatures (if any) do the two temperature scales have the same value?

5. Two joggers begin to run on a $\frac{1}{4}$-mile track at the same place. One of them runs at the steady pace of 8 minutes per mile, and the other runs at the steady pace of 6 minutes per mile. How far will the slower runner have run when the faster runner catches up?

6. The situation is the same as in Exercise 5, but now the faster runner runs at 5 minutes per mile and the slower at 7 minutes per mile. If the faster runner runs 2.5 miles, how far has the slower runner run when the faster runner stops? How many times has the faster runner passed the slower runner?

7. Penni Saver plans to invest $10,000, part of it in utility bonds paying 9 percent per year and the rest in a savings account paying 6 percent per year. How much should be allocated to each investment if the yearly income from the two investments is to be the same?

8. Miniburg has a population of 50,000 and is growing at the rate of 7500 per year. Maxiburg has a population of 1,000,000 and is decreasing at a rate of 125,000 per year. If these rates continue, how many years will it take before both cities have the same population?

9. A system consisting of the equations of three lines can have one solution, no solution, or infinitely many solutions. Graph the lines in each of the following systems and determine the set of solutions of each system.

   a. $3x - y = 6$
       $x + y = 4$
       $5x + y = 14$

   b. $3x - y = 6$
       $x + y = 4$
       $5x + y = 10$

   c. $3x - y = 6$
       $-6x + 2y = -12$
       $\frac{3}{2}x - \frac{y}{2} = 3$

10. Show that if $DB - AE \neq 0$ in system (5-11), the lines determined by the equations of (5-11) have different slopes or one line has a slope and one does not.

5-5 SYSTEMS OF LINEAR INEQUALITIES IN TWO VARIABLES

Just as the set of points (x,y) which satisfy an *equality* of the form $3x + 2y = 6$ may be important in an application, the set of points (x,y) which satisfy an *inequality* of the form $3x + 2y \leq 6$ may also be important. The symbol $\leq$ means that the quantity $3x + 2y$ is to be *no larger* than 6. The relation $3x + 2y < 6$ is referred to as a *strict inequality*. It means that $3x + 2y$ is to be *less than* 6. The symbols $\geq$ and $>$ have similar meanings. In problems arising in the social, life, and management sciences inequalities occur at least as frequently as equalities.

EXAMPLE 5-14

A hiker determines that on long hikes she needs to obtain at least 600 calories by eating a combination of chocolate and raisins. If chocolate contains 150 calories per ounce and raisins contain 80 calories per ounce, which combinations of the two foods will satisfy her requirement?

Solution: In order to answer the question we let x denote the number of ounces of chocolate and y the number of ounces of raisins. Notice that x and y are necessarily subject to the conditions $x \geq 0$ and $y \geq 0$. It follows from the given data that x ounces of chocolate contains $150x$ calories and that y ounces of raisins contains $80y$ calories. Also, the total calories obtained from the two foods is the sum of the calories obtained from chocolate and the calories obtained from raisins, that is,

$$\text{Total calories} = 150x + 80y$$

Therefore, since she requires at least 600 calories (the total calories must equal 600 or more), we have the condition

$$150x + 80y \geq 600 \tag{5-13}$$

Recall that x and y are also subject to the conditions

$$x \geq 0 \qquad y \geq 0 \tag{5-14}$$

The set of all (x,y) satisfying (5-13) and (5-14), with $x =$ number of ounces of chocolate and $y =$ number of ounces of raisins, fulfills her requirements. □

In the earlier sections of this chapter we studied equations similar to (5-13) with the $\geq$ replaced by $=$. In this section we investigate the sets of points (x,y) which satisfy inequalities like (5-13).

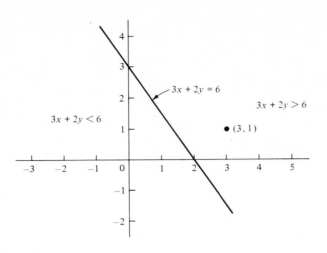

FIGURE 5-17

Geometrically the line determined by $Ax + By = C$ divides the xy plane into three sets: (1) the set of (x,y) such that $Ax + By = C$, that is, the line itself, (2) the set (x,y) such that $Ax + By < C$, one of the half planes bounded by the line, and (3) the set of (x,y) such that $Ax + By > C$, the other half plane bounded by the line. For example, the line determined by $3x + 2y = 6$ divides the plane into the three sets or regions shown in Fig. 5-17. Which of the half planes bounded by the line is described by $3x + 2y > 6$ and which is described by $3x + 2y < 6$ can be determined by testing a single point not on the line. In this example we can test the point $(3,1)$. We have $3x + 2y = 3(3) + 2(1) = 11$, which is larger than 6. Therefore the point $(3,1)$ lies in the half plane described by $3x + 2y > 6$, and the half planes are as labeled in Fig. 5-17. In this example an easier point to test would have been $(0,0)$, the origin of the coordinate system. The point $(0,0)$ is usually the easiest point to test if the line does not pass through the origin.

If we are concerned with the set of points (x,y) satisfying the strict inequality $Ax + By < C$, the line $Ax + By = C$ *does not* belong to the set. If we are concerned with the set of points (x,y) satisfying the inequality $Ax + By \le C$, the line *does* belong to the set. Most of the inequalities in this book will be of the form $Ax + By \le C$ (or $Ax + By \ge C$).

EXAMPLE 5-15

The graph of the set of points satisfying (5-13), $150x + 80y \ge 600$, is shown as the shaded area in Fig. 5-18. The line bounding the set is to be included in the set.   □

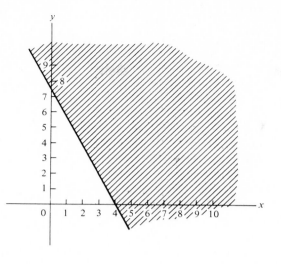

FIGURE 5-18

In Example 5-14 the set of points (x,y) of interest is defined as the set of points satisfying the three inequalities in (5-13) and (5-14). Since these inequalities must be satisfied simultaneously, the desired set is the intersection of the three sets:

$$A = \{(x,y): \quad 150x + 80y \geq 600\}$$

$$B = \{(x,y): \quad x \geq 0\} \quad \text{and} \quad C = \{(x,y): \quad y \geq 0\}$$

The set $A \cap B \cap C$ is the area which is shaded and bounded by the bold line in Fig. 5-19. □

FIGURE 5-19

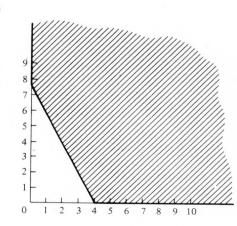

EXAMPLE 5-16

Graph the set of points which satisfy the inequalities

$$x \geq 1$$

$$3x + 4y \leq 12$$

(5-15)

Solution: The graph of the line whose equation is $x = 1$ is the vertical line through the point $(1,0)$. The line $3x + 4y = 12$ is the line through $(4,0)$ and $(0,3)$. The set $A = \{(x,y): \ x \geq 1\}$ is shaded with horizontal shading in Fig. 5-20, and the set $B = \{(x,y): \ 3x + 4y \leq 12\}$ is the set with vertical shading. Thus, the set $A \cap B$ of points satisfying both inequalities of the system (5-15) is the cross-hatched region in Fig. 5-20. The boundary of this set is shown with bold lines in Fig. 5-20. The corner point P of the set $A \cap B$ can be determined by solving the system of equations

$$x = 1$$

$$3x + 4y = 12$$

The system is easily solved, and we find $P = (1, \frac{9}{4})$. □

FIGURE 5-20

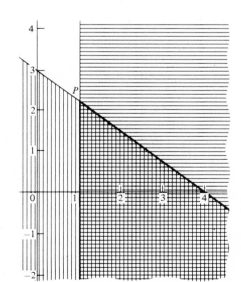

Exercises for Sec. 5-5

1. Graph the sets of points which satisfy each of the following systems of inequalities

    a. $5x + 3y \leq 2$ **b.** $x + 3y \geq 2$

         $x - y \leq 1$ $y \leq 2$

2. Graph the sets of points which satisfy each of the following systems of inequalities

    a. $4x + 5y \leq 10$ **b.** $2x + 3y \geq 6$

           $x \leq 1$ $x - y \geq 2$

           $y \geq -1$ $x + y \leq 3$

3. Find a system of inequalities which determines the shaded area in each of the graphs in Fig. 5-21.

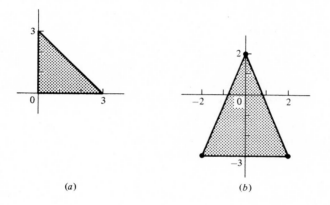

             (*a*)   (*b*)

FIGURE 5-21

4. Find a system of inequalities which determines the shaded area in each of the graphs in Fig. 5-22.

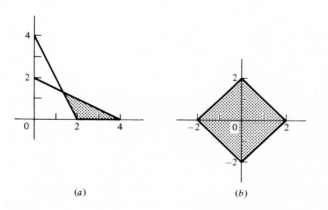

             (*a*)   (*b*)

FIGURE 5-22

5. An investor has $100,000 to invest in common and preferred stock. She determines that no more than three-fourths of the total should be invested in either type and that the amount invested in common stock should be at least as large as the amount invested in preferred stock. Graphically represent the set of choices available to the investor.

6. Find the vertices of the triangles shaded in Fig. 5-21.

7. The points (1,1), (1,2), and (3,2) are the vertices of a triangle. Find a set of inequalities which are satisfied by the points inside the triangle.

8. A rectangle which lies in the first quadrant has one vertex at (1,0) and another vertex at (0,1), and the area of the rectangle is 4. What are the coordinates of the other two vertices of the rectangle?

9. The manager of a paper-box plant is scheduling the work for one production line for a week. He can produce standard and heavy-duty boxes. Each standard box requires 2 pounds of kraft paper and each heavy-duty box requires 4 pounds. Also it requires 8 man-hours to produce 100 standard boxes and 3 man-hours to produce 100 heavy-duty boxes. (The machine which produces heavy-duty boxes is more efficient.) There are 50 tons of kraft paper and 2400 man-hours available during the week. Finally, the manager has a contract which requires him to deliver 10,000 heavy-duty boxes at the end of the week. Represent the set of choices available to the manager graphically.

10. A biologist is planning to conduct two experiments in a salt marsh. He has a total of 1 acre, and he plans to conduct one experiment on a portion of the marsh and another experiment on another portion. He can collect data for the first experiment at such a rate that it would require 2 days to collect data from the entire acre. For the second experiment the data collection is much slower, and it would require 10 days to collect data for the entire acre. Data processing for the first experiment would require 10 minutes for data from the entire acre, and processing the data from the second experiment would require 3 minutes for the data from the whole acre. The biologist has 6 days to collect data, and his budget allows for 8 minutes of computer data processing time. Graphically represent the set of alternatives available to the biologist.

11. A teamster about retire has $100,000 to invest for income during her retirement. She has decided to divide the money between high-risk corporate notes paying 12 percent per year and government securities paying 6 percent per year. If she wants at least $8000 per year in income from this money, what allocations of her funds are open to her? Graph the set of possible investments.

5-6 FORMULATION OF LINEAR PROGRAMMING PROBLEMS

One of the major uses of mathematics in the management sciences, and to a lesser extent in the social and life sciences, is in the formulation and solution of optimization problems. In general, an *optimization problem* is the problem of selecting values of variables in such a way that some quantity which depends upon the variables is made as large or as small as possible. Normally the permissible choices are constrained by conditions which the variables, either individually or jointly, must satisfy. Linear optimization problems are those in which the relations between variables are given by linear equations or inequalities. We discuss linear optimization problems involving more than two variables in Chap. 7. Linear optimization problems of the form considered here are commonly referred to as *linear programming problems.*

Frequently a linear optimization problem is not posed in the form of equations and other symbolic relations but in words, the words which describe the situation as it is viewed by one who first meets the problem. The translation of such problems from verbal to mathematical form is the topic of this section. The technique is best described by illustrating it. In each case our task is to identify and formulate a mathematical version of the problem which is posed in verbal form. This is the process of model building as it was described in Chap. 1. Thus, the topic of this section can be described more accurately as the construction of *linear models* for optimization problems involving two variables.

EXAMPLE 5-17

A plastics manufacturer has 1200 barrels of crude oil to use in the production of two distillates, *A* and *B*. It takes 2 barrels of crude oil and 25 kilowatthours of electric energy to produce 1 barrel of distillate *A*, and 3 barrels of crude oil and 15 kilowatthours of electric energy to produce 1 barrel of distillate *B*. The manufacturer has 7500 kilowatthours to devote to this production process. Also, due to restrictions on emissions of by-products into the atmosphere, at most 200 barrels of distillate *A* can be produced. If the net profit on 1 barrel of distillate *A* is $2 and on 1 barrel of distillate *B* is $1, formulate a mathematical problem whose solution gives the number of barrels of each distillate to be produced to yield a maximum net profit.

Solution: To begin our formulation of a mathematical problem we determine the variables in the situation described above. We let x denote the number of barrels of distillate *A* produced, and we let y denote the number of barrels of distillate *B*. Immediately from the meaning of x and y we have the inequalities

$$x \geq 0 \qquad y \geq 0$$

The constraint imposed by the restriction on emissions gives the inequality

$$x \leq 200$$

Next, $2x$ is the amount (in barrels) of crude oil needed to produce x barrels of distillate *A*, and $3y$ is the amount (in barrels) of crude oil needed to produce y barrels of distillate *B*. The constraint on the total amount of crude oil available leads to the inequality

$$2x + 3y \leq 1200$$

Likewise the constraint on the total amount of electric energy available leads to the inequality

$$25x + 15y \leq 7500$$

Finally, the net profit resulting from the production of x barrels of distillate A is $2x$ dollars, and the net profit resulting from the production of y barrels of distillate B is y dollars ($1 per barrel). Consequently, the total net profit is $2x + y$ dollars.

The mathematical problem is the following:

Find the pair (or pairs) of values x and y satisfying

$$x \geq 0$$
$$y \geq 0$$
$$x \leq 200 \qquad\qquad (5\text{-}16)$$
$$2x + 3y \leq 1200$$
$$25x + 15y \leq 7500$$

for which $2x + y$ is as large as possible.   ☐

It is useful to introduce additional terminology to use in discussing linear optimization problems.

Definition   The inequalities which identify permissible values of the variables are the *constraints* of the problem. The inequality constraints considered in this book will always be *inclusive*, i.e., involving $\leq$ or $\geq$ (instead of $<$ and $>$).

Definition   The set of points (x,y) satisfying the constraints of the problem is known as the *feasible set* for the problem. The function to be maximized or minimized is known as the *objective function* for the problem.

The formulation of a linear optimization problem involves finding the equalities or inequalities which describe the feasible set and specifying the objective function. For instance, in Example 5-17 the constraints are the inequalities (5-16), the feasible set is the set described by these inequalities, and the objective function is the function whose value at (x,y) is $2x + y$.

EXAMPLE 5-18

Tone Down Sounds, Inc., purchases a portion of the plastic used in its record-manufacturing division from an independent supplier and the remainder from a wholly owned subsidiary. As a result of an antitrust agreement, Tone Down must purchase at least $\frac{1}{3}$ of the plastic used from the independent supplier. Furthermore, the independent supplier will not accept orders for less than 15,000 pounds of plastic. The marketing division has determined that Tone Down will need to manufacture 480,000 records in the next quarter, for which 30,000 pounds of plastic will be needed. The company must pay its subsidiary 80¢ per pound for the plastic, and the independent supplier charges $1 per pound. Formulate a mathematical problem whose solution provides the company with an acquisition policy which minimizes its costs for purchasing plastic.

Solution: Let x and y denote the number of pounds of plastic purchased from the subsidiary and the independent supplier, respectively. It is clear from this definition that

$$x \geq 0 \quad \text{and} \quad y \geq 0$$

The requirement that at least 30,000 pounds of plastic be purchased gives the inequality

$$x + y \geq 30{,}000$$

Purchases from the independent supplier are subject to two provisions. First, at least $\frac{1}{3}$ of the total plastic purchased must come from the independent supplier. This is the condition

$$y \geq \tfrac{1}{3}(x + y) \quad \text{or} \quad y \geq \tfrac{1}{3}x + \tfrac{1}{3}y$$

After subtracting y from each side of the last inequality we have*

$$\tfrac{1}{3}x - \tfrac{2}{3}y \leq 0$$

which is equivalent to

$$x - 2y \leq 0$$

---

 * Here and in the remaining sections of this chapter we freely use the following properties of inequalities

  If $a \leq b$ and $c \leq d$, then $a + c \leq b + d$.
  If $a \leq b$ and $\lambda > 0$, then $\lambda a \leq \lambda b$.
  If $a \leq b$ and $\lambda < 0$, then $\lambda a \geq \lambda b$.

There are corresponding properties for strict inequalities $(a < b)$, which we shall also use as necessary.

Purchases from the independent supplier are also subject to the provision that if any plastic is purchased, an order of at least 15,000 pounds must be placed. This gives the inequality

$$y \geq 15,000$$

The feasible set for this problem is

$$x \geq 0$$
$$y \geq 0$$
$$x + y \geq 30,000 \qquad (5\text{-}17)$$
$$x - 2y \leq 0$$
$$y \geq 15,000$$

The cost of purchasing x pounds of plastic from the subsidiary is $.8x$ (measured in dollars), and the cost of purchasing y pounds of plastic from the supplier is y (in dollars). Therefore, the acquisition cost is $.8x + y$. The mathematical problem is

Find (x,y) satisfying (5-17) for which $.8x + y$ is as small as possible   □

EXAMPLE 5-19   (CONTINUATION OF EXAMPLE 1-3)

The firm Public Opinion Survey Enterprises (POSE) has been hired to sample public opinion in Metroburg by conducting a survey. Metroburg consists of an inner city and several suburbs. The contract with POSE specifies the following conditions:

a. At most 2000 people are to be sampled.
b. At least 400 people from the suburbs must be sampled.
c. At least half of those sampled must be from the inner city.
d. POSE is to be compensated with a fee of $2000 plus $10 for each response obtained.

Since POSE operates with the intention of making a profit, it is interested in conducting the survey in a manner that produces maximum profits after expenses. The following information is available about the Metroburg area:

1. Of the people in the city 10 percent will respond to a mail questionnaire.
2. Of the people in the suburbs 20 percent will respond to a mail questionnaire.
3. The postage and processing of each mail questionnaire costs $1.50.

4. Of those sought for an interview in the suburbs 60 percent will be interviewed.
5. Of those sought for an interview in the city 50 percent will be interviewed.
6. Each interview attempt costs $7.25 in the city and $9 in the suburbs.

Formulate a precise mathematical problem for this situation and graphically represent the set of choices available to POSE.

Solution: Let x and y denote the number of responses to be obtained from individuals living in the city and in the suburbs, respectively. The cost data provide the following comparisons between mail surveys and personal interviews. All costs are given in dollars.

In view of information items 1 and 3, we see that to obtain x responses by mail in the city a mailing of $10x$ items at a cost of $10(1.50x)$ is required. Thus each response obtained through the mail costs $15. Likewise, in view of items 5 and 6, to obtain x responses in the city through personal interviews requires $2x$ interview attempts at a cost of $2(7.25x)$. Thus each response obtained through a personal interview costs $14.50. We conclude that in the city it is less expensive to obtain responses through personal interviews.

Similarly, to obtain y responses by mail from the suburbs requires the mailing of $5y$ items at a cost of $5(1.50y) = 7.50y$. Consequently, each response obtained by mail costs $7.50. Finally, to obtain y responses through personal interviews in the suburbs requires $\frac{5}{3}y$ interview attempts at a cost of $\frac{5}{3}(9.00y) = 15.00y$. Thus each response obtained through a personal interview costs $15. We conclude that in the suburbs it is best to obtain responses by mail.

The conditions of the contract with POSE give the following inequalities on x and y:

$$x + y \leq 2000$$
$$y \geq 400$$
$$x - y \geq 0 \qquad\qquad (5\text{-}18)$$
$$x \geq 0$$
$$y \geq 0$$

The first three inequalities in (5-18) arise (in order) from conditions **a, b** and **c** of the contract. The last two inequalities are obvious constraints.

The gross profit is the amount paid to the company. In terms of x and y and using condition **d**, this is $2000 + 10x + 10y$. The net profit is the gross profit minus the costs incurred in conducting the survey. This cost is $14.50 per response in the city and $7.50 per response in the suburbs. The net profit to POSE is

$$2000 + (-4.50x) + 2.50y$$

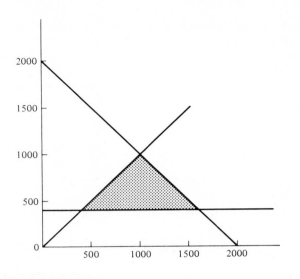

FIGURE 5-23

The mathematical problem is to maximize net profit subject to (5-18). The feasible set for this problem is shown in Fig. 5-23. □

In Example 5-19 a point arises that we have not considered before: it does not make sense to consider less than a complete response, and consequently we should restrict our consideration to integer-valued x and y. Constraints of this sort result in integer programming problems, and there are methods for extending the ideas of this chapter and Chap. 7 to such problems. In this discussion we shall not take into account integer constraints on the variables.

Exercises for Sec. 5-6

In each case formulate a mathematical problem resulting from the use of a linear programming model of the situation described.

1. Suppose that in Example 5-17 the constraint that at most 200 barrels of distillate A can be produced is replaced by the constraint that at most 500 barrels of crude oil and 4000 kilowatthours of electric energy can be devoted to the production of distillate A. Find the appropriate system of inequalities to replace (5-16).

2. Suppose that in Example 5-18 Tone Down Sounds is required to purchase at least $\frac{1}{4}$ of the total plastic used from an independent supplier and that the supplier allocates its production so that no customer recieves more than 20,000 pounds per quarter. If all other conditions are the same, formulate a mathematical problem using a linear programming model.

3. An office-furniture manufacturer has available 9 tons of sheet steel which is to be used to make desks and filing cabinets. It requires 50 pounds of steel and 3 hours of labor to make a filing cabinet and 75 pounds of steel and 2 hours of labor to make a desk. There are 750 hours of labor available but only enough tops for 200 desks. The net profit is $20 on each desk and $15 on each filing cabinet. Formulate a mathematical problem whose solution provides the manufacturer with a production schedule which maximizes total net profit.

4. A purchasing agent for a college finds it necessary to decide how much hard and soft chalk should be purchased for each month. He knows that a typical instructor will use 2 pieces of soft chalk or 1 piece of hard chalk for each class. In addition, he has observed that hard chalk always amounts to at least $\frac{1}{4}$ of the total used. Finally, his supplier has limited him to a purchase of at most 60 gross (8640 pieces) of soft chalk. There are 3600 classes to be taught each month. If soft chalk is $1.50 per gross and hard chalk is $3.50 per gross, how much of each should be purchased to meet the needs and to keep costs to a minimum?

5. The Slapumup Construction Company makes and sells both single-family homes and condominium units. They must decide how many of each to build in the coming year. The company has done some market research, and they believe that they can sell at most 30 single-family homes and at most 70 condominium units. The assembly operations for the projects are divided into three steps: foundation work F, exterior work E, and interior work I. The company's time estimates for each step are given in crew-days per unit as follows:

	F	E	I
Single-family units	3	8	9
Condominium units	1	5	15

Due to equipment and labor limitations the company can work at most 60 crew-days on F work, 200 on E work, and 300 on I work. If the profit on a single family home is $8000 and on a condominium unit is $10,000, how many units of each should be built?

6. The Trendy Toy Company makes two skateboard models, called the Zip and Zap. Both models require two operations in production: (1) finishing the frame and (2) installing and balancing the wheels. These operations require the following amounts of time (in minutes):

	Frame	Wheels
Zip	15	5
Zap	10	20

There are 60 hours of labor per day available for each operation; the profit per Zip model is $5 and per Zap model $8. How should daily production be scheduled to maximize profits?

7. The Fragrant Fertilizer Company makes two types of fertilizer: 20-5-5 for lawns and 10-15-10 for gardens. The numbers in each case refer to the percent by weight of nitrate, phosphate, and potash, respectively, in a 100-pound sack of fertilizer. The company has the following amounts of raw materials on hand: 3.5 tons of nitrate, 2 tons of phosphate, and 2 tons of potash. The profit per 100 pounds of lawn fertilizer is $6, and the profit per 100 pounds of garden fertilizer is $4. The company has a contract to supply at least 1 ton of garden fertilizer. How much of each fertilizer should the company produce to maximize profits?

8. The following mathematical problem describes a production problem of the Fragrant Fertilizer Company (Exercise 7). How should the data of Exercise 7 be changed to yield this mathematical problem? Find the value of x and y so that $x + y$ is a maximum and

$$x \geq 0$$

$$y \geq 0$$

$$20x + 10y \leq 6000$$

$$5x + 15y \leq 5000$$

$$5x + 10y \leq 4000$$

$$x \geq 30$$

9. The government has mobilized to inoculate the student population against sleeping sickness. There are 200 doctors available and 450 nurses. An inoculation team can consist of either 1 doctor and 3 nurses (called a *full team*) or 1 doctor and 2 nurses (called a *half team*). On the average a full team can inoculate 180 people per hour while a half team can inoculate only 100 people per hour. How many teams of each type should be formed in order to maximize the number of inoculations per hour?

5-7 GRAPHICAL SOLUTION OF LINEAR PROGRAMMING PROBLEMS

In this section we turn to the solution of linear programming problems similar to those formulated in the preceding section. We begin with an example and then present a general method.

EXAMPLE 5-20 (CONTINUATION OF EXAMPLE 5-17).

The problem formulated in Example 5-17 is to find the maximum value of $2x + y$ for (x,y) satisfying

$$x \geq 0$$
$$y \geq 0$$
$$x \leq 200$$
$$2x + 3y \leq 1200 \qquad \text{(5-19)}$$
$$25x + 15y \leq 7500$$

Solution: The feasible set, i.e., the set of (x,y) satisfying the inequalities (5-19), is shown as the shaded region in Fig. 5-24. To solve the maximization problem we begin by considering the sets on which the function $2x + y$ is constant. For example, notice that the points $Q = (100,300)$ and $R = (150,200)$ have the property that the objective function $2x + y$ has the same value, namely 500, at both points. Indeed at Q we have

$$2x + y = 2(100) + 300 = 500$$

and at R we also have

$$2x + y = 2(150) + 200 = 500$$

The points $(0,500)$ and $(250,0)$ have the same property, and in fact *every point* (x,y) on the line $2x + y = 500$ has that property. That is, the set of points at which the objective function takes the value 500 is a line. The set of points at which the objective function takes the value 200 is also a line but a different line. A similar statement holds in general: for any constant c the set of points (x,y) for

FIGURE 5-24

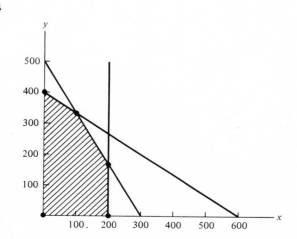

which the objective function $2x + y$ takes the value c, that is, the set $\{(x,y): 2x + y = c\}$, is a line. Some of these lines intersect the feasible set of the problem, and some do not. For example, the point Q on the line $2x + y = 500$ belongs to the feasible set while the point $S = (50,400)$ on the same line does not. No point on the line $2x + y = 800$ belongs to the feasible set. The lines corresponding to several different values of c and the points Q, R, and S are shown in Fig. 5-25. Since we are interested in finding the point (x,y) in the feasible set which makes $2x + y$ as large as possible, we are interested in determining the largest value of c such that the line $2x + y = c$ intersects the feasible set. We denote that value of c by c^*. The boundary of the feasible set is shown by a dotted polygonal line in Fig. 5-25. In this example the values of c corresponding to different lines increase as one considers lines with larger x intercepts. It is clear from Fig. 5-25 that the value c^* corresponds to the line whose intersection with the feasible set is the point labeled P. The coordinates of the point P can be obtained by solving the system of equations

$$x = 200$$

$$25x + 15y = 7500$$

The solution of this system is $x = 200$, $y = \frac{500}{3}$, and therefore the value c^* is

$$2(200) + \tfrac{500}{3} = 400 + \tfrac{500}{3} = \tfrac{1700}{3}$$

FIGURE 5-25

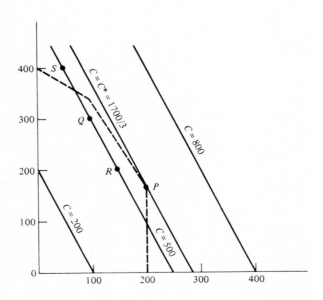

The problem is now solved. The maximum value of $2x + y$ for (x,y) satisfying the conditions (5-19) is $\frac{1700}{3}$, and this value is attained for $x = 200$, $y = \frac{500}{3}$. Stated differently, for any (x,y) satisfying (5-19) we have $2x + y \leq \frac{1700}{3}$, and for the point $(200, \frac{500}{3})$ which satisfies (5-19) we have $2x + y = \frac{1700}{3}$.

In terms of the original problem, the manufacturer maximizes net profit with the production schedule: 200 barrels of distillate A and $\frac{500}{3}$ barrels of distillate B. The net profit with this schedule is $\frac{1700}{3}$ dollars. Notice that this schedule uses

$$2(200) + 3(\tfrac{500}{3}) = 900$$

barrels of crude oil. Since there were 1200 barrels of crude oil available, the optimal production schedule is one which does not use all the crude oil. However, all the available electricity is used; since

$$25(200) + 15(\tfrac{500}{3}) = 7500$$

$\square$

EXAMPLE 5-21

Suppose that in Example 5-17 the net profit on 1 barrel of distillate A is \$1 and on 1 barrel of distillate B is \$2. If all other conditions in the example remain the same, how should production be allocated between the two distillates if net profit is to be as large as possible?

Solution: If x and y are as defined in Example 5-17, the constraints of the situation are again given by the system of inequalities (5-19). However, in this case the objective function is different. In this example it is $x + 2y$. The problem, therefore, is to maximize the function $x + 2y$ subject to the conditions (5-19). If we construct a figure analogous to 5-25, the dotted polygonal line which forms the boundary of the feasible set is exactly as in Fig. 5-25, but the slope of the lines $x + 2y = c$ for various values of c are different from the slope of the corresponding lines on that figure. Using the same method as before, we conclude that the largest value of c for which the set $\{(x,y): x + 2y = c\}$ has a point in common with the feasible set is $c = 800$, and the common point is $(0,400)$. $\square$

Notice that in both Examples 5-20 and 5-21 the maximum value of the objective function was attained at one of the corner points of the feasible set. This is true for a very general class of problems which includes all linear programming problems considered in this book. This statement is not one which is obviously true, and the reader has a legitimate right to ask its basis. It is situations like this which demonstrate the utility of mathematical theorems. We observed a fact in two special cases. If it were true in all cases of interest, it

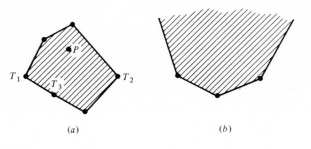

(a)   *(b)*

FIGURE 5-26

would greatly simplify the problem-solving process. Its truth under certain care-fully specified hypotheses is asserted in a *theorem* and demonstrated in a *proof*. Although a proof for the general case is beyond the scope of this book, we shall carefully state the conditions under which the result is valid. We hope that Examples 5-20 and 5-21 provide some intuitive support for the result.

In the discussion which follows we shall be concerned only with sets which arise as feasible sets for linear programming problems. Examples of such sets are shown in Fig. 5-26.

Definition   A set of points in the plane is said to be *bounded* if it is contained in some circle centered at (0,0). Otherwise it is said to be *unbounded*.

Set (*a*) in Fig. 5-26 is bounded; set (*b*) is unbounded.

Definition   A point *T* is a *corner point* of a feasible set for a linear programming problem if every line segment which is contained in the set and which contains *T* has *T* as one of its endpoints.

In Fig. 5-26*a* the points T_1 and T_2 are corner points, but the points *P* and T_3 are not. This definition makes precise the intuitive notion of a corner point used in Examples 5-20 and 5-21. As in those examples, corner points can always be determined by solving a system of linear equations.

Using these definitions, we can formulate a mathematical theorem which provides the basis for the method of solving the linear programming problems introduced in this book.

Theorem   Let $\mathscr{F}$ be the feasible set for a linear programming problem and let *f* be a linear function defined on $\mathscr{F}$.

1. If $\mathscr{F}$ is bounded, then *f* assumes its maximum (and minimum) value at a corner point of the set $\mathscr{F}$.

2. If $\mathscr{F}$ is unbounded, then either f assumes arbitrarily large positive (negative) values or else it assumes its maximum (minimum) value at a corner point of the set.

The first assertion can be restated as follows: given a linear programming problem with bounded feasible set $\mathscr{F}$ and a linear function f defined on $\mathscr{F}$, there are corner points P and Q belonging to $\mathscr{F}$ such that the value of the function at P is at least as large as its value at any other point in $\mathscr{F}$ and its value at Q is no larger than its value at any other point in $\mathscr{F}$. That is, for any point R in $\mathscr{F}$ we have

$$f(Q) \leq f(R) \leq f(P)$$

The assertion regarding unbounded sets is necessarily phrased somewhat differently from the assertion regarding bounded sets. Indeed, a linear function need not assume its maximum value on an unbounded feasible set $\mathscr{F}$. For example, the linear function $x + y$ does not assume a maximum value on the feasible set $\mathscr{F} = \{(x,y): \; x \geq 0\}$. In fact, the value of the function $x + y$ can be made as large as one pleases simply by taking the point $(n,0)$ with n sufficiently large (see Fig. 5-27). However, if a linear function does not assume arbitrarily large values, it must assume its maximum value at a corner point.

FIGURE 5-27

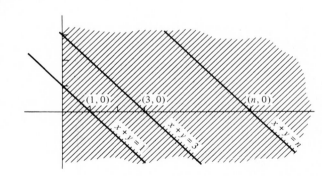

This theorem provides the justification for the following method of solving linear programming problems in two variables.

1. **Graph the feasible set for the problem.**
2. **Determine the coordinates of each of the corner points of the feasible set. Since each corner point is the intersection of two lines, it can be determined by solving a pair of linear equations.**
3. **Evaluate the objective function at each corner point. If the problem is to maximize the objective function, select the largest of these values; if the problem is to minimize the objective function, select the smallest of these values.**

a. **If the feasible set is bounded, the value selected is the solution of the problem.**

b. **If the feasible set is unbounded and if the problem has a solution, the value selected is the solution.**

EXAMPLE 5-22

Solve the problem formulated in Example 5-18.

Solution: The equations defining the feasible set are

$$x + y \geq 30{,}000$$

$$y \geq 15{,}000$$

$$2y - x \geq 0$$

$$x \geq 0$$

$$y \geq 0$$

and the objective function is $.8x + y$. The feasible set, which is graphed in Fig. 5-28, has three corner points, denoted by P, Q, and R in Fig. 5-28. The co-ordinates of each of the corner points can be determined by simultaneously solving the pair of equations corresponding to the lines which intersect at the corner point. Thus, to determine the coordinates of P, Q, and R we solve the respective pairs of equations

	P	Q	R
	$x = 0$	$x + y = 30{,}000$	$y = 30{,}000$
	$x + y = 30{,}000$	$y = 15{,}000$	$2y - x = 0$

FIGURE 5-28

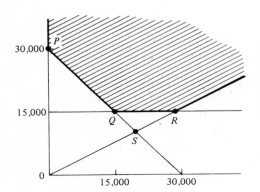

TABLE 5-3

Corner Point	Value of Objective Function
(0, 30,000)	30,000
(15,000, 15,000)	27,000
(30,000, 15,000)	39,000

We conclude that $P = (0, 30{,}000)$, $Q = (15{,}000, 15{,}000)$, and $R = (30{,}000, 15{,}000)$. Evaluating the objective function at each of these points, we obtain the results recorded in Table 5-3. For example, at the corner point $P = (0, 30{,}000)$, the objective function $.8x + y$ has the value $.8(0) + 1(30{,}000) = 30{,}000$.

The last step is to examine the values of the objective function at the corner points and select the smallest (recall that the original problem was to minimize acquisition costs). This is 27,000 which is attained for (15,000 15,000). Therefore, if there is a smallest value of the objective function, it is achieved for $x = 15{,}000$ and $y = 15{,}000$ and it is equal to 27,000. In fact, this is the minimum value of the objective function on the feasible set. This follows since the objective function $.8x + y$ necessarily assumes only positive values (recall $x \geq 0$, $y \geq 0$), and consequently this function cannot assume arbitrarily large negative values.

Step 1 of the method outlined above helps us determine that the point S, which is the intersection of two of the lines which bound half planes arising in the constraints, is not a corner point of the feasible set. Without graphing the feasible set it is sometimes difficult to determine that points like S are not corner points.

In terms of the original problem, the optimal acquisition policy for Tone Down is to purchase 15,000 pounds of plastic from an independent supplier and 15,000 pounds from its subsidiary. In this case its acquisition cost is $27,000, and this is the minimum such cost.   □

EXAMPLE 5-23

Find the largest value of $x + y$ for x and y satisfying the constraints

$$x \geq 0$$
$$y \geq 0$$
$$x + 2y \geq 6$$
$$x - y \geq -4$$
$$2x + y \leq 8$$

Solution: We begin with step 1, graphing the feasible set. This set is shown as the shaded area in Fig. 5-29. Note that the constraint $y \geq 0$ is redundant in this problem. That is, the feasible set is the same whether or not the constraint $y \geq 0$ is used to determine it.

Next we determine the corner points, which are $(0,4)$, $(\frac{4}{3},\frac{16}{3})$, $(\frac{10}{3},\frac{4}{3})$ and $(0,3)$. Evaluating the objective function $x + y$ at these corner points, we obtain the information given in Table 5-4. Examination of the table shows that the largest value assumed by the objective function is $\frac{20}{3}$ and that this value is assumed at the corner point $(\frac{4}{3},\frac{16}{3})$. It follows from the theorem that this is the largest value that the objective function assumes on the entire feasible set. $\square$

FIGURE 5-29

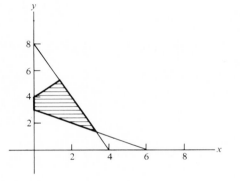

TABLE 5-4

Corner Point	Value of Objective Function
$(0,4)$	4
$\left(\dfrac{4}{3}, \dfrac{16}{3}\right)$	$\dfrac{20}{3}$
$\left(\dfrac{10}{3}, \dfrac{4}{3}\right)$	$\dfrac{14}{3}$
$(0,3)$	3

Exercises for Sec. 5-7

1. Each of the following systems of inequalities describes a set. Graph that set and find its corner points.

   a. $x - y \leq 2$
   $x + 2y \leq 2$
   $2x + y \geq -2$

   b. $x + y \leq 2$
   $x - y \leq 0$
   $x - y \geq -3$

2. Each of the following systems of inequalities describes a set. Graph that set and find its corner points.

a.
$$x \geq 0$$
$$y \geq 0$$
$$y \leq 2$$
$$x + y \leq 3$$
$$x - 3y \leq 1$$

b. $3x + 2y \leq 10$
$$y \leq 3$$
$$x + y \geq 1$$
$$x - 2y \leq 1$$
$$x \geq 0$$

3. Find the maximum value of the function $3x + 5y$ on the feasible set defined by

$$x + 3y \leq 50 \qquad x \geq 0 \qquad x - 3y \geq -25 \qquad y \geq 0$$

4. Find the maximum value of the function $2x - 5y$ on the feasible set defined by

$$x + 3y \leq 30 \qquad x \geq 0 \qquad 3x - 4y \geq -24 \qquad y \geq 0$$

5. Find the minimum value of the function $5x - 3y$ on the feasible set defined in Exercise 3.

6. Find the minimum value of the function $4x - 2y$ on the feasible set defined in Exercise 4.

7. Maximize $2x - y$ over the set of points (x,y) defined by

$$2x + y \leq 50 \qquad 6x - 2y \leq 50 \qquad x \geq 0 \qquad y \geq 0$$

8. Complete the solution of the survey problem posed in Example 1-3 as formulated precisely in Example 5-19.

Solve the linear programming problem formulated in:

9. Exercise 3 of Sec. 5-6.

10. Exercise 4 of Sec. 5-6.   **11.** Exercise 5 of Sec. 5-6.

12. Exercise 6 of Sec. 5-6.   **13.** Exercise 7 of Sec. 5-6.

14. Exercise 9 of Sec. 5-6.

IMPORTANT TERMS

You should be able to describe, define, or give examples of each of the following:

Cartesian coordinate system
Line
General equation of a line
Slope intercept equation of a line
Two-point equation of a line
Intercepts
Slope

Parallel lines
Linear programming problem
Constraints
Feasible set
Objective function
Bounded (unbounded) set
Corner point

REVIEW EXERCISES

1. On a cartesian coordinate system graph the pairs of points (x,y), where x has the values -10, -5, 0, 5, 10 and y is related to x by the formula $y = 1 - x + x^2$.

2. Find the slope, the y intercept and the x intercept for each of the lines:
   a. $5y = 10$
   b. $3x + 6y + 9 = 0$
   c. $\dfrac{x}{2} = 10$

3. Find the equation of the line through the point $(5,10)$ which is parallel to the line through the points $(1,2)$ and $(-1,8)$.

Find all solutions of the system of equations:

4. $\begin{aligned} -2x - 3y &= 6 \\ 6x + 6y &= -3 \end{aligned}$

5. $\begin{aligned} x + y &= 16 \\ -x + 3y &= -6 \\ 2x + 4y &= 12 \end{aligned}$

6. $\begin{aligned} x + 2y &= 16 \\ -x + 3y &= -6 \\ x - y &= 7 \end{aligned}$

7. Graph the set of points which satisfy the inequalities:

   $x \le 5 \qquad x + y \le 10 \qquad x \ge 1 \qquad x - y \ge 10$

8. Find the inequalities which describe the set of points which lie inside a triangle with vertices at the points $(-5,-5)$, $(1,1)$, $(3,0)$.

9. For each of the lines described below, find the equation of the line, the slope of the line, and the intercepts. Then graph the lines on the same coordinate system.
   a. A horizontal line through $(3,6)$
   b. A line through $(2,6)$ with slope 1.5
   c. A line with undefined slope through $(-3,-2)$
   d. A line through $(6,-4)$ and through the origin
   e. A line with x intercept $-17/3$ and y intercept $17/2$

10. Find the largest and smallest values of $x - y$ (if they exist) for x and y which satisfy the constraints:

    $x \ge 0 \qquad y \ge 0 \qquad x - y \ge -4 \qquad x + 2y \ge 6$

11. A car which cost \$3400 when new is linearly depreciated. The book value of the car decreases \$525 each year until it reaches \$250 (the salvage value of the car), after which it no longer changes.
    a. What is the book value of the car at the end of 3 years? At the end of 7 years?
    b. Find the two linear equations which describe the relationship between v, the book value of the car, and t, the number of years since the the car was new. For which values of t is each equation valid?
    c. Find the slopes of the lines in part b.

12. A car is linearly depreciated over 10 years. Its original book value is \$4800, and after 4 years its book value is \$3028.
    a. Find the equations of the lines relating the book value of the car to its age. (Assume that the book value of the car is constant after it becomes 10 years old.)
    b. How old will the car be when its book value first drops below \$2000?
    c. What will be the book value of the car at the end of 5 years? At the end of 15 years?

13. Assume that the cars described in Exercises 11 and 12 were manufactured in the same year.
    a. When, if ever, do they have the same book values?
    b. Graph the lines representing the book values of the cars on the same coordinate system.

14. a. Graph the set of points which satisfy the inequalities:

    $$3y \ge -2x$$
    $$2y - 3x \le 17$$
    $$-3 \le x \le 0$$
    $$y \le 6$$
    $$2y \ge 3x + 6$$

    b. Find the corner points of the set in part a.
    c. Find the maximum of $2x - y$ over this set.
    d. Find the minimum of $3y + x$ over this set.

15. a. Graph the set of points which satisfy the inequalities:

$$y - 3x \leq 2$$
$$2y + 3x \geq 12$$
$$y \leq 8$$
$$y + x \leq 14$$
$$2x - y \leq 16$$
$$2x - 5y \leq 8$$

b. Find the corner points of the set in part **a**.
c. Find the maximum of $2x + 3y$ over this set.
d. Find the minimum of $4y - 3x$ over this set.

16. Formulate a mathematical problem for the following situation. Mr. Smith produces widgets. For each 100 left-handed widgets he uses 1 pound of metal and 5 pounds of fiber glass. For each 100 right-handed widgets he uses 2 pounds of metal and 3 pounds of fiber glass. Each week Mr. Smith has 65 pounds of metal and 150 pounds of fiber glass available. He makes a profit of $2.50 on each left-handed widget and a profit of $2 on each right-handed widget. How many of each kind of widget should Mr. Smith produce (per week) to maximize his profit?

17. Solve the problem posed in Exercise 16. What is Mr. Smith's maximum profit if he produces only one kind of widget?

18. Suppose that in Exercise 16 the production of 100 widgets (either model) requires 10 kilowatt-hours of electricity. Also suppose 350 kilowatt-hours are available per week for this process. How many widgets of each type should be made (per week) to yield the maximum profit? What is the maximum profit?

SYSTEMS OF LINEAR EQUATIONS AND MATRICES

SIX

6-1   INTRODUCTION

Systems of linear equations arise in many different ways, some direct, others indirect. Several of the situations discussed in Chap. 5 illustrate how systems of equations in two variables originate. A situation which leads immediately to a system of equations in more than two variables is described in Example 6-1. In Sec. 6-7, where we study a Leontief input-output model for an economy, we shall see how systems of equations arise somewhat less directly, and in Chap. 8 we shall encounter (rather unexpectedly) systems of equations in the study of Markov-chain models.

EXAMPLE 6-1

The Zeeoff Oil Company has access to supplies of three different types of crude oil, known as California (C), Texas (T), and Venezuela (V) crude, respectively. The company refines the crude oil and produces gasoline and heating oil together with several by-products. The various types of crude oil yield different amounts of gasoline and heating oil, as shown in Table 6-1, where the number of barrels of the two products which can be refined from 100 barrels of crude oil of each type are given. If the company plans to produce 10,000 barrels of gasoline and 20,000 barrels of heating oil, what mixes of crude oil (if any) will yield these amounts of refined products?

TABLE 6-1

Type of Crude Oil	Number of Barrels per 100 Barrels of Crude Oil		
	Gasoline	Heating Oil	Other
California	37	36	27
Texas	45	23	32
Venezuela	15	77	8

Solution: A mix of crude oils is determined by specifying the number of barrels of each type that are to be refined to meet certain specifications. Let x_1, x_2, and x_3 denote the number of barrels of crude oil of types C, T, and V, respectively, which are to be refined. Each barrel of type C crude oil yields .37 barrels of gasoline, and consequently x_1 barrels of type C crude oil yield .37x_1 barrels of gasoline. Likewise, x_2 barrels of type T crude oil yield .45x_2 barrels of gasoline, and x_3 barrels of type V crude oil yield .15x_3 barrels of gasoline. Thus, a mixture of x_1, x_2, and x_3 barrels of types C, T, and V crude oils, respectively, yields

$$.37x_1 + .45x_2 + .15x_3$$

barrels of gasoline. A similar argument shows that the same mix yields

$$.36x_1 + .23x_2 + .77x_3$$

barrels of heating oil. The desired mixes of types C, T, and V crude oils which produce 10,000 barrels of gasoline and 20,000 barrels of heating oil are specified by the triples x_1, x_2, x_3 of numbers which satisfy the system of equations

$$.37x_1 + .45x_2 + .15x_3 = 10,000$$
$$.36x_1 + .23x_2 + .77x_3 = 20,000$$

(6-1)
□

One of the primary goals of this chapter is to develop methods of solving problems like that posed in Example 6-1. To this end it is useful to introduce the terminology and notation of vectors and matrices. We shall see (in Sec. 6-6) that matrices also arise in situations other than those involving systems of linear equations. Some of these situations will be studied in this chapter, and others will be studied in Chaps. 8 to 10.

6-2   SYSTEMS OF LINEAR EQUATIONS AND MATRIX NOTATION

A system of m linear equations in n variables can be written in the form

$$
\begin{aligned}
a_{11}x_1 + a_{12}x_2 + \cdots + a_{1n}x_n &= b_1 \\
a_{21}x_1 + a_{22}x_2 + \cdots + a_{2n}x_n &= b_2 \\
&\ \ \vdots \\
a_{m1}x_1 + a_{m2}x_2 + \cdots + a_{mn}x_n &= b_m
\end{aligned}
\tag{6-2}
$$

It is assumed that the a_{ij}'s and b_j's are known numbers. The x_i's are unknown numbers which are to be determined.

Definition   A *solution* of the system (6-2) is an ordered set of numbers $x_1^0, x_2^0, \ldots, x_n^0$ which when substituted into Eqs. (6-2) for $x_1, x_2, \ldots, x_n$, respectively, result in numerical equalities.

It is convenient to introduce notation to simplify the representation of systems such as (6-2).

Definition   A *matrix* is a rectangular array of numbers. If the array has m (horizontal) rows and n (vertical) columns, the matrix is said to be $m \times n$, or, equivalently, to have size $m \times n$. The number in the ith row and the jth column of the matrix is said to be the (i,j) *entry* in the matrix. A matrix with one column is a *column vector*, and a matrix with one row is a *row vector*. The entries in vectors are called *coordinates*. The matrix with entries a_{ij} is denoted (a_{ij}).

Thus, a vector is an ordered set of numbers. A column vector with n coordinates, an n-vector for short, can be represented as

$$
\begin{bmatrix}
x_1 \\
x_2 \\
\vdots \\
x_n
\end{bmatrix}
$$

and a row vector with m coordinates can be represented as

$$
\begin{bmatrix} x_1 & x_2 & \cdots & x_m \end{bmatrix}
$$

We use the symbol $\mathbf{x}$ to denote a column vector:

$$\mathbf{x} = \begin{bmatrix} x_1 \\ x_2 \\ \vdots \\ x_n \end{bmatrix} \tag{6-3}$$

The size of the vector $\mathbf{x}$ will always be clear from the way it arises. For instance, we can represent a mix of crude oils in Example 6-1 by a vector $\mathbf{x}$ with coordinates x_1, x_2, x_3. We introduce a vector with three coordinates since there are three types of oil available.

The matrix

$$\begin{bmatrix} a_{11} & a_{12} & \cdots & a_{1n} \\ a_{21} & a_{22} & \cdots & a_{2n} \\ \vdots & \vdots & & \vdots \\ a_{m1} & a_{m2} & \cdots & a_{mn} \end{bmatrix} \tag{6-4}$$

in which the entries are the coefficients which appear in the system (6-2), is called the *coefficient matrix* of that system. The (i,j) element of the coefficient matrix is a_{ij}. Note that the system must be written in the form (6-2). In particular, the variables must appear in the same order in every equation, and every variable must have a coefficient, possibly zero, in each equation. The matrix (6-4) will be denoted by the outline letter $\mathbb{A}$. In general we use outline letters $\mathbb{A}$, $\mathbb{B}$, $\mathbb{C}$, ... to denote matrices. For example, the coefficient matrix above is denoted by $\mathbb{A}$ or by (a_{ij}).

Matrices and vectors are useful both as notational devices and as computational tools. Matrices can be manipulated like numbers (with certain crucial restrictions), and there is a rich theory associated with them. At the moment we concentrate on the use of matrices in solving systems of linear equations.

Definition   If $\mathbb{A}$ is an $m \times n$ matrix and $\mathbf{x}$ is an n-vector, then the product $\mathbb{A}\mathbf{x}$ is the vector given by

$$\mathbb{A}\mathbf{x} = \begin{bmatrix} a_{11}x_1 + a_{12}x_2 + \cdots + a_{1n}x_n \\ a_{21}x_1 + a_{22}x_2 + \cdots + a_{2n}x_n \\ \vdots \\ a_{m1}x_1 + a_{m2}x_2 + \cdots + a_{mn}x_n \end{bmatrix}$$

It is clear from the definition that the product $\mathbb{A}\mathbf{x}$ of an $m \times n$ matrix $\mathbb{A}$ and an n-vector $\mathbf{x}$ is an m-vector. The ith coordinate of this m-vector is

$$a_{i1}x_1 + a_{i2}x_2 + \cdots + a_{in}x_n$$

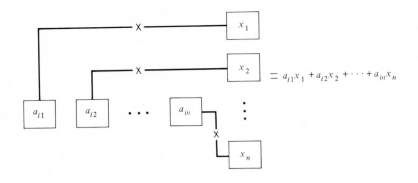

FIGURE 6-1

This last expression can be obtained by multiplying the elements in the ith row of the matrix A (there are n of them) by the corresponding elements in $\mathbf{x}$ and adding the resulting products. This way of viewing the ith entry in the product $A\mathbf{x}$ is pictured in Fig. 6-1. It is described as *row-by-column* multiplication.

The product $A\mathbf{x}$ of an $m \times n$ matrix A and a k-vector $\mathbf{x}$ with $k \neq n$ is not defined.

EXAMPLE 6-2

Let A be the 3×4 matrix

$$\begin{bmatrix} 2 & 5 & 1 & -1 \\ 1 & -1 & 0 & -2 \\ 3 & 2 & -1 & 1 \end{bmatrix}$$

(*a*) Find the product $A\mathbf{x}$, where $\mathbf{x}$ is a 4-vector as in (6-3).
(*b*) Find the product $A\mathbf{y}$, where

$$\mathbf{y} = \begin{bmatrix} 1 \\ 0 \\ -2 \\ -1 \end{bmatrix}$$

Solution: (*a*) If $\mathbf{x}$ is a 4-vector as in (6-3), then

$$A\mathbf{x} = \begin{bmatrix} 2 & 5 & 1 & -1 \\ 1 & -1 & 0 & -2 \\ 3 & 2 & -1 & 1 \end{bmatrix}\begin{bmatrix} x_1 \\ x_2 \\ x_3 \\ x_4 \end{bmatrix} = \begin{bmatrix} 2x_1 + 5x_2 + x_3 - x_4 \\ x_1 - x_2 - 2x_4 \\ 3x_1 + 2x_2 - x_3 + x_4 \end{bmatrix}$$

The product $\mathbb{A}\mathbf{y}$ is

$$\mathbb{A}\mathbf{y} = \begin{bmatrix} 2 & 5 & 1 & -1 \\ 1 & -1 & 0 & -2 \\ 3 & 2 & -1 & 1 \end{bmatrix} \begin{bmatrix} 1 \\ 0 \\ -2 \\ -1 \end{bmatrix}$$

$$= \begin{bmatrix} 2(1) + 5(0) + 1(-2) + (-1)(-1) \\ 1(1) + (-1)(0) + 0(-2) + (-2)(-1) \\ 3(1) + 2(0) + (-1)(-2) + 1(-1) \end{bmatrix} = \begin{bmatrix} 1 \\ 3 \\ 4 \end{bmatrix}$$

$\square$

Definition

Two column n-vectors

$$\mathbf{x} = \begin{bmatrix} x_1 \\ x_2 \\ \vdots \\ x_n \end{bmatrix} \qquad \text{and} \qquad \mathbf{y} = \begin{bmatrix} y_1 \\ y_2 \\ \vdots \\ y_n \end{bmatrix}$$

are *equal*, written $\mathbf{x} = \mathbf{y}$, if $x_1 = y_1$, $x_2 = y_2$, ..., $x_n = y_n$. That is, vectors of the same size are equal if all of the pairs of corresponding coordinates are equal. Vectors of different sizes are never equal.

With these definitions of matrix, vector, matrix-vector product, and equality of vectors the system (6-2) can be represented in a much more convenient form. In fact, with $\mathbf{x}$ and $\mathbb{A}$ as defined in (6-3) and (6-4), respectively, and with $\mathbf{b}$ defined by

$$\mathbf{b} = \begin{bmatrix} b_1 \\ b_2 \\ \vdots \\ b_m \end{bmatrix}$$

the system (6-2) can be written

$$\mathbb{A}\mathbf{x} = \mathbf{b} \tag{6-5}$$

That is, the systems (6-2) and (6-5) are simply different representations for the same set of equations. In the general case the sizes of $\mathbb{A}$, $\mathbf{x}$, and $\mathbf{b}$ are related to the number of equations and the number of variables as follows:

Number of equations = number of rows in $\mathbb{A}$

= number of coordinates in $\mathbf{b}$

Number of variables = number of columns in $\mathbb{A}$

= number of coordinates in $\mathbf{x}$

We introduce the symbol **0** to denote the vector each of whose coordinates is the number zero. We do not distinguish notationally between the n-vector **0** and the m-vector **0**, with $m \neq n$. The size of the vector will usually be clear from the discussion; otherwise we shall specifically comment on the matter.

EXAMPLE 6-3

Find the matrix $\mathbb{A}$ and the vector **b** which arise when the system

$$
\begin{aligned}
2x_1 - 3x_2 + x_3 &= -9 \\
4x_1 + 2x_2 - 2x_3 &= 0 \\
-3x_1 + x_2 - 3x_3 &= 4
\end{aligned}
\tag{6-6}
$$

is written in the form $\mathbb{A}\mathbf{x} = \mathbf{b}$ and verify that

$$
\mathbf{x} = \begin{bmatrix} -1 \\ \frac{5}{2} \\ \frac{1}{2} \end{bmatrix}
$$

is a solution of (6-6).

Solution: The matrix $\mathbb{A}$ is the coefficient matrix of (6-6), and the vector **b** is the vector whose coordinates are the numbers on the right-hand side of (6-6). Thus

$$
\mathbb{A} = \begin{bmatrix} 2 & -3 & 1 \\ 4 & 2 & -2 \\ -3 & 1 & -3 \end{bmatrix} \quad \text{and} \quad \mathbf{b} = \begin{bmatrix} -9 \\ 0 \\ 4 \end{bmatrix}
$$

In order to verify that $\begin{bmatrix} -1 \\ \frac{5}{2} \\ \frac{1}{2} \end{bmatrix}$ is a solution of (6-6) we compute the product

$\mathbb{A} \begin{bmatrix} -1 \\ \frac{5}{2} \\ \frac{1}{2} \end{bmatrix}$ and show that it is **b**. We have

$$
\mathbb{A} \begin{bmatrix} -1 \\ \frac{5}{2} \\ \frac{1}{2} \end{bmatrix} = \begin{bmatrix} 2 & -3 & 1 \\ 4 & 2 & -2 \\ -3 & 1 & -3 \end{bmatrix} \begin{bmatrix} -1 \\ \frac{5}{2} \\ \frac{1}{2} \end{bmatrix} = \begin{bmatrix} -9 \\ 0 \\ 4 \end{bmatrix} = \mathbf{b} \qquad \square
$$

Exercises for Sec. 6-2

1. For each of the following pairs A and x form the product Ax.

a. $A = \begin{bmatrix} 1 & 2 \\ 3 & 4 \end{bmatrix}$ $\quad x = \begin{bmatrix} 5 \\ 6 \end{bmatrix}$ $\qquad$ **b.** $A = \begin{bmatrix} 1 & 0 & 1 \\ 0 & 1 & 0 \\ 1 & 1 & 1 \end{bmatrix}$ $\quad x = \begin{bmatrix} 1 \\ 2 \\ -3 \end{bmatrix}$

2. Form the product Ax for

$$A = \begin{bmatrix} 1 & -1 & 2 & -3 \\ 4 & 0 & -1 & 3 \end{bmatrix} \qquad x = \begin{bmatrix} -1 \\ 0 \\ 3 \\ -4 \end{bmatrix}$$

3. Find the matrix A and the vector b when each of the following systems is written in the form (6-5):

a. $\begin{aligned} 3x_1 + 2x_2 - x_3 &= 4 \\ x_1 + x_2 + x_3 &= 1 \\ 2x_1 + x_2 - x_3 &= \tfrac{5}{2} \end{aligned}$
$\qquad$ **b.** $\begin{aligned} 3x_1 - 2x_2 + 4x_3 - x_4 &= 1 \\ 2x_1 + x_2 + 5x_3 + 2x_4 &= 3 \\ -x_1 + x_2 + 3x_3 \quad\quad &= 2 \end{aligned}$

4. Find the matrix A and the vector b when each of the following systems is written in the form (6-5):

a. $\begin{aligned} x_1 + 2x_2 + x_3 &= 5 \\ 2x_1 + 3x_2 - 2x_3 &= 2 \end{aligned}$
$\qquad$ **b.** $\begin{aligned} x_1 + x_2 &= -2x_1 + 1 \\ 2x_1 - x_2 &= 3x_1 + 4 \end{aligned}$

5. For each pair of matrix A and vector b given below, write the system of equations $Ax = b$ in the form (6-2):

a. $A = \begin{bmatrix} 2 & 3 & 1 \\ 6 & -2 & -8 \\ 8 & 5 & -3 \end{bmatrix}$ $\quad b = \begin{bmatrix} 6 \\ 7 \\ 17 \end{bmatrix}$

b. $A = \begin{bmatrix} 2 & 2 & \tfrac{3}{2} \\ 4 & \tfrac{5}{2} & 2 \\ 2 & -\tfrac{3}{2} & -1 \end{bmatrix}$ $\quad b = \begin{bmatrix} 2 \\ 8 \\ 10 \end{bmatrix}$

6. Write the system $Ax = b$ in the form (6-2) when

$$A = \begin{bmatrix} 4 & 5 & 2 & -1 \\ 3 & 2 & 2 & 2 \\ 2 & 1 & 3 & 3 \end{bmatrix} \qquad b = \begin{bmatrix} 46 \\ 29 \\ 23 \end{bmatrix}$$

7. Verify that $x = \begin{bmatrix} \tfrac{1}{2} \\ 1 \\ -\tfrac{1}{2} \end{bmatrix}$ is a solution of part **a** of Exercise 3.

8. Verify that $x = \begin{bmatrix} \tfrac{5}{2} \\ 0 \\ 1 \end{bmatrix}$ and $x = \begin{bmatrix} \tfrac{3}{2} \\ 1 \\ 0 \end{bmatrix}$ are both solutions of $Ax = b$ with A and b as in part **a** of Exercise 5.

9. Verify that $\mathbf{x} = \begin{bmatrix} 5 \\ 4 \\ 3 \\ 0 \end{bmatrix}$ is a solution of $\mathbb{A}\mathbf{x} = \mathbf{b}$ with $\mathbb{A}$ and $\mathbf{b}$ as in Exercise 6.

10. Define

$$\mathbb{A} = \begin{bmatrix} 1 & 3 \\ -2 & -6 \end{bmatrix} \qquad \mathbf{b} = \begin{bmatrix} 5 \\ -10 \end{bmatrix}$$

Show that $\mathbf{x} = \begin{bmatrix} 5 - 3t \\ t \end{bmatrix}$ is a solution of $\mathbb{A}\mathbf{x} = \mathbf{b}$ for every number t.

11. Verify that $\mathbf{x} = \begin{bmatrix} 7t - 11 \\ 8 - 4t \\ t \end{bmatrix}$ is a solution of part **a** of Exercise 4 for every value of t.

12. For Exercise 10 show that the second equation in $\mathbb{A}\mathbf{x} = \mathbf{b}$ is -2 times the first equation.

13. In part **a** of Exercise 5 show that the third equation is equal to $\frac{23}{11}$ times the first equation plus $\frac{7}{11}$ times the second equation, i.e., show that

$$\tfrac{23}{11}(\text{equation 1}) + \tfrac{7}{11}(\text{equation 2}) = (\text{equation 3})$$

6-3 SOLVING SYSTEMS OF LINEAR EQUATIONS

Once a problem has been reduced to a system of equations, one is usually interested in finding solutions of the system. In this section we develop an algorithm, i.e., a well-defined method of proceeding in a step-by-step manner, which will produce all the solutions of a system of linear equations when there are solutions.

There are of course systems for which there are no solutions. For example,

$$3x - y = 4$$
$$6x - 2y = 3$$

is a system which has no solutions. Indeed, if x and y are any numbers such that $3x - y = 4$, then $2(3x - y) = 6x - 2y$ must be equal to $2(4) = 8$. Consequently $6x - 2y$ cannot equal 3. There are no values for x and y such that both $3x - y = 4$ and $6x - 2y = 3$.

Definition   A system of equations which has at least one solution is said to be *consistent*, and a system which has no solutions is said to be *inconsistent*.

There are consistent systems with infinitely many solutions. For example, the system

$$x_1 - 2x_2 + x_3 = 4$$
$$2x_1 + 3x_2 - 2x_3 = -1$$
$$8x_1 + 5x_2 - 4x_3 = 5$$

has infinitely many solutions. The vectors

$$\mathbf{w} = \begin{bmatrix} 0 \\ -7 \\ -10 \end{bmatrix} \qquad \mathbf{y} = \begin{bmatrix} 1 \\ -3 \\ -3 \end{bmatrix} \qquad \text{and} \qquad \mathbf{z} = \begin{bmatrix} 2 \\ 1 \\ 4 \end{bmatrix}$$

are solutions, as can be verified by substitution. In fact, the vector $\mathbf{x}$ with $x_1 = t, x_2 = 4t - 7$, and $x_3 = 7t - 10$ is a solution for any number t. This solution can be conveniently represented in vector notation as

$$\mathbf{x}(t) = \begin{bmatrix} t \\ 4t - 7 \\ 7t - 10 \end{bmatrix}$$

The specific solutions $\mathbf{w}$, $\mathbf{y}$, and $\mathbf{z}$ given above are special cases of the solution $\mathbf{x}(t)$ corresponding to $t = 0$, 1, and 2, respectively. That is, $\mathbf{x}(0) = \mathbf{w}$, $\mathbf{x}(1) = \mathbf{y}$, and $\mathbf{x}(2) = \mathbf{z}$.

Some systems have exactly one solution. For example, the vector $\mathbf{x} = \begin{bmatrix} 1 \\ 0 \\ -1 \end{bmatrix}$

is the only solution of the system

$$\mathbb{A}\mathbf{x} = \mathbf{b} \text{ with } \mathbb{A} = \begin{bmatrix} 2 & 1 & 0 \\ 0 & 3 & -1 \\ 1 & 2 & 1 \end{bmatrix} \text{ and } \mathbf{b} = \begin{bmatrix} 2 \\ 1 \\ 0 \end{bmatrix}$$

The solution method introduced here is one in which the original system is successively transformed into simpler and simpler systems *without changing the set of solutions* until the system can be solved by inspection. This is the technique introduced in Sec. 5-4 to solve systems of two equations in two variables. Although the details of the method developed here differ from those of Chap. 5, the basic idea is the same. The method rests on the following result.

Theorem   If a system of equations is transformed into a new system by any one of the following operations:

1. The interchange of two equations
2. The multiplication of any equation by a nonzero constant
3. The replacement of any equation by the sum of that equation and a constant multiple of any other equation

then the set of solutions of the transformed system is the same as the set of solutions of the original system.

This theorem can be proved by directly verifying: (*i*) that every solution of the original system is also a solution of the transformed system and (*ii*) that every solution of the transformed system is also a solution of the original system.

Our method of solving systems of equations is to use operations of types 1, 2, and 3 to successively replace the original system by other systems until a system which has a coefficient matrix of a particular form is obtained. According to the theorem, the transformed systems all have the same solutions as the original system. The particular form of the coefficient matrix we seek will be such that the solutions of the transformed system can easily be determined. The coefficient matrix we seek has the following properties:

A. The first nonzero entry in each row is 1.
B. If a_{ij} is the first nonzero entry in the ith row, it is the only nonzero entry in the submatrix of $\mathbb{A}$ with a_{ij} at the upper right-hand corner. This submatrix is shown as the shaded area in Fig. 6-2.

FIGURE 6-2

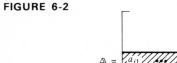

Definition   A matrix which has the form described by conditions A and B is said to be in *echelon form*.

The following matrix is in echelon form

$$\begin{bmatrix} 1 & 3 & -1 & 0 & -2 & -1 \\ 0 & 0 & 1 & -2 & 3 & 0 \\ 0 & 0 & 0 & 1 & -1 & 4 \\ 0 & 0 & 0 & 0 & 0 & 1 \end{bmatrix}$$

We illustrate the solution method with an example.

EXAMPLE 6-4

Solve the system of equations

$$\begin{aligned} 2x_1 + x_2 \qquad &= 2 \\ 3x_2 - x_3 &= 1 \\ x_1 + 2x_2 + x_3 &= 0 \end{aligned} \qquad (6\text{-}7)$$

Solution: We begin by transforming the system into one in which the coefficient of x_1 in the first equation is 1, so that condition A is satisfied for the first row. To accomplish this we interchange the first and third equations (an operation of type 1). We obtain

$$\begin{aligned} x_1 + 2x_2 + x_3 &= 0 \\ 3x_2 - x_3 &= 1 \\ 2x_1 + x_2 \qquad &= 2 \end{aligned}$$

In order to satisfy condition B we must transform the system into one in which the coefficients of x_1 in the second and third equations are zero. Since $a_{21} = 0$, we need to consider only the third equation. Multiplying the first equation by -2, adding it to the third, and replacing the third equation by this sum (an operation of type 3), we have

$$\begin{aligned} x_1 + 2x_2 + x_3 &= 0 \\ 3x_2 - x_3 &= 1 \\ -3x_2 - 2x_3 &= 2 \end{aligned}$$

We next seek to satisfy condition A for the second row in the coefficient matrix. Multiplying the second equation by $\frac{1}{3}$, we have

$$\begin{aligned} x_1 + 2x_2 + x_3 &= 0 \\ x_2 - \tfrac{1}{3}x_3 &= \tfrac{1}{3} \\ -3x_2 - 2x_3 &= 2 \end{aligned}$$

To continue, in order to satisfy condition B for the second row, it is necessary to transform the system into one in which the coefficient of x_2 in the third equation is zero. To accomplish this we multiply the second equation by 3, add it to the third equation, and replace the third equation by this sum. We have

$$x_1 + 2x_2 + x_3 = 0$$
$$x_2 - \tfrac{1}{3}x_3 = \tfrac{1}{3}$$
$$-3x_3 = 3$$

In order to have the coefficient matrix in the desired form it remains only to satisfy condition A for the third row. To this end we multiply the third row by $-\tfrac{1}{3}$. The transformed system which satisfies conditions A and B is

$$x_1 + 2x_2 + x_3 = 0$$
$$x_2 - \tfrac{1}{3}x_3 = \tfrac{1}{3} \qquad\qquad (6\text{-}8)$$
$$x_3 = -1$$

with coefficient matrix $\mathbb{A} = \begin{bmatrix} 1 & 2 & 1 \\ 0 & 1 & -\tfrac{1}{3} \\ 0 & 0 & 1 \end{bmatrix}$

Each of the transformed systems, and in particular the last, has the same solutions as the original system since we used only operations of types 1 to 3 to change one system into another.

It is clear from (6-8) that $x_3 = -1$. Substituting this into the second equation of the system (6-8), we have

$$x_2 - \tfrac{1}{3}(-1) = \tfrac{1}{3} \qquad \text{or} \qquad x_2 = \tfrac{1}{3} - \tfrac{1}{3} = 0$$

Substituting $x_3 = -1$ and $x_2 = 0$ into the first equation, we have

$$x_1 + 2(0) + 1(-1) = 0 \qquad \text{or} \qquad x_1 = 1$$

A solution (in fact the only solution) of (6-8), and therefore also a solution of (6-7), is $x_1 = 1$, $x_2 = 0$, $x_3 = -1$ or, equivalently,

$$\mathbf{x} = \begin{bmatrix} 1 \\ 0 \\ -1 \end{bmatrix} \qquad\qquad \square$$

The technique of using x_3 to determine x_2 and then x_3 and x_2 to determine x_1 is known as *back substitution*. The entire process introduced here is known as *reduction to echelon form and back substitution*.

It should be noted that in Example 6-4 the variables x_1, x_2, and x_3 were just "carried along for the ride." That is, the entire process could be carried out by working only with the coefficient matrix $\mathbb{A}$ and the vector $\mathbf{b}$. To take advantage of this simplification we introduce additional notation.

Definition

The *augmented matrix* of a system of equations $\mathbb{A}\mathbf{x} = \mathbf{b}$ is the matrix formed by juxtaposing the vector $\mathbf{b}$ on the right of the matrix $\mathbb{A}$. A vertical line between $\mathbb{A}$ and $\mathbf{b}$ indicates the origin of the entries in the augmented matrix.

The augmented matrix for the problem of Example 6-4 is

$$\left[\begin{array}{ccc|c} 2 & 1 & 0 & 2 \\ 0 & 3 & -1 & 1 \\ 1 & 2 & 1 & 0 \end{array}\right]$$

In working the problem of the next example we shall show each step in both the form in which the variables are displayed and the form in which the augmented matrix is utilized.

EXAMPLE 6-5

Solve the system of equations

$$\begin{array}{rcl} x_2 + 3x_3 &=& -1 \\ x_1 - x_2 - x_3 &=& 1 \\ 3x_1 \qquad + 2x_3 &=& 4 \end{array} \tag{6-9}$$

Solution: The augmented matrix for the system (6-9) is

$$\left[\begin{array}{ccc|c} 0 & 1 & 3 & -1 \\ 1 & -1 & -1 & 1 \\ 3 & 0 & 2 & 4 \end{array}\right]$$

We begin by interchanging the first two equations or, in matrix terms, the first two rows (an operation of type 1). The new system and augmented matrix are

$$\begin{array}{rcl} x_1 - x_2 - x_3 &=& 1 \\ x_2 + 3x_3 &=& -1 \\ 3x_1 \qquad + 2x_3 &=& 4 \end{array} \qquad \left[\begin{array}{ccc|c} 1 & -1 & -1 & 1 \\ 0 & 1 & 3 & -1 \\ 3 & 0 & 2 & 4 \end{array}\right]$$

Next, we multiply the first equation (row) by -3 and add it to the third equation (row) to obtain

$$
\begin{aligned}
x_1 - x_2 - x_3 &= 1 \\
x_2 + 3x_3 &= -1 \\
3x_2 + 5x_3 &= 1
\end{aligned}
\qquad
\left[
\begin{array}{ccc|c}
1 & -1 & -1 & 1 \\
0 & 1 & 3 & -1 \\
0 & 3 & 5 & 1
\end{array}
\right]
$$

Next, we multiply the second equation (row) by -3 and add it to the third equation (row). This gives

$$
\begin{aligned}
x_1 - x_2 - x_3 &= 1 \\
x_2 + 3x_3 &= -1 \\
- 4x_3 &= 4
\end{aligned}
\qquad
\left[
\begin{array}{ccc|c}
1 & -1 & -1 & 1 \\
0 & 1 & 3 & -1 \\
0 & 0 & -4 & 4
\end{array}
\right]
$$

Finally, we multiply the third equation (row) by $-\frac{1}{4}$ and obtain

$$
\begin{aligned}
x_1 - x_2 - x_3 &= 1 \\
x_2 + 3x_3 &= -1 \\
x_3 &= -1
\end{aligned}
\qquad
\left[
\begin{array}{ccc|c}
1 & -1 & -1 & 1 \\
0 & 1 & 3 & -1 \\
0 & 0 & 1 & -1
\end{array}
\right]
$$

Obviously, $x_3 = -1$. Substituting this value for x_3 in the second equation, we find $x_2 + 3(-1) = -1$ or $x_2 = -1 + 3 = 2$. Using these values for x_2 and x_3 in the first equation, we have

$$
x_1 - 2 - (-1) = 1 \qquad \text{or} \qquad x_1 = 1 + 2 - 1 = 2
$$

Therefore, a solution of the system of equations (6-9) is

$$
\mathbf{x} =
\begin{bmatrix}
2 \\
2 \\
-1
\end{bmatrix}
\qquad \square
$$

The technique illustrated in Example 6-5 does not always lead to a solution of the system of equations:

If in using this technique it ever happens that the transformed system has an equation with zero on the left-hand side and a nonzero number on the right, the original system has no solution.

In terms of the augmented matrix (or its transform) this condition is that there is a row with only zeros to the left of the vertical line and a nonzero entry in that row to the right of the vertical line.

EXAMPLE 6-6

Solve the system of equations

$$3x_1 - x_2 = 4$$
$$6x_1 - 2x_2 = 3$$

Solution: To illustrate the method we again proceed by using both the equations and the augmented matrix. The augmented matrix for the given system is

$$\begin{bmatrix} 3 & -1 & | & 4 \\ 6 & -2 & | & 3 \end{bmatrix}$$

Multiplying the first equation by -2 and adding it to the second, we obtain

$$3x_1 - x_2 = 4 \qquad\qquad \begin{bmatrix} 3 & -1 & | & 4 \\ 0 & 0 & | & -5 \end{bmatrix}$$
$$0 = -5$$

We conclude that the original system has no solutions. ☐

The next example illustrates the technique to be used when a system has more than one solution. A system of linear equations has either *no solutions, exactly one solution, or infinitely many solutions*. There are no other possibilities. In Chap. 5 we commented on the geometry associated with two linear equations in two variables, and it is also useful to consider the geometry of linear equations in three variables. The solution set of a linear equation in three variables is a plane in ordinary three-dimensional space. The set of points which are simultaneously the solution of two linear equations in three variables will be empty if the planes associated with the two equations are parallel, it will be a line if the planes intersect but are not coincident, and it will be a plane if the planes are coincident. The set of points which are simultaneously the solution of three linear equations in three variables can be empty, a single point, or a line or a plane. In the first case the system has no solution, in the second case a unique solution, and in the third case infinitely many solutions—either all points on a line or all points in a plane. Since there are infinitely many solutions associated with intersections which are lines and with those which are planes, we see that there can be different kinds of solution sets which have infinitely many elements. We continue our examination of this situation after the next example, and we make the ideas precise in a theorem.

EXAMPLE 6-7

Find all solutions of the system

$$
\begin{aligned}
x_1 + 2x_2 - x_3 &= 5 \\
2x_1 + 3x_2 - 3x_3 &= 8 \\
x_2 + x_3 &= 2
\end{aligned}
\qquad (6\text{-}10)
$$

Solution: In this example we shall work only with the augmented matrix. We begin by interchanging the second and third rows. Recall that this corresponds to interchanging the second and third equations, an operation of type 1. We have

$$
\begin{bmatrix}
1 & 2 & -1 & 5 \\
0 & 1 & 1 & 2 \\
2 & 3 & -3 & 8
\end{bmatrix}
$$

Multiplying the first row by -2 and adding it to the third row, we obtain

$$
\begin{bmatrix}
1 & 2 & -1 & 5 \\
0 & 1 & 1 & 2 \\
0 & -1 & -1 & -2
\end{bmatrix}
$$

Next we add the second row to the third row, and we have

$$
\begin{bmatrix}
1 & 2 & -1 & 5 \\
0 & 1 & 1 & 2 \\
0 & 0 & 0 & 0
\end{bmatrix}
\qquad (6\text{-}11)
$$

The augmented matrix is now in echelon form.

The system of equations which is associated with this augmented matrix is

$$
\begin{aligned}
x_1 + 2x_2 - x_3 &= 5 \\
x_2 + x_3 &= 2
\end{aligned}
$$

Suppose that $x_3 = 0$. The system then becomes

$$
\begin{aligned}
x_1 + 2x_2 &= 5 \\
x_2 &= 2
\end{aligned}
$$

from which we conclude at once that $x_2 = 2$ and $x_1 = 5 - 2(2) = 1$. Thus one solution of the original system is $x_1 = 1$, $x_2 = 2$, and $x_3 = 0$. Likewise, if $x_3 = -1$, the system becomes

$$x_1 + 2x_2 - (-1) = 5$$
$$x_2 + (-1) = 2$$

from which we conclude that $x_2 = 3$ and $x_1 = -2$. Consequently, another solution of the original system is $x_1 = -2$, $x_2 = 3$, and $x_3 = -1$. The same method works when x_3 is set equal to any number. That is, for any choice of x_3 we can determine x_2 and x_1. To keep our work as general as possible, we set $x_3 = t$. Here t stands for a number whose value is not determined by the problem, a *parameter*. With $x_3 = t$ the equations resulting from the augmented matrix (6-11) become

$$x_1 + 2x_2 - t = 5$$
$$x_2 + t = 2$$

From the second of these equations we have $x_2 = 2 - t$. Substituting this value into the first equation, we have

$$x_1 + 2(2 - t) - t = 5$$
or
$$x_1 = 3t + 1$$

We conclude that for each number t a solution of system (6-10) is given by $x_1 = 3t + 1$, $x_2 = 2 - t$, $x_3 = t$. In vector terms, for each number t the vector

$$\begin{bmatrix} 3t + 1 \\ 2 - t \\ t \end{bmatrix}$$ is a solution. Since there are infinitely many choices for t, any number

will do, and since each choice of t gives a different solution of the system, there are infinitely many solutions of the system (6-10). Every solution can be written in this form for some value of t. $\square$

In Example 6-7 a single parameter t was introduced. The geometry in this case is that the solution set of system (6-10) is a line in 3 space. It may happen that more than one parameter can be introduced. For example, if the solution set of a system of three equations in three variables is a plane in 3-space, then two independent parameters can be introduced. There is a means of determining the number of independent parameters which can be introduced.

Theorem   If a consistent system of linear equations in n variables is reduced to echelon form with k nonzero rows, then there are solutions which depend on $n - k$ independent parameters. These $n - k$ parameters can be identified with $n - k$ of the variables. If $n > k$, then there are infinitely many solutions.

Note: Although the echelon form for a system of equations is not unique, the number k of nonzero rows is the same for each echelon form.

EXAMPLE 6-8

Find all solutions of the system

$$
\begin{aligned}
x_1 - x_2 + 2x_3 - x_4 &= 7 \\
x_1 + x_2 - 4x_3 + x_4 &= 3 \\
2x_1 - 2x_3 &= 10 \\
-2x_2 + 6x_3 - 2x_4 &= 4
\end{aligned}
$$

Solution: Again we work only with the augmented matrix. The original augmented matrix and the matrix obtained by multiplying row one by -2 and adding it to row 3 are

$$
\left[\begin{array}{rrrr|r}
1 & -1 & 2 & -1 & 7 \\
1 & 1 & -4 & 1 & 3 \\
2 & 0 & -2 & 0 & 10 \\
0 & -2 & 6 & -2 & 4
\end{array}\right]
\qquad
\left[\begin{array}{rrrr|r}
1 & -1 & 2 & -1 & 7 \\
1 & 1 & -4 & 1 & 3 \\
0 & 2 & -6 & 2 & -4 \\
0 & -2 & 6 & -2 & 4
\end{array}\right]
$$

Now we add row 3 to row 4 and obtain

$$
\left[\begin{array}{rrrr|r}
1 & -1 & 2 & -1 & 7 \\
1 & 1 & -4 & 1 & 3 \\
0 & 2 & -6 & 2 & -4 \\
0 & 0 & 0 & 0 & 0
\end{array}\right]
$$

Next we multiply row one by -1 and add it to row 2 to obtain

$$
\left[\begin{array}{rrrr|r}
1 & -1 & 2 & -1 & 7 \\
0 & 2 & -6 & 2 & -4 \\
0 & 2 & -6 & 2 & -4 \\
0 & 0 & 0 & 0 & 0
\end{array}\right]
$$

To continue, we multiply row two by -1 and add it to row 3 to obtain a matrix which corresponds to a system of two equations. Finally, we multiply row two by $\frac{1}{2}$. The resulting matrix has echelon form

$$
\left[\begin{array}{rrrr|r}
1 & -1 & 2 & -1 & 7 \\
0 & 1 & -3 & 1 & -2 \\
0 & 0 & 0 & 0 & 0 \\
0 & 0 & 0 & 0 & 0
\end{array}\right]
$$

It follows from this augmented matrix that the two equations are

$$x_1 - x_2 + 2x_3 - x_4 = 7$$
$$x_2 - 3x_3 + x_4 = -2$$

Thus the original system has been reduced to one with four variables and two nonzero equations. Using the theorem quoted above, we know that we can choose two of the variables to be parameters. We let $x_3 = s$ and $x_4 = t$, where s and t represent arbitrary numbers. We can immediately solve for x_2 in terms of s and t. Back substitution then yields x_1 in terms of s and t. We obtain

$$x_2 = 3s - t - 2$$
$$x_1 = 5 + s$$

We conclude that all solutions of the original system are given by $x_1 = 5 + s$, $x_2 = 3s - t - 2$, $x_3 = s$, and $x_4 = t$, where s and t are any numbers. As vectors, all solutions of the original system are given by

$$\mathbf{x} = \mathbf{x}(s,t) = \begin{bmatrix} 5 + s \\ 3s - t - 2 \\ s \\ t \end{bmatrix}$$

where s and t are arbitrary numbers.

Two specific solutions can be obtained by setting, say, $s = 0$, $t = 1$ and $s = 2$, $t = -1$. They are

$$\mathbf{x}(0,1) = \begin{bmatrix} 5 \\ -3 \\ 0 \\ 1 \end{bmatrix} \quad \text{and} \quad \mathbf{x}(2,-1) = \begin{bmatrix} 7 \\ 5 \\ 2 \\ -1 \end{bmatrix}$$

respectively.   □

Exercises for Sec. 6-3

Find all solutions of each of the following systems:

1. a. $2x_1 + 4x_2 = 8$
$\quad\ \ 4x_1 - 2x_2 = 4$

b. $2x_1 + 4x_2 = 8$
$\quad\ 4x_1 - 2x_2 = 4$
$\quad\ 6x_1 + 2x_2 = 12$

2. a. $2x_1 + 4x_2 = 8$
$\quad\ \ 4x_1 - 2x_2 = 4$
$\quad\ \ 6x_1 + 2x_2 = 10$

b. $2x_1 + 4x_2 = 8$
$\quad\ 4x_1 + 8x_2 = 16$

3. a.
$$3x_1 \qquad + 2x_3 = 9$$
$$x_1 - \;\; x_2 - 3x_3 = -3$$
$$-x_1 + 2x_2 + 4x_3 = 5$$

b.
$$2x_1 + 2x_2 + \;\; x_3 = -2$$
$$3x_1 \qquad + \;\; x_3 = 4$$
$$2x_2 + 3x_3 = -2$$

4. a.
$$2x_3 + 3x_4 = 5$$
$$x_1 + \;\; x_2 - \;\; x_3 \qquad = 2$$
$$2x_2 - 2x_3 - \;\; x_4 = 1$$

b.
$$2x_1 - x_2 - 3x_3 = 6$$
$$x_1 \qquad + 2x_3 = 1$$
$$x_1 + x_2 + 3x_3 = -3$$

5. Find the solutions of part **a** of Exercise 4 which have $x_2 = 0$. Repeat for $x_2 = 2$.

6. Find all solutions of the following system:

$$3x_1 - 2x_2 + \;\; x_3 + 2x_4 = 6$$
$$2x_1 + 4x_2 - 2x_3 + 3x_4 = 2$$

(*Hint:* You will need to introduce two parameters.)

7. Find all solutions of the system in Exercise 6 which have $x_2 = 2$.

8. A sum of $100,000 is to be invested, partly in bonds yielding 10 percent per year, partly in preferred stocks yielding 8 percent per year, and the remainder in common stocks yielding 5 percent per year. The total investment is to yield $6500 per year, and the amount invested in common stocks must equal the amount invested in preferred stocks. How much should be invested in each area?

9. A contractor has a supply of kitchen cabinets on hand when he learns that the style has been discontinued. He would like to use as many of the cabinets as possible to avoid loss. Since all cabinets in a house must be of the same style, the contractor is interested in determining how many houses of each type should be furnished with the discontinued style of cabinets. There are 4 types of cabinets, base, base corner, wall, and sink base. The contractor builds 3 styles of house, ranch, bilevel, and colonial. The number of cabinets of each type used in each style house and the number of each type available are given in Table 6-2. Using this information, determine whether the contractor can use all his cabinets and if all cabinets cannot be used, decide how many houses of each type should be built to use the maximum total number of cabinets. (*Hint:* The second part of the problem requires techniques other that those introduced in this chapter. An elementary trial-and-error approach can be used.)

TABLE 6-2

House Style	Cabinet Type			
	Base	**Base Corner**	**Sink Base**	**Wall**
Ranch	6	1	1	4
Bilevel	6	1	1	8
Colonial	7	2	1	5
Number available	75	15	10	53

6-4   MATRIX ALGEBRA

In Secs. 6-2 and 6-3 we used the notation and terminology of matrices to facilitate our discussion of systems of linear equations. Although this is an important use of matrices, there are also many other problems whose solution is aided by using matrices. We consider some of these problems in detail in later chapters, and we include examples of them together with the definitions of this section. The primary purpose of this and the following section is to develop mathematical tools, e.g., the addition and multiplication of matrices, which will be applied later in the book.

EXAMPLE 6-9

The Ampsunwatts Electric Automobile Company records the sales made by each person on its sales staff on a large bulletin board. Since the company has a sales staff of 4 and manufactures a two-door and a four-door automobile and a light truck, the bulletin board, represented as a matrix, appears as in Table 6-3.

TABLE 6-3

Type of Vehicle	Salesperson			
	A	**B**	**C**	**D**
Two-door automobiles	3	2	0	4
Four-door automobiles	1	1	5	1
Light trucks	4	3	1	3

The entries are given for the month of October. If we agree to remember the meaning of the entry in each location, we can represent the sales figures for October by a 3×4 matrix, call it $\mathbb{A}$. The corresponding matrix for another month would be viewed as the same as the one for October only if its entries were exactly the same as those for October. If the corresponding matrix for November, call it $\mathbb{B}$, is

$$\begin{bmatrix} 1 & 1 & 2 & 3 \\ 2 & 2 & 3 & 1 \\ 2 & 1 & 2 & 3 \end{bmatrix}$$

the matrix which represents the combined sales for October and November is

$$\begin{bmatrix} 4 & 3 & 2 & 7 \\ 3 & 3 & 8 & 2 \\ 6 & 4 & 3 & 6 \end{bmatrix}$$

This new matrix represents a sum of $\mathbb{A}$ and $\mathbb{B}$. If the salesmanager sets a goal for December for each salesperson which is twice the November sales in each category, then the goal can be represented by the matrix

$$\begin{bmatrix} 2 & 2 & 4 & 6 \\ 4 & 4 & 6 & 2 \\ 4 & 2 & 4 & 6 \end{bmatrix}$$

Since the November sales matrix is denoted by $\mathbb{B}$, it is reasonable to view the matrix which is the December goal as $2\mathbb{B}$. $\qquad\qquad\square$

This example leads to the following definitions of *equality of matrices*, *matrix addition* and *scalar multiplication*.

Definition

Let $\mathbb{A} = (a_{ij})$ and $\mathbb{B} = (b_{ij})$ be two $m \times n$ matrices. $\mathbb{A}$ is said to be *equal* to $\mathbb{B}$, written $\mathbb{A} = \mathbb{B}$, if $a_{ij} = b_{ij}$ for $1 \leq i \leq m$, $1 \leq j \leq n$.

That is, two matrices $\mathbb{A}$ and $\mathbb{B}$ are equal if they are the same size and their corresponding entries are all equal.

Definition

With $\mathbb{A}$ and $\mathbb{B}$ as above, the *sum*, $\mathbb{A} + \mathbb{B}$, of the matrices $\mathbb{A}$ and $\mathbb{B}$ is defined to be the matrix $\mathbb{C} = (c_{ij})$, where $c_{ij} = a_{ij} + b_{ij}$, $1 \leq i \leq m$, $1 \leq j \leq n$.

That is, the sum of two $m \times n$ matrices is the $m \times n$ matrix each of whose entries is the sum of the corresponding entries of $\mathbb{A}$ and $\mathbb{B}$. Addition of matrices is defined only for matrices with the same size, i.e., the same number of rows and the same number of columns. The sum of three or more $m \times n$ matrices, $\mathbb{A}$, $\mathbb{B}$, $\mathbb{C}$, ..., can be determined by computing the sum $\mathbb{A} + \mathbb{B}$, then $(\mathbb{A} + \mathbb{B}) + \mathbb{C}$, and so on. In the expression $(\mathbb{A} + \mathbb{B}) + \mathbb{C}$ the term $\mathbb{A} + \mathbb{B}$ is a single matrix and consequently the sum $(\mathbb{A} + \mathbb{B}) + \mathbb{C}$ is the sum of two matrices and is covered by the definition. In fact, the parentheses are unnecessary: the sum $(\mathbb{A} + \mathbb{B}) + \mathbb{C}$ is the same matrix as the sum $\mathbb{A} + (\mathbb{B} + \mathbb{C})$. Therefore, there is no ambiguity in writing simply $\mathbb{A} + \mathbb{B} + \mathbb{C}$ for the sum, and we shall do so in the future.

Definition

If c is a number and $\mathbb{A}$ is an $m \times n$ matrix, then the *scalar multiple* $c\mathbb{A}$ is defined to be the matrix $\mathbb{D} = (d_{ij})$, where $d_{ij} = ca_{ij}$, $1 \leq i \leq m$, $1 \leq j \leq n$.

That is, the scalar multiple of a matrix $\mathbb{A}$ is the matrix each of whose entries is the corresponding entry of $\mathbb{A}$ multiplied by the scalar.

EXAMPLE 6-10

Let

$$A = \begin{bmatrix} 3 & 2 \\ 1 & -1 \\ 0 & 2 \end{bmatrix} \quad B = \begin{bmatrix} 2 & 1 \\ 0 & 2 \\ -1 & -3 \end{bmatrix} \quad C = \begin{bmatrix} a & 2 \\ b & -1 \\ 0 & c \end{bmatrix}$$

Find $A + B$, $2A$, $(-1)B$, and conditions on a, b, and c such that $A = C$.

Solution:

$$A + B = \begin{bmatrix} 3 & 2 \\ 1 & -1 \\ 0 & 2 \end{bmatrix} + \begin{bmatrix} 2 & 1 \\ 0 & 2 \\ -1 & -3 \end{bmatrix} = \begin{bmatrix} 5 & 3 \\ 1 & 1 \\ -1 & -1 \end{bmatrix}$$

$$2A = 2\begin{bmatrix} 3 & 2 \\ 1 & -1 \\ 0 & 2 \end{bmatrix} = \begin{bmatrix} 6 & 4 \\ 2 & -2 \\ 0 & 4 \end{bmatrix}$$

$$(-1)B = (-1)\begin{bmatrix} 2 & 1 \\ 0 & 2 \\ -1 & -3 \end{bmatrix} = \begin{bmatrix} -2 & -1 \\ 0 & -2 \\ 1 & 3 \end{bmatrix}$$

$$A = C \quad \text{or} \quad \begin{bmatrix} 3 & 2 \\ 1 & -1 \\ 0 & 2 \end{bmatrix} = \begin{bmatrix} a & 2 \\ b & -1 \\ 0 & c \end{bmatrix}$$

if and only if $a = 3$, $b = 1$, and $c = 2$. □

There are several features of matrix addition and scalar multiplication which will be useful to us:

If A and B are $m \times n$ matrices and c and d are numbers, then

$$A + B = B + A$$

$$cA + dA = (c + d)A$$

$$cA + cB = c(A + B)$$

$$(cd)A = c(dA)$$

In the work which follows in this and later chapters we shall use these properties freely without specifically mentioning them. At times more than one of them may be combined into a single step in a computation. We write $-A$ for the matrix $(-1)A$ and $A - B$ for $A + (-B)$.

In Sec. 6-2 we defined a matrix-vector product $A\mathbf{x}$, where A is an $m \times n$ matrix and $\mathbf{x}$ is a (column) n-vector. Clearly, $\mathbf{x}$ may also be considered as an $n \times 1$ matrix, and from this point of view we have defined a certain type of matrix product. We now extend the definition.

Definition

If $A = (a_{ij})$ is an $m \times n$ matrix and $B = (b_{ij})$ is an $n \times k$ matrix, then the *matrix product* AB is defined to be the $m \times k$ matrix $C = (c_{ij})$, where

$$c_{ij} = a_{i1}b_{1j} + a_{i2}b_{2j} + \cdots + a_{in}b_{nj} \qquad \begin{aligned} 1 \le i \le m \\ 1 \le j \le k \end{aligned}$$

Notice that the matrix product AB is only defined for matrices A and B with the number of columns of A equal to the number of rows of B. Since the (i,j) entry of the product AB is determined by the ith row of the matrix A and the jth column of the matrix B, the (i,j) entry is said to be obtained through *row-by-column* multiplication.

EXAMPLE 6-11

Find the matrix product AB, where

$$A = \begin{bmatrix} 2 & 3 & -1 \\ -1 & 0 & 2 \end{bmatrix} \qquad \text{and} \qquad B = \begin{bmatrix} 1 & -1 \\ -1 & 0 \\ 1 & 2 \end{bmatrix}$$

Solution: Since A is 2×3 and B is 3×2, the product AB is defined and is a 2×2 matrix.

$$AB = \begin{bmatrix} 2(1) + 3(-1) + (-1)(1) & 2(-1) + 3(0) + (-1)(2) \\ (-1)(1) + 0(-1) + 2(1) & (-1)(-1) + 0(0) + 2(2) \end{bmatrix}$$

$$= \begin{bmatrix} -2 & -4 \\ 1 & 5 \end{bmatrix} \qquad\qquad \square$$

Multiplication of three or more matrices can be defined by grouping the matrices and computing the product in steps by multiplying two matrices at a time. For example, the product of an $m \times n$ matrix A, an $n \times r$ matrix B, and an $r \times s$ matrix C (in that order) can be defined as $(AB)C$. This is to be interpreted as the result of multiplying the $m \times r$ matrix AB (which is defined by the definition of multiplication of two matrices) by the $r \times s$ matrix C. It can be shown that this is the same as the product of the $m \times n$ matrix A and the $n \times s$ matrix BC. That is, $(AB)C = A(BC)$, and in the future we write simply ABC since there is no possible ambiguity.

In contrast to the freedom we have in grouping matrices in a product in different ways, in general we cannot change the order of the factors in a matrix product. If the product $\mathbb{AB}$ of the matrices $\mathbb{A}$ and $\mathbb{B}$ in that order is defined, it may be that the product of these matrices in the reverse order is not defined. Even if $\mathbb{AB}$ and $\mathbb{BA}$ are both defined, they may not be equal.

EXAMPLE 6-12

If $\mathbb{A} = \begin{bmatrix} 2 & 0 \\ 1 & 3 \\ 0 & -1 \end{bmatrix}$ and $\mathbb{B} = \begin{bmatrix} 1 & 1 \\ -1 & -1 \end{bmatrix}$, $\mathbb{AB}$ is defined but the product of $\mathbb{B}$ and $\mathbb{A}$ is not defined.

With $\mathbb{B}$ as above and $\mathbb{A} = \begin{bmatrix} 2 & 2 \\ 1 & 1 \end{bmatrix}$ we have

$$\mathbb{AB} = \begin{bmatrix} 0 & 0 \\ 0 & 0 \end{bmatrix} \quad \text{and} \quad \mathbb{BA} = \begin{bmatrix} 3 & 3 \\ -3 & -3 \end{bmatrix}$$

This shows that in general $\mathbb{AB} \neq \mathbb{BA}$. It also illustrates that one can have a matrix $\mathbb{A}$ which is not all zeros and a matrix $\mathbb{B}$ which is not all zeros whose product is a matrix with all zeros. Thus, the familiar result "if $ab = 0$, then either $a = 0$ or $b = 0$" which holds for numbers, does not hold for matrices. □

There is a matrix $\mathbb{I}$ which plays the same role in matrix multiplication that the number 1 plays with respect to multiplication of numbers.

Definition   The $n \times n$ matrix $\mathbb{I}$ defined by

$$\mathbb{I} = \left.\begin{bmatrix} 1 & 0 & 0 & \cdots & 0 \\ 0 & 1 & 0 & \cdots & 0 \\ 0 & 0 & 1 & \cdots & 0 \\ \vdots & & & & \vdots \\ 0 & 0 & 0 & \cdots & 1 \end{bmatrix}\right\} n \text{ Rows}$$

$$\underbrace{}_{n \text{ Columns}}$$

is said to be the $n \times n$ *identity* matrix.

We shall not distinguish notationally between identity matrices of different sizes. When we write $\mathbb{I}\mathbb{A}$ we assume that $\mathbb{I}$ has the correct size for the product to be defined.

It is a useful fact for every matrix A and the identity matrices I for which the products AI and IA are defined we have

$$AI = A \quad \text{and} \quad IA = A$$

Exercises for Sec. 6-4

1. Let

$$A = \begin{bmatrix} 2 & 1 & 0 \\ 3 & -1 & 2 \end{bmatrix} \qquad B = \begin{bmatrix} 1 & 0 & 1 \\ -1 & 2 & -2 \\ 3 & 1 & 1 \end{bmatrix} \qquad C = \begin{bmatrix} -1 & -1 & 3 \\ 2 & 2 & 0 \end{bmatrix}$$

Decide which of the following operations are defined and carry out those which are defined:

a. $A + B$ **b.** AB **c.** $(A + C)B$

2. Repeat Exercise 1 for the operations

a. $2A$ **b.** AC **c.** $AB + C$

3. Let

$$A = \begin{bmatrix} 2 & 2 \\ 3 & 1 \end{bmatrix} \qquad B = \begin{bmatrix} 3 & 1 \\ -1 & 0 \\ 4 & 2 \end{bmatrix} \qquad C = \begin{bmatrix} 4 & 0 & 1 \\ 1 & -1 & 2 \\ 3 & 1 & 1 \end{bmatrix} \qquad D = \begin{bmatrix} -1 & 0 \\ 2 & 1 \end{bmatrix}$$

Find:

a. $2B$ **b.** BA **c.** CBA

4. With matrices A, B, C, and D defined as in Exercise 3, find

a. $3A - 2D$ **b.** CB **c.** $CBA - B$

5. Let $A = \begin{bmatrix} 3 & 1 \\ 2 & 1 \end{bmatrix}$. For what value of c does the matrix $B = \begin{bmatrix} 1 & -1 \\ -2 & c \end{bmatrix}$ satisfy $AB = I$?

6. For what values of c and d does the matrix equation

$$c \begin{bmatrix} 3 \\ 1 \end{bmatrix} + d \begin{bmatrix} -1 \\ 2 \end{bmatrix} = \begin{bmatrix} 7 \\ 0 \end{bmatrix}$$

hold?

7. Find three 2×2 matrices A, B, and C (C not all zeros) such that $AC = BC$ but $A \neq B$.

8. Find two 3×3 matrices A and B such that $AB = BA = I$ but $A \neq I$ and $B \neq I$.

9. For a square matrix A (same number of rows and columns) the product AA is always defined. We call this product A^2. Likewise, $A^3 = AAA$, and for any positive integer n, $A^n = AA \cdots A$ (n factors). For $A = \begin{bmatrix} 1 & 0 \\ \frac{1}{2} & \frac{1}{2} \end{bmatrix}$ find A^2 and A^3.

10. Let $\mathbb{P} = \begin{bmatrix} \frac{1}{3} & \frac{2}{3} \\ \frac{2}{3} & \frac{1}{3} \end{bmatrix}$. Using the definitions of Exercise 9, find $\mathbb{P}^2$, $\mathbb{P}^4$, and $\mathbb{P}^8$.

11. Let $\mathbb{P}$ be as in Exercise 10 and let $\mathbf{x} = \begin{bmatrix} \frac{1}{2} \\ \frac{1}{2} \end{bmatrix}$. Show that $\mathbb{P}\mathbf{x} = \mathbf{x}$ and $\mathbb{P}^n\mathbf{x} = \mathbf{x}$ for any positive integer n.

6-5 MATRIX INVERSES

Linear equations in a single scalar variable are especially easy to solve. In order to find x for which $5x = 3$ one multiplies both sides of the equation by $\frac{1}{5}$ (the reciprocal of 5) to obtain

$$\tfrac{1}{5}(5x) = x = \tfrac{1}{5}(3) = \tfrac{3}{5}$$

or $x = \frac{3}{5}$. Superficially a system of equations written in matrix form as $\mathbb{A}\mathbf{x} = \mathbf{b}$ appears much as the above scalar equation $5x = 3$. One hopes that there is a reciprocal of the matrix $\mathbb{A}$ which can be used to solve the equation $\mathbb{A}\mathbf{x} = \mathbf{b}$ in the same way as $\frac{1}{5}$ was used to solve the scalar equation $5x = 3$. Unfortunately, in general a matrix $\mathbb{A}$ will not have a reciprocal which is analogous to the reciprocal of a nonzero number. There is, however, a special case in which the method described above can be applied to matrix equations.

Definition   An $n \times n$ matrix $\mathbb{A}$ is said to be *invertible* if there is an $n \times n$ matrix $\mathbb{B}$ such that $\mathbb{B}\mathbb{A} = \mathbb{A}\mathbb{B} = \mathbb{I}$. The matrix $\mathbb{B}$ with this property is said to be the *inverse* of $\mathbb{A}$ and is denoted $\mathbb{A}^{-1}$.

We do not discuss the concept of an inverse for a nonsquare matrix, i.e., for an $m \times n$ matrix with $m \neq n$. Not all square matrices are invertible; some are and some are not. If a matrix is invertible, there is only one matrix $\mathbb{B}$ such that $\mathbb{B}\mathbb{A} = \mathbb{I} = \mathbb{A}\mathbb{B}$; that is, the inverse of a square matrix is unique.

If $\mathbb{A}$ is invertible, the solution of the system of equations $\mathbb{A}\mathbf{x} = \mathbf{b}$ is straightforward. Indeed, if $\mathbb{A}^{-1}$ is the inverse of $\mathbb{A}$, then

$$\mathbb{A}^{-1}\mathbb{A}\mathbf{x} = \mathbb{I}\mathbf{x} = \mathbf{x}$$

Therefore, if we multiply the equation $\mathbb{A}\mathbf{x} = \mathbf{b}$ on the left by $\mathbb{A}^{-1}$, we obtain

$$\mathbb{A}^{-1}\mathbb{A}\mathbf{x} = \mathbb{A}^{-1}\mathbf{b} \qquad \text{or} \qquad \mathbf{x} = \mathbb{A}^{-1}\mathbf{b}$$

Since $\mathbb{A}^{-1}$ and $\mathbf{b}$ are known matrices, this provides a vector $\mathbf{x}$ which satisfies the equation $\mathbb{A}\mathbf{x} = \mathbf{b}$.

In general one would not solve a single system $\mathbb{A}\mathbf{x} = \mathbf{b}$ by computing the inverse of $\mathbb{A}$ since this computation frequently involves more labor than the direct solution of $\mathbb{A}\mathbf{x} = \mathbf{b}$ by the methods described in Sec. 6-3. However, if one is asked to solve several systems with the same coefficient matrix, for example,

$$\mathbb{A}\mathbf{x} = \mathbf{b} \qquad \mathbb{A}\mathbf{x} = \mathbf{b}' \qquad \mathbb{A}\mathbf{x} = \mathbf{b}''$$

it may be desirable to compute $\mathbb{A}^{-1}$ since one can then obtain the solutions of these equations immediately as $\mathbb{A}^{-1}\mathbf{b}$, $\mathbb{A}^{-1}\mathbf{b}'$, and $\mathbb{A}^{-1}\mathbf{b}''$, respectively. Also, there are important uses of the inverse matrix other than its use in solving systems of equations. One such use will be discussed in our study of Markov chains (Chap. 8).

The method of computing the inverse matrix presented here is a direct extension of the method introduced in Sec. 6-3. In reducing a matrix to the echelon form described by conditions A and B of Sec. 6-3, we used the first nonzero entry in each row to eliminate all nonzero entries with the same column and higher row index. For example, if the first nonzero entry in the ith row is the (i,j) entry, we use this entry to transform the coefficient matrix into one in which the (k,j) entries for $k > i$ are all zero. Here we continue the same idea and transform the matrix into one in which there is exactly one nonzero entry in each row and each column. This version of the reduction process is known as *gaussian elimination*. We use gaussian elimination only for the computation of inverse matrices.

Given a matrix $\mathbb{A}$, our problem is to find a matrix $\mathbb{B}$ such that $\mathbb{A}\mathbb{B} = \mathbb{I}$. If we write $\mathbf{b}^1$, $\mathbf{b}^2$, ..., $\mathbf{b}^n$ for the columns of $\mathbb{B}$ in order, the equation $\mathbb{A}\mathbb{B} = \mathbb{I}$ is equivalent to n vector equations

$$\mathbb{A}\mathbf{b}^1 = \begin{bmatrix} 1 \\ 0 \\ \vdots \\ 0 \end{bmatrix}, \; \mathbb{A}\mathbf{b}^2 = \begin{bmatrix} 0 \\ 1 \\ \vdots \\ 0 \end{bmatrix}, \; \ldots, \; \mathbb{A}\mathbf{b}^n = \begin{bmatrix} 0 \\ 0 \\ \vdots \\ 1 \end{bmatrix} \tag{6-12}$$

For example, given the matrix

$$\mathbb{A} = \begin{bmatrix} 2 & 1 \\ 5 & 3 \end{bmatrix}$$

the problem of finding the inverse $\mathbb{B} = [\mathbf{b}^1 \quad \mathbf{b}^2]$ of $\mathbb{A}$ is equivalent to solving

$$\mathbb{A}\mathbb{B} = \mathbb{I} = \begin{bmatrix} 1 & 0 \\ 0 & 1 \end{bmatrix}$$

or

$$\mathbb{A}\mathbf{b}^1 = \begin{bmatrix} 1 \\ 0 \end{bmatrix} \qquad \mathbb{A}\mathbf{b}^2 = \begin{bmatrix} 0 \\ 1 \end{bmatrix}$$

In the general case if we solve the n systems (6-12) for the vectors $\mathbf{b}^1, \mathbf{b}^2, \dots, \mathbf{b}^n$, we have the matrix $\mathbb{B}$, that is, the inverse of $\mathbb{A}$. If $\mathbb{A}$ is invertible, then each of the systems (6-12) can be solved by the method of gaussian elimination. In particular, we can use the operations of types 1 to 3 introduced in Sec. 6-3 on the augmented matrices:

$$\left[\mathbb{A} \ \middle| \ \begin{matrix} 1 \\ 0 \\ \vdots \\ 0 \end{matrix}\right], \left[\mathbb{A} \ \middle| \ \begin{matrix} 0 \\ 1 \\ \vdots \\ 0 \end{matrix}\right], \dots, \left[\mathbb{A} \ \middle| \ \begin{matrix} 0 \\ 0 \\ \vdots \\ 1 \end{matrix}\right]$$

However, since the coefficient matrices are the same in each case, we can solve all n systems simultaneously! We simply set up the augmented matrix

$$\left[\mathbb{A} \ \middle| \ \begin{matrix} 1 & 0 & \cdots & 0 \\ 0 & 1 & \cdots & 0 \\ \hdotsfor{4} \\ 0 & 0 & \cdots & 1 \end{matrix}\right] \quad \text{or} \quad [\mathbb{A} \mid \mathbb{I}] \tag{6-13}$$

and then use the operations of types 1 to 3 to successively replace this by simpler systems until the $n \times n$ matrix on the left of (6-13) is the identity matrix. Then the matrix on the right of the new augmented matrix is $\mathbb{A}^{-1}$.

Method of Determining Invertibility and Computing the Inverse of $\mathbb{A}$

1. Form the augmented matrix $[\mathbb{A} \mid \mathbb{I}]$.
2. Use operations of types 1 to 3 to transform this augmented matrix into the augmented matrix of a system with the same solutions.
   a. If $[\mathbb{A} \mid \mathbb{I}]$ can be so transformed into $[\mathbb{I} \mid \mathbb{B}]$, then $\mathbb{A}$ is invertible and $\mathbb{A}^{-1} = \mathbb{B}$.
   b. If $[\mathbb{A} \mid \mathbb{I}]$ cannot be transformed into the form $[\mathbb{I} \mid \mathbb{B}]$, then $\mathbb{A}$ is not invertible. $\mathbb{A}$ is not invertible if and only if a row consisting entirely of zeros is obtained on the left-hand side of one of the augmented matrices obtained from $[\mathbb{A} \mid \mathbb{I}]$ by row-reduction operations.

EXAMPLE 6-13

Find the inverse (if it exists) of the matrix

$$\mathbb{A} = \begin{bmatrix} 2 & 0 & 1 \\ 2 & 1 & -1 \\ 3 & 1 & -1 \end{bmatrix}$$

Solution: We form the augmented matrix

$$\left[\begin{array}{rrr|rrr} 2 & 0 & 1 & 1 & 0 & 0 \\ 2 & 1 & -1 & 0 & 1 & 0 \\ 3 & 1 & -1 & 0 & 0 & 1 \end{array}\right]$$

First, we multiply the first row by $\frac{1}{2}$ to obtain

$$\left[\begin{array}{rrr|rrr} 1 & 0 & \frac{1}{2} & \frac{1}{2} & 0 & 0 \\ 2 & 1 & -1 & 0 & 1 & 0 \\ 3 & 1 & -1 & 0 & 0 & 1 \end{array}\right]$$

Next we multiply the first row by -2 and add it to the second row and by -3 and add it to the third row. The new matrix is

$$\left[\begin{array}{rrr|rrr} 1 & 0 & \frac{1}{2} & \frac{1}{2} & 0 & 0 \\ 0 & 1 & -2 & -1 & 1 & 0 \\ 0 & 1 & -\frac{5}{2} & -\frac{3}{2} & 0 & 1 \end{array}\right]$$

The next step is to multiply the second row by -1 and add it to the third row. The resulting matrix is

$$\left[\begin{array}{rrr|rrr} 1 & 0 & \frac{1}{2} & \frac{1}{2} & 0 & 0 \\ 0 & 1 & -2 & -1 & 1 & 0 \\ 0 & 0 & -\frac{1}{2} & -\frac{1}{2} & -1 & 1 \end{array}\right]$$

Multiplying the third row by -2, we obtain

$$\left[\begin{array}{rrr|rrr} 1 & 0 & \frac{1}{2} & \frac{1}{2} & 0 & 0 \\ 0 & 1 & -2 & -1 & 1 & 0 \\ 0 & 0 & 1 & 1 & 2 & -2 \end{array}\right]$$

Finally, multiplying the third row by 2 and adding it to the second row and multiplying the third row by $-\frac{1}{2}$ and adding it to the first row, we obtain

$$\left[\begin{array}{rrr|rrr} 1 & 0 & 0 & 0 & -1 & 1 \\ 0 & 1 & 0 & 1 & 5 & -4 \\ 0 & 0 & 1 & 1 & 2 & -2 \end{array}\right]$$

We conclude that the matrix $\mathbb{A}$ is invertible and that

$$\mathbb{A}^{-1} = \left[\begin{array}{rrr} 0 & -1 & 1 \\ 1 & 5 & -4 \\ 1 & 2 & -2 \end{array}\right]$$

Although it is not a necessary part of the method, we verify our work in this example by showing that $A^{-1}A = I$. We have

$$\begin{bmatrix} 0 & -1 & 1 \\ 1 & 5 & -4 \\ 1 & 2 & -2 \end{bmatrix}\begin{bmatrix} 2 & 0 & 1 \\ 2 & 1 & -1 \\ 3 & 1 & -1 \end{bmatrix}$$

$$= \begin{bmatrix} 0(2) + (-1)(2) + 1(3) & 0(0) + (-1)(1) + 1(1) & 0(1) + (-1)(-1) + 1(-1) \\ 1(2) + 5(2) + (-4)(3) & 1(0) + 5(1) + (-4)(1) & 1(1) + 5(-1) + (-4)(-1) \\ 1(2) + 2(2) + (-2)(3) & 1(0) + 2(1) + (-2)(1) & 1(1) + 2(-1) + (-2)(-1) \end{bmatrix}$$

$$= \begin{bmatrix} 1 & 0 & 0 \\ 0 & 1 & 0 \\ 0 & 0 & 1 \end{bmatrix}$$

We leave it as an exercise for the reader (Exercise 1) to verify that $AA^{-1} = I$.

$\square$

It is worthwhile to consider an example which illustrates how one can infer from the failure of the method proposed here that a matrix is not invertible.

EXAMPLE 6-14

Find the inverse (if it exists) of the matrix

$$A = \begin{bmatrix} 2 & 0 & 1 \\ 3 & 1 & -1 \\ -2 & -2 & 4 \end{bmatrix}$$

Solution: We form the augmented matrix

$$\begin{bmatrix} 2 & 0 & 1 & 1 & 0 & 0 \\ 3 & 1 & -1 & 0 & 1 & 0 \\ -2 & -2 & 4 & 0 & 0 & 1 \end{bmatrix}$$

First, we multiply the first row by $\frac{1}{2}$ to obtain

$$\begin{bmatrix} 1 & 0 & \frac{1}{2} & \frac{1}{2} & 0 & 0 \\ 3 & 1 & -1 & 0 & 1 & 0 \\ -2 & -2 & 4 & 0 & 0 & 1 \end{bmatrix}$$

Next we multiply the first row by -3 and add it to the second row, and we multiply the first row by 2 and add it to the third row. The new matrix is

$$\left[\begin{array}{ccc|ccc} 1 & 0 & \frac{1}{2} & \frac{1}{2} & 0 & 0 \\ 0 & 1 & -\frac{5}{2} & -\frac{3}{2} & 1 & 0 \\ 0 & -2 & 5 & 1 & 0 & 1 \end{array}\right]$$

We continue by multiplying the second row by 2 and adding it to the third row. We have

$$\left[\begin{array}{ccc|ccc} 1 & 0 & \frac{1}{2} & \frac{1}{2} & 0 & 0 \\ 0 & 1 & -\frac{5}{2} & -\frac{3}{2} & 1 & 0 \\ 0 & 0 & 0 & -2 & 2 & 1 \end{array}\right]$$

Notice the third row of this matrix. It has only zeros to the left of the vertical line. Therefore $\mathbb{A}$ is not invertible. In particular, since there is a nonzero entry in the (3,4) spot of the augmented matrix, we conclude that the system of equations

$$\mathbb{A}\mathbf{x} = \begin{bmatrix} 1 \\ 0 \\ 0 \end{bmatrix} \tag{6-14}$$

has no solutions. In fact, in this case neither do the systems

$$\mathbb{A}\mathbf{x} = \begin{bmatrix} 0 \\ 1 \\ 0 \end{bmatrix} \quad \text{and} \quad \mathbb{A}\mathbf{x} = \begin{bmatrix} 0 \\ 0 \\ 1 \end{bmatrix} \qquad \square$$

We remind the reader that the noninvertibility of $\mathbb{A}$ does not mean that every equation of the form $\mathbb{A}\mathbf{x} = \mathbf{b}$ is unsolvable. For example, with the matrix $\mathbb{A}$ of Example 6-14, which we have just shown to be noninvertible, the equation

$$\mathbb{A}\mathbf{x} = \begin{bmatrix} 1 \\ 2 \\ -2 \end{bmatrix}$$

is solvable, and the vector

$$\mathbf{x} = \begin{bmatrix} 1 \\ -2 \\ -1 \end{bmatrix}$$

is a solution.

EXAMPLE 6-15

Let $A = \begin{bmatrix} 2 & 1 \\ 3 & -1 \end{bmatrix}$, and use the inverse of A to solve the equations

$$A\mathbf{x} = \begin{bmatrix} 1 \\ -1 \end{bmatrix} \qquad A\mathbf{x} = \begin{bmatrix} 2 \\ 1 \end{bmatrix} \qquad \text{and} \qquad A\mathbf{x} = \begin{bmatrix} 1 \\ 3 \end{bmatrix}$$

Solution: We begin by computing A^{-1}. First we form the augmented matrix

$$\left[\begin{array}{cc|cc} 2 & 1 & 1 & 0 \\ 3 & -1 & 0 & 1 \end{array}\right]$$

and we proceed to use row reduction operations to determine A^{-1}:

$$\left[\begin{array}{cc|cc} 1 & \frac{1}{2} & \frac{1}{2} & 0 \\ 3 & -1 & 0 & 1 \end{array}\right]$$

$$\left[\begin{array}{cc|cc} 1 & \frac{1}{2} & \frac{1}{2} & 0 \\ 0 & -\frac{5}{2} & -\frac{3}{2} & 1 \end{array}\right]$$

$$\left[\begin{array}{cc|cc} 1 & \frac{1}{2} & \frac{1}{2} & 0 \\ 0 & 1 & \frac{3}{5} & -\frac{2}{5} \end{array}\right]$$

$$\left[\begin{array}{cc|cc} 1 & 0 & \frac{1}{5} & \frac{1}{5} \\ 0 & 1 & \frac{3}{5} & -\frac{2}{5} \end{array}\right]$$

Therefore

$$A^{-1} = \begin{bmatrix} \frac{1}{5} & \frac{1}{5} \\ \frac{3}{5} & -\frac{2}{5} \end{bmatrix}$$

The solution of $A\mathbf{x} = \begin{bmatrix} 1 \\ -1 \end{bmatrix}$ is

$$\mathbf{x} = A^{-1} \begin{bmatrix} 1 \\ -1 \end{bmatrix} = \begin{bmatrix} \frac{1}{5} & \frac{1}{5} \\ \frac{3}{5} & -\frac{2}{5} \end{bmatrix} \begin{bmatrix} 1 \\ -1 \end{bmatrix} = \begin{bmatrix} 0 \\ 1 \end{bmatrix}$$

The solution of $A\mathbf{x} = \begin{bmatrix} 2 \\ 1 \end{bmatrix}$ is

$$\mathbf{x} = A^{-1} \begin{bmatrix} 2 \\ 1 \end{bmatrix} = \begin{bmatrix} \frac{3}{5} \\ \frac{4}{5} \end{bmatrix}$$

and the solution of $A\mathbf{x} = \begin{bmatrix} 1 \\ 3 \end{bmatrix}$ is

$$\mathbf{x} = A^{-1}\begin{bmatrix} 1 \\ 3 \end{bmatrix} = \begin{bmatrix} \frac{4}{5} \\ -\frac{3}{5} \end{bmatrix}$$

□

Exercises for Sec. 6-5

1. In Example 6-13 show that $AA^{-1} = I$.

2. Find the inverses of each of the following matrices:
   a. $\begin{bmatrix} 3 & 2 \\ 2 & 1 \end{bmatrix}$ b. $\begin{bmatrix} .4 & .6 \\ .1 & .9 \end{bmatrix}$

3. Find the inverses (if they exist) of the following matrices:
   a. $\begin{bmatrix} 0 & 1 & 2 & 1 \\ 1 & 0 & -1 & -1 \\ 1 & 2 & 0 & 0 \\ 1 & 0 & -2 & 0 \end{bmatrix}$ b. $\begin{bmatrix} 1 & -\frac{1}{4} & 0 \\ -\frac{1}{4} & 1 & -\frac{1}{4} \\ 0 & -\frac{1}{4} & 1 \end{bmatrix}$

4. Find the inverse of the matrix A and use it to solve the following equations:
   $$A\mathbf{x} = \begin{bmatrix} 1 \\ 3 \\ 1 \end{bmatrix} \quad \text{and} \quad A\mathbf{x} = \begin{bmatrix} 2 \\ 1 \\ 0 \end{bmatrix} \quad \text{where } A = \begin{bmatrix} 2 & 1 & 0 \\ -1 & 1 & 1 \\ 0 & -1 & 1 \end{bmatrix}$$

5. Let $A = \begin{bmatrix} 5 & 2 \\ 2 & 1 \end{bmatrix}$. Find A^{-1} and $(A^{-1})^{-1}$ by the method of this section.

6. Let $A = \begin{bmatrix} 5 & 2 \\ 2 & 1 \end{bmatrix}$ and $B = \begin{bmatrix} 1 & 3 \\ 3 & 5 \end{bmatrix}$. Find A^{-1}, B^{-1}, $(AB)^{-1}$, and $B^{-1}A^{-1}$.

7. Let A, B, and C be 2×2 matrices and suppose
   $$A = \begin{bmatrix} 2 & 1 \\ -3 & 1 \end{bmatrix} \quad \text{and} \quad C = \begin{bmatrix} 2 & 6 \\ 1 & -3 \end{bmatrix}.$$
   Find B such that
   a. $AB = C$ b. $BA = C$
   (*Hint:* Use A^{-1}.)

8. Use the concept of a matrix inverse to solve the following systems of equations:
   a. $x + y + z = 15$ b. $x + y + z = 9$ c. $x + y + z = 0$
      $y - z = 1$ $y - z = 0$ $y - z = 1$
      $x \quad + z = 10$ $x \quad + z = 6$ $x \quad + z = 0$

9. Let $A = \begin{bmatrix} 1 & 3 \\ 1 & 5 \end{bmatrix}$. Compute $AA = A^2$, A^{-1}, $A^{-1}A^{-1}$, and $(AA)^{-1}$. Is $(A^2)^{-1}$ the same as $(A^{-1})^2$?

10. (Continuation of Exercise 9) Use the definition of a matrix inverse to show that for any square matrix A, if A has an inverse, then
    $$(A^n)^{-1} = (A^{-1})^n \quad \text{for } n = 1, 2 \ldots$$

6-6   MATRICES AND THE CHAIN OF COMMAND

In the earlier sections of this chapter we commented on the usefulness of matrices for notational, computational, and theoretical purposes. This section, which is in essence a single example, illustrates the use of matrices as both a notational and computational tool.

In most organizations there is a chain of command which dictates the flow of directives between members of the organization. In military units this flow follows the basic command structure given by the various ranks of officers and the positions they hold. In business a command structure is often given in terms of an organization chart. In such organizations it often happens that between some individuals directives flow in both directions while between other individuals directives always flow in only one direction. An example of the former is provided by two corporate officers with the same rank but different areas of responsibility, and an example of the latter is provided by a general and a colonel. In any specific setting the command structure used may be a simple one or a complicated one. When the structure is large, and perhaps complicated as well, matrices can often be used to help study the structure. We illustrate this use of matrices by the following example.

Suppose a business has six officers, a president P, a vice president for finance VPF, a vice president for production VPP, a treasurer T, a director of labor relations DLR, and a director of research DR. Also suppose that the organization chart of the company is as shown in Fig. 6-3.

The arrows which connect the various units of the chart in Fig. 6-3 indicate the flow of directives between those units. For example, the president gives direct orders to both vice presidents and to the treasurer. No one gives direct orders to the president, and accordingly no arrows are directed toward the unit labeled P.

The organization chart of a large company may contain many more units than this example, and it may well be difficult to use a chart of the type shown

FIGURE 6-3

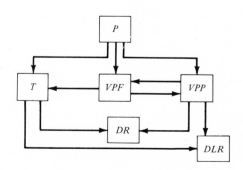

in Fig. 6-3 to follow the flow of directives through the organization. An alternative means of presenting the information contained in Fig. 6-3 is to use a communication matrix. Instead of arrows on a figure, a communication matrix uses matrix entries to indicate the flow of directives. In particular, the rows and columns of a matrix are identified with the units in an organization, and the entries are determined by the following rules:

If u issues directives to v, then the (u, v) entry in the matrix is 1. If u does not issue directives to v, then the (u, v) entry in the matrix is 0.

The resulting matrix is known as a communication matrix.

Definition   The $n \times n$ matrix $\mathbb{C} = (c_{ij})$ is called a *communication matrix* if each entry c_{ij} is either 1 or 0 and $c_{ii} = 0$ for $i = 1, 2, \ldots, n$.

A communication matrix which contains the same information as Fig. 6-3 is a 6×6 matrix obtained by associating the rows and columns of the matrix with the units of the chart in such a way that if a unit is associated with the ith row, the unit is also associated with the ith column. The (i,j) entry of the matrix is set equal to 1 if the unit associated with row i gives directives to the unit associated with column j. Otherwise the (i,j) entry is set equal to 0. One communication matrix for Fig. 6-3 is shown below, where the units which are to be identified with the rows and columns are indicated above and to the left of the matrix.

$$
\mathbb{C} =
\begin{array}{c}
 \\
P \\
VPF \\
VPP \\
T \\
DLR \\
DR
\end{array}
\begin{array}{c}
\begin{array}{cccccc}
P & VPF & VPP & T & DLR & DR
\end{array} \\
\left[
\begin{array}{cccccc}
0 & 1 & 1 & 1 & 0 & 0 \\
0 & 0 & 1 & 1 & 0 & 0 \\
0 & 1 & 0 & 0 & 1 & 1 \\
0 & 0 & 0 & 0 & 1 & 1 \\
0 & 0 & 0 & 0 & 0 & 0 \\
0 & 0 & 0 & 0 & 0 & 0
\end{array}
\right]
\end{array}
$$

The matrix $\mathbb{C}$ is just one possible communication matrix for Fig. 6-3 because the association of the units with the rows and columns is not unique. A different association would result in a different matrix; however, the information contained in the matrix would be the same.

At first glance it might seem that we have gained very little by converting Fig. 6-3 into the matrix $\mathbb{C}$. In fact an argument could be made that we have lost ground since it seems to be easier to evaluate the command structure from Fig. 6-3 than it is from the associated communication matrix $\mathbb{C}$. However, as we noted earlier, this is an extremely simple organization chart. The chart for a large company may have well over 100 entries, and it may be difficult to obtain useful information about the command structure by simply looking at the chart. On the other hand, such information can be easily obtained from the related communication matrix and the repeated products of this matrix with itself.

Definition

If A is an $n \times n$ matrix, then the kth *power* of A is the $n \times n$ matrix obtained by multiplying A by itself k times:

$$A^k = \underbrace{AA \ldots A}_{k \text{ Factors}}$$

The powers of a matrix can be evaluated by a computer, and the necessity of having to work with large matrices (100×100 or larger) is not a serious obstacle. Thus, the use of a communication matrix and its powers may be the only feasible way of analyzing the structure of an organization. The powers of a communication matrix C have the following interpretation. Consider $C^2 = CC$. The entry in the (i,j) position of C^2 is obtained by multiplying the ith row of C by the jth column of C. If $C = (c_{ij})$ and $CC = C^2 = (c_{ij}(2))$, then

$$c_{ij}(2) = c_{i1}c_{1j} + c_{i2}c_{2j} + \cdots + c_{in}c_{nj}$$

Since each entry c_{ij} is either 1 or 0, we see that each of the products in the sum defining $c_{ij}(2)$ is either 1 or 0. Also, a typical product in the sum, say $c_{ik}c_{kj}$, will be 1 only if c_{ik} and c_{kj} are both 1. This will occur only if unit i gives directives to unit k and unit k gives directives to unit j. Thus, the term $c_{ij}(2)$ is nonzero only if directives can flow from unit i to unit j in two steps (first to some unit k and then from k to j). In fact, since each of the terms $c_{ik}c_{kj}$ in the sum can be analyzed in this way, we see that each time we have a unit k such that $c_{jk} = 1$ and $c_{kj} = 1$ we have a command chain of length 2 from unit i to unit j. Therefore, the number of nonzero terms in the sum gives the number of such command chains. Moreover, since each nonzero term is in fact 1, the number of command chains is equal to the sum, or

$$c_{ij}(2) = \text{number of command chains of length 2 from unit } i \text{ to unit } j$$

This information provides us with an interpretation of the entries in C^2. If the entry in the (i,j) position, $c_{ij}(2)$, is a zero, then there is no two-step path for directives to flow from unit i to unit j. If $c_{ij}(2)$ is not zero, then there is a two-step path for directives to flow from i to j; and in fact, $c_{ij}(2)$ is the number of such paths. Thus the square of the matrix C, $C^2 = CC$, contains information on the paths of flow of directives which are two steps long. Using the special 6×6 matrix C given above, we have

$$C^2 = CC = \begin{bmatrix} 0 & 1 & 1 & 1 & 2 & 2 \\ 0 & 1 & 0 & 0 & 2 & 2 \\ 0 & 0 & 1 & 1 & 0 & 0 \\ 0 & 0 & 0 & 0 & 0 & 0 \\ 0 & 0 & 0 & 0 & 0 & 0 \\ 0 & 0 & 0 & 0 & 0 & 0 \end{bmatrix}$$

The entries in the matrix $\mathbb{C}^2$ indicate that the president (row 1) can send directives to everyone (other than himself) in just two steps. On the other hand, the treasurer, the director of labor relations, and the director of research cannot send directives to anyone in exactly two steps.

The reasoning used to interpret the entries in $\mathbb{C}^2$ can also be applied to $\mathbb{C}^3$ and higher powers of the matrix $\mathbb{C}$. If the (i,j) entry in $\mathbb{C}^3$ is nonzero, directives can flow in exactly *three* steps from the unit associated with row i to the unit associated with row j. We deduce this by examining the sum of products used to form the (i,j) entry in $\mathbb{C}^3$, which we label $c_{ij}(3)$, and by noting that this sum is nonzero only if one of the products is nonzero. Since $\mathbb{C}^3 = \mathbb{C}^2\mathbb{C}$, we see that

$$c_{ij}(3) = \sum_{k=1}^{n} c_{ik}(2)c_{kj}$$

and thus $c_{ij}(3) \neq 0$ means there is a k such that $c_{ik}(2) \neq 0$ and $c_{kj} \neq 0$. Equivalently, directives can flow from i to k in two steps, from k to j in one step, and hence from i to j in three steps.

In our example the matrix $\mathbb{C}^3$ is given by

$$\mathbb{C}^3 = \mathbb{C}^2\mathbb{C} = \begin{bmatrix} 0 & 1 & 1 & 1 & 2 & 2 \\ 0 & 0 & 1 & 1 & 0 & 0 \\ 0 & 1 & 0 & 0 & \boxed{2} & 2 \\ 0 & 0 & 0 & 0 & 0 & 0 \\ 0 & 0 & 0 & 0 & 0 & 0 \\ 0 & 0 & 0 & 0 & 0 & 0 \end{bmatrix}$$

As an example of the interpretation which can be given to the entries of $\mathbb{C}^3$ we note that the entry in position $(3,5)$ is 2. Since row 3 corresponds to the vice president for production and column 5 corresponds to the director for labor relations we see that there are two paths of length 3 from VPP to DLR. In fact these paths are $VPP \rightarrow VPF \rightarrow VPP \rightarrow DLR$ and $VPP \rightarrow VPF \rightarrow T \rightarrow DLR$. In this simple setting there is a path of length 1, that is, direct communication, from VPP to DLR, and hence the knowledge that there are two paths of length 3 is not especially significant. However, in a large setting it may be difficult to determine all paths of directives of length 3 and to determine who can give orders to whom in three or fewer steps. The powers of the communication matrix provide an automatic method of determining the relationship between the units of the command structure, especially the role each plays in initiating directives. In fact, a measure of the power or influence of each unit can be given in terms of the number of other units which receive directives from the first unit in a specific number of steps. The more units which receive directives, the more the power or influence of the originating unit. To determine the number of units which receive directives in k steps from unit i (row i of the communication matrix) one simply counts the number of nonzero entries in the ith row of the *k*th power of the matrix. Each nonzero entry in this row corresponds to a unit which receives directives from unit *i* in *k* steps. Naturally, some of these may also receive directives from unit *i* in fewer than *k* steps.

We note that in our specific example the president (row 1) may send directives to all other units of the structure in two or fewer steps. No other unit has this property.

Exercises for Sec. 6-6

1. Find a communication matrix for the organization chart in Fig. 6-4. Be sure to identify the rows and columns of your matrix with units of the organization.

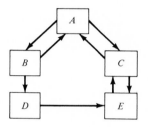

FIGURE 6-4

2. Repeat Exercise 1 for the organization chart in Fig. 6-5.

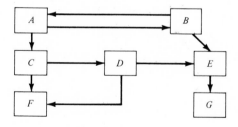

FIGURE 6-5

3. Repeat Exercise 1 for the organization chart in Fig. 6-6.

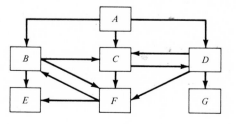

FIGURE 6-6

Find an organization chart for the communication matrix:

4. $\begin{bmatrix} 0 & 1 & 0 & 0 \\ 0 & 0 & 1 & 0 \\ 0 & 0 & 0 & 1 \\ 1 & 0 & 0 & 0 \end{bmatrix}$
5. $\begin{bmatrix} 0 & 1 & 0 & 0 & 0 \\ 0 & 1 & 0 & 0 & 0 \\ 0 & 0 & 0 & 0 & 1 \\ 0 & 0 & 0 & 0 & 1 \\ 0 & 0 & 0 & 0 & 0 \end{bmatrix}$
6. $\begin{bmatrix} 0 & 0 & 0 & 0 & 0 & 0 \\ 1 & 0 & 0 & 0 & 0 & 0 \\ 1 & 1 & 0 & 0 & 0 & 0 \\ 1 & 1 & 1 & 0 & 0 & 0 \\ 1 & 1 & 1 & 1 & 0 & 0 \\ 1 & 1 & 1 & 1 & 1 & 0 \end{bmatrix}$

7. In Exercise 2 find the unit (or units) which gives directives to the largest number of other units in two or fewer steps.

8. In Exercise 5 find the unit (or units) which gives directives to the largest number of other units in three or fewer steps.

9. In Exercise 6 find the unit (or units) which gives directives to the largest number of other units in two or fewer steps.

10. Six teenagers (who know each other) are asked: Who among the other five do you trust the most? The teenagers' names are Abel, Bob, Cathy, Diane, Edith, and Frank (A, B, C, D, E, F for short), and they respond

> A trusts F the most
> B trusts F the most
> C trusts D the most
> D trusts E the most
> E trusts C the most
> F trusts D the most

Represent the students' responses in a matrix similar to a communication matrix, and compute the square of this matrix.

11. (Exercise 10 continued.) Give an interpretation to the powers of the matrix used in Exercise 10. Develop a method for determining the most trustworthy members of this group in the opinion of the others in the group.

6-7   A LINEAR ECONOMIC MODEL

In 1973 Professor Wassily Leontief was awarded the Nobel prize for economics in recognition of his development of mathematical methods to study interdependencies in an economy. In their simplest form Leontief input-output models are very straightforward. In practice these models are quite complicated because of the very large number of variables involved and because of modifications that have been made to increase the accuracy of the predictions. Here we shall consider only one of the simplest forms of the model.

Suppose that we have an economy in which there are n industries, each of which produces a single good. We assume that each industry utilizes the goods of some or all of the industries in the production process. Thus, some of the output of an industry is input for other industries, and some is required in its own productive processes. For example, some lumber is utilized in the production of lumber, and steel is used to produce more steel. Let x_i denote the amount of the ith good (measured in appropriate units) produced in unit time. We suppose that this output is used in the production of the ith good, in the production of other goods, and in the satisfaction of the demand external to the economic system being considered. Thus

$$x_i = \text{amount of } i\text{th good used to produce goods}$$
$$+ \text{ amount used to satisfy external demand} \quad (6\text{-}15)$$

The amount of the ith good used in the production processes can be further subdivided into the amounts used to produce goods 1, 2, 3, ..., n, respectively. We repeat for emphasis that the situation in which good i is used to produce good i is included in the model.

We set a_{ij} equal to the amount of good i used to produce 1 unit of good j, and we assume that the amount of good i needed to produce x_j units of good j is $a_{ij}x_j$. We also assume that the various production processes are unrelated in the sense that the amount of good i used to produce x_1 units of good 1, and x_2 units of good 2, ... and x_n units of good n is the sum of the amounts of good i needed to produce those quantities of the goods individually:

$$a_{i1}x_1 + a_{i2}x_2 + \cdots + a_{in}x_n$$

It is these assumptions which make the model linear. Finally, we let d_i denote the external demand for good i. Therefore, the representation of x_i given in Eq. (6-15) has the form

$$x_i = a_{i1}x_1 + a_{i2}x_2 + \cdots + a_{in}x_n + d_i$$

We assume that all the goods produced in the economy can be described by similar equations. It follows that we have a system of linear equations in the variables $x_1, x_2, \ldots, x_n$:

$$x_1 = a_{11}x_1 + a_{12}x_2 + \cdots + a_{1n}x_n + d_1$$
$$x_2 = a_{21}x_1 + a_{22}x_2 + \cdots + a_{2n}x_n + d_2$$
$$\vdots$$
$$x_n = a_{n1}x_1 + a_{n2}x_2 + \cdots + a_{nn}x_n + d_n$$

or in matrix vector notation

$$\mathbf{x} = \mathbb{A}\mathbf{x} + \mathbf{d}$$

where $\mathbb{A}$ is the $n \times n$ matrix whose (i,j) entry is a_{ij} and $\mathbf{x}$ and $\mathbf{d}$ are n-vectors whose coordinates are $x_1, \ldots, x_n$ and $d_1, \ldots, d_n$, respectively.

Definition   In a Leontief input-output model represented by the equation $\mathbf{x} = \mathbb{A}\mathbf{x} + \mathbf{d}$, the matrix $\mathbb{A}$ is the *technology* or *input-output* matrix, the vector $\mathbf{d}$ is the *external demand* vector, and the vector $\mathbf{x}$ is a *production schedule*.

Since the coordinates of a production schedule are the amounts of various goods produced, each of these coordinates must be nonnegative. Also, since most production processes utilize labor as one of their inputs, we shall assume that one of the industries, say the nth, is the "household" whose output is labor. Therefore, the nth column of the technology matrix $\mathbb{A}$ should be interpreted as a vector of behavior parameters rather than technology coefficients.

EXAMPLE 6-16

Consider a Leontief input-output model with two goods, lumber and labor. Suppose that it requires .1 unit of lumber and 2 units of labor to produce 1 unit of lumber, and .3 unit of lumber to produce 1 unit of labor. Find the technology matrix $\mathbb{A}$.

Solution: Let x_1 and x_2 denote the number of units of lumber and labor, respectively, produced in the economy. The amount of lumber necessary to produce 1 unit of lumber is .1 and the amount of lumber required to produce 1 unit of labor is .3. Therefore we have

$$x_1 = .1x_1 + .3x_2$$

Likewise $$x_2 = 2x_1$$

The technology matrix $\mathbb{A}$ is the matrix which arises when these equations are written in matrix form. We have

$$\begin{bmatrix} x_1 \\ x_2 \end{bmatrix} = \begin{bmatrix} .1 & .3 \\ 2 & 0 \end{bmatrix} \begin{bmatrix} x_1 \\ x_2 \end{bmatrix}$$

and consequently the technology matrix $\mathbb{A}$ is

$$\mathbb{A} = \begin{bmatrix} .1 & .3 \\ 2 & 0 \end{bmatrix} \qquad \square$$

Next, since we have assumed that each industry produces a single good, the jth industry produces the jth good, it follows that $a_{jk} \geq 0$ for $j \neq k$. Indeed, if there were industries j and k with $a_{jk} < 0$, the effect of producing x_k units of good k would be to reduce the amount of good j required by an amount $|a_{jk} x_k|$. Thus, if this were the case, the production of good k would be equivalent to the production of good j. We assume that this is not the case.

Finally, we assume that all the coordinates of the demand vector $\mathbf{d}$ are nonnegative. This assumption reflects the fact that we are assuming $\mathbf{d}$ to be a genuine demand vector and demands ought to increase the amount of a good required.

The problem we consider is that of finding a production schedule $\mathbf{x}$ which satisfies the demand $\mathbf{d}$ with a given technology $\mathbb{A}$. Thus we seek $\mathbf{x}$ with nonnegative coordinates such that

$$\mathbf{x} = \mathbb{A}\mathbf{x} + \mathbf{d} \qquad \text{or} \qquad (\mathbb{I} - \mathbb{A})\mathbf{x} = \mathbf{d}$$

We assume that $\mathbb{I} - \mathbb{A}$ is invertible. This assumption can be given economic justification. Indeed, the failure of $\mathbb{I} - \mathbb{A}$ to be invertible would imply a particular and exact relationship between the coefficients a_{ij} of $\mathbb{A}$. Since the coefficients of $\mathbb{A}$ are usually estimated and consequently not known precisely, it is not unduly restrictive to assume that $\mathbb{I} - \mathbb{A}$ is invertible.

For the remainder of this discussion we shall assume that $\mathbb{A}$ has nonnegative entries, that $\mathbb{I} - \mathbb{A}$ is invertible, and that $(\mathbb{I} - \mathbb{A})^{-1}$ has nonnegative entries. One can make additional assumptions about the nature of the economic system which will guarantee that $(\mathbb{I} - \mathbb{A})^{-1}$ has nonnegative entries, but we shall not discuss this assumption in economic terms. With these assumptions, the equation $(\mathbb{I} - \mathbb{A})\mathbf{x} = \mathbf{d}$ can be solved for the production schedule $\mathbf{x}$ by multiplying both sides by $(\mathbb{I} - \mathbb{A})^{-1}$. We conclude that

$$\mathbf{x} = (\mathbb{I} - \mathbb{A})^{-1}\mathbf{d} \qquad (6\text{-}16)$$

Since $\mathbf{d}$ and $(\mathbb{I} - \mathbb{A})^{-1}$ contain only nonnegative entries, it follows that $\mathbf{x}$ has only nonnegative entries, and consequently, $\mathbf{x}$ is a legitimate production schedule.

EXAMPLE 6-17

Consider a Leontief input-output model for an economy with two goods, lumber and labor, and the technology matrix

$$\mathbb{A} = \begin{bmatrix} .1 & .3 \\ 2 & 0 \end{bmatrix}$$

Here the first row (and column) is identified with lumber and the second row (and column) is identified with labor. Suppose that the external demand vector is $\begin{bmatrix} 12 \\ 3 \end{bmatrix}$. What is the required production schedule?

Solution: We propose to use Eq. (6-16), and consequently we need to determine $(\mathbb{I} - \mathbb{A})^{-1}$.

$$\mathbb{I} - \mathbb{A} = \begin{bmatrix} 1 & 0 \\ 0 & 1 \end{bmatrix} - \begin{bmatrix} .1 & .3 \\ 2 & 0 \end{bmatrix} = \begin{bmatrix} .9 & -.3 \\ -2 & 1 \end{bmatrix}$$

and if we use the methods of Sec. 6-5, we find

$$(\mathbb{I} - \mathbb{A})^{-1} = \begin{bmatrix} \frac{10}{3} & 1 \\ \frac{20}{3} & 3 \end{bmatrix}$$

Therefore, since the entries of $(\mathbb{I} - \mathbb{A})^{-1}$ are all nonnegative, the assumptions are all fulfilled and the desired production schedule is

$$\mathbf{x} = \begin{bmatrix} \frac{10}{3} & 1 \\ \frac{20}{3} & 3 \end{bmatrix} \begin{bmatrix} 12 \\ 3 \end{bmatrix} = \begin{bmatrix} 43 \\ 89 \end{bmatrix}$$

The economy requires 43 units of lumber and 89 units of labor to meet the external demand. Notice that $31 \ (= 43 - 12)$ units of lumber and $86 \ (= 89 - 3)$ units of labor are utilized within the economy.   □

EXAMPLE 6-18

Suppose that an economy has three goods, lumber, coal, and labor, and that the technology matrix is

$$\mathbb{A} = \begin{bmatrix} .5 & 0 & .2 \\ .2 & .8 & .12 \\ 1 & .4 & 0 \end{bmatrix}$$

where lumber, coal, and labor are identified with the first, second, and third rows (and columns), respectively. What is the production schedule which will satisfy the external demand given by the vector $\begin{bmatrix} 5 \\ 3 \\ 4 \end{bmatrix}$?

Solution: Again, we proceed by determining $(\mathbb{I} - \mathbb{A})^{-1}$ and using Eq. (6-16). In this case

$$\mathbb{I} - \mathbb{A} = \begin{bmatrix} .5 & 0 & -.2 \\ -.2 & .2 & -.12 \\ -1 & -.4 & 1 \end{bmatrix} \quad \text{and} \quad (\mathbb{I} - \mathbb{A})^{-1} = \begin{bmatrix} 7.6 & 4 & 2 \\ 16 & 15 & 5 \\ 14 & 10 & 5 \end{bmatrix}$$

The production schedule $\mathbf{x}$ can be obtained from Eq. (6-16):

$$\mathbf{x} = (\mathbb{I} - \mathbb{A})^{-1}\mathbf{d} = \begin{bmatrix} 7.6 & 4 & 2 \\ 16 & 15 & 5 \\ 14 & 10 & 5 \end{bmatrix}\begin{bmatrix} 5 \\ 3 \\ 4 \end{bmatrix} = \begin{bmatrix} 58 \\ 145 \\ 120 \end{bmatrix}$$

To satisfy the external demand, the economy needs 58 units of lumber, 145 units of coal, and 120 units of labor. ☐

Exercises for Sec. 6-7

1. In Example 6-17 compute the production schedule for the following external-demand vectors:

   a. $\mathbf{d} = \begin{bmatrix} 15 \\ 5 \end{bmatrix}$

   b. $\mathbf{d} = \begin{bmatrix} 1000 \\ 125 \end{bmatrix}$

2. In Example 6-18 compute the production schedule for the following external-demand vectors:

   a. $\mathbf{d} = \begin{bmatrix} 30 \\ 10 \\ 20 \end{bmatrix}$

   b. $\mathbf{d} = \begin{bmatrix} 100 \\ 200 \\ 300 \end{bmatrix}$

   Find the production schedule for the technology matrix and demand vector:

3. $\mathbb{A} = \begin{bmatrix} .5 & .4 \\ .25 & .2 \end{bmatrix}$ $\mathbf{d} = \begin{bmatrix} 2 \\ 1 \end{bmatrix}$

4. $\mathbb{A} = \begin{bmatrix} .2 & .04 \\ .6 & .05 \end{bmatrix}$ $\mathbf{d} = \begin{bmatrix} 2 \\ 10 \end{bmatrix}$

5. $\mathbb{A} = \begin{bmatrix} .5 & 0 & .3 \\ 0 & .8 & .4 \\ 0 & .2 & .4 \end{bmatrix}$ $\mathbf{d} = \begin{bmatrix} 3 \\ 1 \\ 2 \end{bmatrix}$

6. $\mathbb{A} = \begin{bmatrix} .2 & .4 & 0 \\ 0 & .4 & .5 \\ .8 & .1 & .5 \end{bmatrix}$ $\mathbf{d} = \begin{bmatrix} 8 \\ 2 \\ 3 \end{bmatrix}$

7. $\mathbb{A} = \begin{bmatrix} .10 & .01 & 1.5 \\ .05 & .20 & 1.0 \\ .30 & .30 & 0 \end{bmatrix}$ $\mathbf{d} = \begin{bmatrix} 6 \\ 3 \\ 4 \end{bmatrix}$

8. Consider an economy in which the goods are called agriculture, manufacturing, and labor. Also suppose that the units of measurement are such that Table 6-4 describes the relationships between the units.

   TABLE 6-4

	Number of Units Needed to Produce 1 Unit of		
Good	**Agriculture**	**Manufacturing**	**Labor**
Agriculture	.2	.1	.2
Manufacturing	.5	.6	.1
Labor	.1	.2	.1

   Find the associated technology matrix and decide whether this matrix satisfies the assumptions of this section.

IMPORTANT TERMS

You should be able to describe, define, or give examples of each of the following;

Matrix

Column vector

Row vector

System of linear equations

Echelon form

Back substitution

Augmented matrix

Inverse of a matrix

Communication matrix

Leontief model

Technology matrix

External demand vector

Production schedule

REVIEW EXERCISES

1. Write the following systems of equations in matrix form:

   a. $2x_1 + 2x_2 = 4$
   $6x_1 - x_2 = 6$

   b. $2x_1 - 2x_2 + 2x_3 = 7$
   $4x_1 - 9x_2 - 7x_3 = 4$
   $-11x_1 + x_2 + x_3 = 15$

2. Write the following systems of equations in matrix form:

   a. $x_1 + x_2 + 3x_4 = 7$
   $x_2 = 10$
   $x_1 - x_3 = 20$

   b. $x_1 + x_2 = 5 - x_2$
   $3x_1 - 3x_2 = 2x_1$

3. Form the product Ax for the following pairs A and x:

   a.
   $$A = \begin{bmatrix} 1 & 2 \\ 5 & 4 \end{bmatrix} \qquad x = \begin{bmatrix} 3 \\ 6 \end{bmatrix}$$

   b.
   $$A = \begin{bmatrix} 1 & 0 & -1 \\ 0 & 2 & 3 \\ -1 & -2 & 4 \end{bmatrix} \qquad x = \begin{bmatrix} 3 \\ -1 \\ 2 \end{bmatrix}$$

4. Verify that the vector $x = \begin{bmatrix} 3 \\ 1 \end{bmatrix}$ is a solution of the system of equations in part **b** of Exercise 2.

5. Verify that the vector $x = \begin{bmatrix} 0 \\ 10 \\ -20 \\ -1 \end{bmatrix}$ is a solution of the system in part **a** of Exercise 2.

Find all solutions of the system of equations:

6. $2x_1 + 4x_2 = 8$
   $4x_1 - 2x_2 = 4$
   $6x_1 + 2x_2 = 12$

7. $2x_1 + 2x_2 - 4x_3 = 12$
   $x_1 + x_2 + x_3 = 6$
   $3x_1 + 3x_2 - 3x_3 = 3$

8. $x_1 - x_2 + x_3 - x_4 = 10$
   $2x_1 + x_2 - 2x_3 + 2x_4 = 20$
   $x_1 + 2x_2 - 3x_3 + 3x_4 = 10$
   $4x_1 - x_2 = 40$

9. Let $A = \begin{bmatrix} 1 & 2 \\ -2 & 1 \end{bmatrix}$ and $B = \begin{bmatrix} 1 & -1 & 2 \\ 2 & 0 & -1 \end{bmatrix}$; find:

   a. AB

   b. $A(AB)$

10. Let $A = \begin{bmatrix} 5 & 3 \\ 3 & 2 \end{bmatrix}$. Find A^{-1}.

11. Let $A = \begin{bmatrix} 1 & 1 & 1 \\ 2 & 3 & 2 \\ 3 & 3 & 4 \end{bmatrix}$. Find A^{-1}.

12. Decide whether the following technology matrices A and B satisfy the assumptions of the Leontief model as given in Sec. 6-7:

$$A = \begin{bmatrix} .4 & .6 \\ .6 & .4 \end{bmatrix} \qquad B = \begin{bmatrix} .8 & .4 \\ .2 & .8 \end{bmatrix}$$

13. Find the production schedule for the following technology matrix A and demand vector $\mathbf{d}$:

$$A = \begin{bmatrix} .5 & .1 \\ .2 & .6 \end{bmatrix} \qquad \mathbf{d} = \begin{bmatrix} 4 \\ 11 \end{bmatrix}$$

14. Solve the following systems of equations:

$$A\mathbf{x} = \begin{bmatrix} -1 \\ 2 \\ 3 \end{bmatrix} \qquad A\mathbf{x} = \begin{bmatrix} -1 \\ 0 \\ 1 \end{bmatrix} \qquad A\mathbf{x} = \begin{bmatrix} 1 \\ 2 \\ 3 \end{bmatrix} \qquad \text{where} \qquad A = \begin{bmatrix} 1 & 1 & 1 \\ 2 & 3 & 2 \\ 3 & 3 & 4 \end{bmatrix}$$

15. Represent the organization chart shown in Fig. 6-7 as a communication matrix and use the matrix to decide which unit (or units) gives directives to the largest number of other units in two or fewer steps.

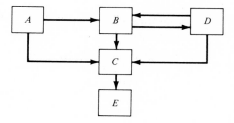

FIGURE 6-7

16. Draw an organization chart to represent the flow of directives determined by the communication matrix

$$\begin{bmatrix} 0 & 0 & 0 & 0 & 0 \\ 0 & 0 & 1 & 1 & 0 \\ 1 & 0 & 0 & 0 & 0 \\ 0 & 1 & 1 & 0 & 0 \\ 0 & 0 & 1 & 1 & 0 \end{bmatrix}$$

COMPUTATIONAL METHODS FOR LINEAR PROGRAMMING

7-1 INTRODUCTION

Linear programming problems commonly involve more than two variables. For such problems it is generally impossible to use the geometric methods of Chap. 5 (which depend on graphical representations in the plane) or any reasonable extension of these methods. In this chapter we describe a method based only on algebra which can be used both for problems with two variables and for those with more than two variables. This method, called the *simplex method*, is related to the geometric technique of Chap. 5 and the gaussian elimination process of Chap. 6.

The simplex method is an algorithm for computing solutions to linear programming problems. This algorithm is important because it applies to a very general class of linear programming problems and because it works well in practice. Linear programming models can be formulated in many different situations, and much of the usefulness of these models rests on the fact that there are effective means of solving the mathematical problems. In practice one usually needs to obtain numbers, and the simplex method provides these numbers. Other algorithms which apply to special types of programming problems, as well as a more general version of the simplex method, are part of the subject of operations research. In this chapter it is our intention to illustrate these important algorithms by using the simplex method to solve linear programming problems of a special form.

7-2 FORMULATION OF PROBLEMS

We concentrate on linear programming problems which have a rather special character. Such concentration can be justified since the problems selected form an important class of all linear programming problems, and moreover many other problems can be converted to this form. We begin with an example and then describe the class of problems to be considered in the remainder of the chapter.

EXAMPLE 7-1

The Scarumsilly Toy Company makes three monster dolls, Scary Harry, Horrible Harriet, and The Glob. The manufacturing of these dolls is a three-step process: (1) the body is molded from plastic; (2) clothes are put on; and (3) special monster features are added. The amounts of time and material for each step vary from doll to doll, and consequently each doll has its own production cost and associated profit. Data of the manufacturing process are shown in Table 7-1.

Problem: Determine a production schedule which meets the restrictions on the amounts of material and time and which provides the maximum profit; i.e., determine how many dolls of each type should be manufactured each hour.

Mathematical Formulation: The variables in this problem are the number of dolls of each type to be manufactured each hour. Thus we begin the mathematical formulation by letting

x = number of Scary Harry dolls manufactured per hour

y = number of Horrible Harriet dolls manufactured per hour

z = number of The Glob dolls manufactured per hour

TABLE 7-1

Doll	Plastic (ounces)	Time for Clothes (minutes)	Time for Special Features (minutes)	Profit per Doll
Scary Harry	4	3	2	$1.00
Horrible Harriet	3	4	4	1.25
The Glob	9	1	3	1.50
Available time or material per hour of operation	160	50	50	

We see from Table 7-1 that the amount (in ounces) of plastic used per hour is $4x + 3y + 9z$. Hence, one restriction or constraint which must be satisfied is

$$4x + 3y + 9z \le 160$$

Two constraints on time also follow from Table 7-1, namely,

$$3x + 4y + z \le 50$$
$$2x + 4y + 3z \le 50$$

Also, it follows from the definition of x, y, and z that each of them is nonnegative. Thus, we have constraints $x \ge 0$, $y \ge 0$, $z \ge 0$. Finally, the profit resulting from each hour of operation is (in dollars):

$$\text{Profit} = x + 1.25y + 1.5z$$

Thus, for the original production problem we have the following mathematical formulation.

Find numbers x, y, and z such that the system of inequalities

$$
\begin{aligned}
x \ge 0 \qquad y \ge 0 \qquad z \ge 0 \\
4x + 3y + 9z \le 160 \\
3x + 4y + z \le 50 \\
2x + 4y + 3z \le 50
\end{aligned}
\tag{7-1}
$$

is satisfied and

$$p = \text{profit} = x + 1.25y + 1.5z$$

is a maximum for all x, y, z satisfying (7-1).   □

The constraints which appear in (7-1) involve linear functions of the variables x, y, and z. This will be true of all the problems studied in this chapter, and in fact all linear programming problems must have linear constraints.

Definition   If a_1, a_2, ..., a_n are n constants, then the function which assigns the number $a_1x_1 + a_2x_2 + \cdots + a_nx_n$ to the vector $(x_1, x_2, \ldots, x_n)$ is a *linear function of n variables*.

Note that system (7-1) is satisfied by $x = y = z = 0$, in which case the profit is 0, and also that it is satisfied by $x = y = z = 2$, in which case the profit is $7.50 per hour. The problem is to find values for x, y, and z so that all inequalities in (7-1) are satisfied and the profit is the maximum possible. Any triple of three numbers (x,y,z) which satisfies (7-1) is said to be a *feasible vector* for this problem. Thus $(0,0,0)$ and $(2,2,2)$ are feasible vectors. The function to be maximized (or minimized) is called the *objective function*. In a linear programming problem the objective function must also be a linear function or a linear function plus a constant. This terminology is completely analogous to that introduced in Chap. 5.

Example 7-1 is typical of the type of problem which will be considered in this chapter. Each problem involves a system of inequalities [such as (7-1)] and an objective function (such as a profit function). The goal is to find a feasible vector which maximizes the objective function over the set of all feasible vectors. In all problems which we consider the inequality constraints will include the conditions that the coordinates of the vector be nonnegative. If the problem under consideration has three variables, we shall write the vectors as (x,y,z) and in cases where there are n variables, $n > 3$, the vectors will be written as $(x_1, x_2, \ldots, x_n)$. Although we concentrate on examples with either three or four variables, the methods of the chapter can be used for much larger values of n. In practical cases problems with n as large as 10,000 have been solved.

The problem in Example 7-1 is an example of a type of problem called a standard maximum problem (SMP). The precise definition of this class of problems is as follows:

Definition   A linear programming problem is written as a *standard maximum problem* (SMP) if:

1. The variables are constrained to be nonnegative.
2. The constraints of the problem (other than the nonnegativity constraints given in condition 1) are all of the form.

$$\text{Linear function of the variables} \leq \text{constant}$$

3. The objective function of the problem is a linear function which is to be maximized.
4. The constants which appear on the right-hand side of the constraints described in condition 2 are positive.

Remark: Condition 4 is not usually part of the definition of an SMP. For our purposes it is convenient to include this condition; however, it is important to note that there are procedures for modifying our techniques to solve problems which obey conditions 1, 2, and 3, but not 4.

EXAMPLE 7-2

Decide which of the following problems are SMPs (including condition 4):

(a)   Find the maximum of $p = 4x - 2y + 3z$ for

$$x \geq 0 \qquad y \geq 0 \qquad z \geq 0$$
$$2x + 9y - 2z \leq 0$$
$$-x + y - z \geq 5$$

(b)   Find the maximum of $p = x_1 + 2x_2 + 3x_3 + 4x_4$ for

$$x_1 \geq 0$$
$$2x_1 + 4x_3 \leq 5$$
$$3x_2 + 5x_4 \leq 6$$

(c)   Find the maximum of $p = 2x + 3y$ for

$$x \geq 0 \qquad y \geq 0$$
$$x + 2y \geq -3$$
$$3x - 4y \geq -1$$

Solution: (a)   Conditions 1 and 2 are satisfied directly, and condition 3 holds since the constraint $-x + y - z \geq 5$ can be written as $x - y + z \leq -5$ by multiplying both sides by -1. However, condition 4 does not hold, and hence problem (a) is *not* an SMP.

(b)   Condition 2 holds, but condition 1 does not hold since all variables are not required to be nonnegative. Hence, this is *not* an SMP.

(c)   All conditions are satisfied and hence this *is* an SMP. To show that conditions 2 and 4 hold it is necessary to multiply each of the constraints by -1. After this multiplication they can be written as

$$-x - 2y \leq 3$$
$$-3x + 4y \leq 1 \qquad\qquad \square$$

Exercises for Sec. 7-2

1. Use the data from Table 7-2 to formulate a linear programming problem similar to that of Example 7-1.

TABLE 7-2

Doll	Plastic (ounces)	Time for Clothes (minutes)	Time for Special Features (minutes)	Profit per Doll
Scary Harry	5	2	3	$1.10
Horrible Harriet	3	4	4	1.30
The Glob	10	1	6	2.00
Available time or material per hour of operation	192	55	45	

2. Formulate the following situation as a linear programming problem. The dietitian at a school cafeteria has 3 foods to serve. Each unit of food I contains 6 ounces of protein, 5 ounces of fat, and 8 ounces of carbohydrates and costs $1.25. Each unit of food II contains 4 ounces of protein, 4 ounces of fat, and 12 ounces of carbohydrates and costs $1.50. Each unit of food III contains 9 ounces of protein, 2 ounces of fat, and 5 ounces of carbohydrates and costs $1.75. A mixture of these foods is to be formed which contains at least 30 ounces of protein, 40 ounces of fat, and 45 ounces of carbohydrates. What is the least expensive mixture?

3. Decide which of the following linear programming problems are SMPs. If a problem is not an SMP, state why not.
   a. Find the maximum of $p = 4x - 3y + 2z$ for

$$x \geq 0 \qquad y \geq 0 \qquad z \geq 0$$
$$2x - y + z \leq 0$$
$$x + y - z \geq 0$$

   b. Find the minimum of $x + y$ for

$$x \geq 0 \qquad y \geq 0$$
$$2x + y \leq 7$$
$$x + 3y \leq 6$$

4. Decide which of the following linear programming problems are SMPs. If a problem is not an SMP, state why not.
   a. Find the maximum of $x + y + z$ for

$$x \geq 0 \qquad y \geq 0 \qquad z \geq 0$$
$$3x - y + z \leq 4$$
$$-x + y + 5z \leq 10$$
$$5x + y - z \leq 5$$

b. Find the maximum of $2x + y + 8z$ for

$$x \geq 1 \qquad y \geq 1 \qquad z \geq 1$$
$$-3x + y - z \geq -3$$
$$x - y - 5z \geq -10$$
$$-5x - y + z \geq -5$$

5. Decide whether the problems of Exercises 3 to 5 of Sec. 5-7 are SMPs. If a problem is not an SMP, state which condition is not satisfied.

7-3 SLACK VARIABLES AND BASIC SOLUTIONS

The constraints in an SMP are inequalities which state that linear functions of the variables are to be less than or equal to nonnegative constants. In Example 7-1 the variables are x, y, and z; these variables must be nonnegative, and they must satisfy the constraints

$$4x + 3y + 9z \leq 160$$
$$3x + 4y + z \leq 50$$
$$2x + 4y + 3z \leq 50$$

To satisfy the first of these inequalities, $4x + 3y + 9z \leq 160$, one must find nonnegative numbers x, y, and z such that either $4x + 3y + 9z = 160$ or $4x + 3y + 9z < 160$. If $4x + 3y + 9z < 160$, then there is a positive number u such that $4x + 3y + 9z + u = 160$. If we consider u to also be a variable, then in order to satisfy this inequality one must find nonnegative numbers x, y, z, and u such that $4x + 3y + 9z + u = 160$. If $u = 0$, we have satisfied the inequality in the form $4x + 3y + 9z = 160$ and if $u > 0$, the inequality is satisfied in the form $4x + 3y + 9z < 160$. The variable u is called the *slack variable* for the first inequality. In terms of the problem of Example 7-1 this variable measures the amount (in ounces) of plastic material which is available per hour and which is *not* used in the production of x Scary Harry dolls, y Horrible Harriet, dolls, and z The Glob dolls. Thus it is the unused or "slack" amount of material.

Similarly, there are slack variables v and w for the second and third inequalities. Thus an equivalent way of representing the constraints of Example 7-1 is as follows: find nonnegative numbers x, y, z, and nonnegative values of the slack variables u, v, w such that

$$4x + 3y + 9z + u = 160$$
$$3x + 4y + z + v = 50 \qquad (7\text{-}2)$$
$$2x + 4y + 3z + w = 50$$

System (7-2) is a system of equations (instead of inequalities) and as such it can be solved for the unknowns x, y, z, u, v, w by using gaussian elimination, as discussed in Sec. 6-3. However, in Example 7-1 and in all linear programming problems it is not enough to just solve the system of equations in (7-2). Instead it is required to find solutions which are nonnegative, i.e., each variable must be zero or positive; and moreover it is necessary to find solutions which maximize the objective function. In Example 7-1 the objective function is $p = x + 1.25y + 1.5z$. Note that the slack variables u, v, and w do not enter directly in the objective function. As we noted before, the slack variables measure unused material and work time, and since profit is determined by the number of dolls produced—and consequently by the material and time actually used—the slack variables are indirectly but not directly related to profit. If the slack variables take on large values, x, y, and z are necessarily small because much of the material and available work time is not used and hence profit is small. Thus the objective function is indirectly affected by the slack variables.

In the discussion of linear programming problems in two variables in Chap. 5 it was noted that the maximum of the objective function always occurred at a corner point of the feasible set (provided, of course, that a maximum existed). Such problems have solutions if the feasible set is bounded and not empty. In two dimensions, the problem can be solved by graphing the feasible set in the plane, finding the corner points, and determining the corner point at which the objective function assumed its maximum value. In higher dimensions the situation is essentially the same. However, it is difficult to determine the corner points of the set by graphing, and an algebraic method is preferable. The algebra and geometry are connected in the concept of a *basic solution*.

Consider an SMP $[\mathbb{A}, \mathbf{b}, \mathbf{c}]$ where $\mathbb{A}$ is $m \times n$. Introduce slack variables to form a system of m equations in $m + n$ variables, and reduce this system to echelon form. A *basic solution* is any nonnegative solution of this system with zero coordinates except possibly those coordinates corresponding to a leading 1 in the echelon form (condition A, p. 230). Different basic solutions arise from different echelon forms.

The identification of basic solutions by using echelon forms is not crucial to what follows. Instead we will obtain basic solutions either by the algorithm known as the simplex method or by geometry. To illustrate how basic solutions are related to corner points we consider an example in two dimensions.

EXAMPLE 7-3

Consider Example 5-20 again. The problem is to find the maximum of $2x + y$ for (x,y) satisfying

$$
\begin{aligned}
x \geq 0 \qquad y \geq 0 \\
x \leq 200 \\
2x + 3y \leq 1200 \\
25x + 15y \leq 7500
\end{aligned}
\tag{7-3}
$$

TABLE 7-3

x	y	u	v	w	Number of Zeros	Basic Solution	Corner Point	Value $2x + y$
0	0	200	1200	7500	2	Yes	Yes	0
100	100	100	700	3500	0	No	No	300
200	0	0	800	2500	2	Yes	Yes	400
200	$\frac{500}{3}$	0	300	0	2	Yes	Yes	$566\frac{2}{3}$
100	200	100	400	2000	0	No	No	400
0	400	200	0	1500	2	Yes	Yes	400
100	$\frac{1000}{3}$	100	0	0	2	Yes	Yes	$533\frac{1}{3}$
0	300	200	300	3000	1	No	No	300

This is clearly an SMP. There are three constraints other than the non-negativity constraints ($x \geq 0$, $y \geq 0$). Hence we add three slack variables u, v, and w and obtain the system

$$x + u = 200$$
$$2x + 3y + v = 1200 \qquad (7\text{-}4)$$
$$25x + 15y + w = 7500$$

This system of equations has the variables x, y, u, v, and w, and each variable is to be nonnegative. Some solutions of this system are shown in Table 7-3, and the graph of the feasible set in the xy plane is shown in Fig. 7-1. □

We note from Table 7-3 and Fig. 7-1 that each basic solution of the system of equations corresponds to a corner point of the feasible set, and conversely each corner point of the feasible set corresponds to a basic solution of the system

FIGURE 7-1

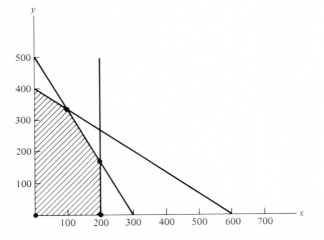

of equations. This holds true in general, and hence a maximum (if it exists) will occur at a basic solution of the system of equations. This important fact provides the connection between the geometric techniques of Chap. 5 and the algebraic techniques of this chapter.

Theorem In any standard maximum problem the basic solutions of the system of equations obtained by adding slack variables correspond to the corner points of the feasible set. Moreover, if a maximum exists, it will occur at a basic solution.

It follows from this result that in order to solve a linear programming problem in higher dimensions one can proceed by finding basic solutions of the related system of equations and by evaluating the objective function at these basic solutions. The basic solutions can be found by using a systematic step-by-step procedure called the *simplex method*, which is similar to gaussian elimination. A description of this method is the topic of the next two sections.

Exercises for Sec. 7-3

Convert each of the following problems into a problem with slack variables, graph the feasible set in the plane, and find the basic solutions which correspond to the corner points:

1. a. Find the maximum of $3x + 8y$ for

$$x \geq 0 \qquad y \geq 0$$
$$2x + 2y \leq 10$$
$$3x + 6y \leq 18$$

b. Find the maximum of $x + y$ for

$$x \geq 0 \qquad y \geq 0$$
$$2x + 8y \leq 16$$
$$3x + 3y \leq 15$$
$$10x + 3y \leq 30$$

2. a. Find the maximum of $4x + 6y$ for

$$x \geq 0 \qquad y \geq 0$$
$$x + y \leq 10$$
$$-x - y \leq -5$$
$$x \leq 7$$
$$y \leq 7$$

b. Find the maximum of $4x - 5y$ for

$$x \geq 0 \qquad y \geq 0$$
$$x \geq y - 1$$
$$x + y \leq 20$$

3. Form a table similar to that of Table 7-3 for the system of equations (7-2). Find two solutions which are basic and two solutions which are not basic.

4. Convert the following problem into a problem with slack variables and find two basic solutions of the resulting system of equations. Find the maximum of $2x + 3y + z$ for

$$x \geq 0 \qquad y \geq 0 \qquad z \geq 0$$
$$x + y \leq 2$$
$$y + z \leq 2$$
$$x + z \leq 2$$

7-4 TABLEAUS AND THE PIVOT OPERATION

The simplex method is an algebraic process which solves an SMP by systematically evaluating the objective function at corner points of the feasible set until a point is found where the objective function assumes its maximum value. The method involves considering the corner points, one after another, until one is found at which the maximum value of the objective function is attained. Of course, the maximum value of the objective function is not known before solving the problem. A major advantage of the method is that at each stage of the process the only corner points considered are those at which the objective function has values at least as large as the values at points already considered. Thus, the simplex method is not a random trial-and-error process or even a systematic technique which must find all corner points and evaluate the objective function at each one. Instead, it starts with one corner point and then systematically considers "better" corner points, if such exist. The criterion for one corner point to be better than another is that the objective function be larger at the better point. Suppose that in the application of the algorithm a corner point has been selected. In this section we describe the notation and the basic operation used to select the next corner point. In the next section we discuss the complete simplex method. Each corner point considered in the simplex method is displayed by means of a rectangular array called a *tableau*. We introduce this notion through an example and describe the method of obtaining the first tableau.

Formation of the Initial Tableau

EXAMPLE 7-4

Form the initial tableau for the following SMP.

Find the maximum of $p = 5x + y + 3z$ for

$$x \geq 0 \qquad y \geq 0 \qquad z \geq 0$$
$$x + y + 2z \leq 100$$
$$10x + 2y + 5z \leq 500$$

Solution: We first convert the constraints into a system of equations by adding slack variables. Since there are two linear inequalities (we do not count $x \geq 0$, $y \geq 0$, and $z \geq 0$), we need two slack variables. We call these slack variables u and v. Thus we consider the new problem.

Find the maximum of $p = 5x + y + 3z$ for

$$x \geq 0 \qquad y \geq 0 \qquad z \geq 0$$
$$u \geq 0 \qquad v \geq 0$$
$$x + y + 2z + u = 100$$
$$10x + 2y + 5z + v = 500$$

This is a system of two equations in five variables, and the system can be solved by the methods introduced in Chap. 6. The augmented matrix for the system (see Sec. 6-3) is

$$\begin{bmatrix} 1 & 1 & 2 & 1 & 0 & | & 100 \\ 10 & 2 & 5 & 0 & 1 & | & 500 \end{bmatrix}$$

We form a tableau which consists of the augmented matrix with all variables listed at the top, the slack variables listed at the left, and the right-hand column labeled Basic Solution. The tableau at this stage, shown as Tableau 7-1, is not yet the initial tableau of the simplex method. We continue by adding a final row

TABLEAU 7-1

	x	y	z	u	v	Basic Solution
u	1	1	2	1	0	100
v	10	2	5	0	1	500

to the tableau, and we label this row p to indicate that the entries are determined from the coefficients in the objective function p. The entry under x is the negative of the coefficient of x in p, -5 in this example. Similarly, the entries under y, z, u, and v are the negatives of the coefficients of those variables in p. They are -1, -3, 0, and 0, respectively, since 1, 3, 0, and 0 are the coefficients of y, z, u, and v in p. Finally, the entry under the column labeled Basic Solution is the value of p when the variables in p are given the values shown in this column. For example, in Tableau 7-1 the variable u has the value 100, v has the value 500, and x, y, and z have value 0. Thus, since $p = 5x + y + z$, for these values of the variables, p has value 0 and the entry under p in the last row is 0. The tableau so constructed is the initial tableau of the problem. The initial tableau for Example 7-4 is shown in Tableau 7-2.

TABLEAU 7-2

	x	y	z	u	v	Basic Solution
u	1	1	2	1	0	100
v	10	2	5	0	1	500
p	-5	-1	-3	0	0	0

Remark: It is important to keep in mind that the entries in the last row of the initial tableau are the *negatives* of the coefficients in p. Thus, if one of these coefficients is a negative number, the corresponding entry in the last row of the initial tableau is a positive number.

EXAMPLE 7-5

Form the initial tableau for the following SMP.
   Find the maximum of $p = x + 3y + 2z$ for

$$x \geq 0 \qquad y \geq 0 \qquad z \geq 0$$
$$3x + 9z \leq 54$$
$$x + 3y + 6z \leq 36$$
$$5y + z \leq 15$$

Solution: We convert the three inequalities to equations by introducing the slack variables u, v, and w. This gives the new problem.

Find the maximum of $p = x + 3y + 2z$ for

$$x \geq 0 \qquad y \geq 0 \qquad z \geq 0$$
$$u \geq 0 \qquad v \geq 0 \qquad w \geq 0$$
$$3x + 9z + u = 54$$
$$x + 3y + 6z + v = 36$$
$$5y + z + w = 15$$

The initial tableau for this system of equations is formed in two steps. First, we form the augmented matrix of the system and list all the variables at the top and the slack variables at the left and label the right-hand column Basic Solution. This gives the tableau shown in Tableau 7-3.

TABLEAU 7-3

	x	y	z	u	v	w	**Basic Solution**
u	3	0	9	1	0	0	54
v	1	3	6	0	1	0	36
w	0	5	1	0	0	1	15

TABLEAU 7-4

	x	y	z	u	v	w	**Basic Solution**
u	3	0	9	1	0	0	54
v	1	3	6	0	1	0	36
w	0	5	1	0	0	1	15
p	−1	−3	−2	0	0	0	0

The second and final step in forming the initial tableau (in practice the two steps are carried out together) is to add a row at the bottom of the tableau labeled p, which is obtained by writing in each column the negative of the coefficient of that variable in the objective function p. In this example, the entry under x is −1, the entry under y is −3, and the entry under z is −2 since $p = x + 3y + 2z$. The entries under u, v, and w are 0 since these variables do not appear in p (they have a zero coefficient in p). With this row added the initial tableau has the form shown in Tableau 7-4. The entry under Basic Solution is the value of p for the basic solution of the system of equations given by $x = 0$, $y = 0$, $z = 0$, $u = 54$, $v = 36$, $w = 15$. Since $p = x + 3y + 2z$, the value is $p = 0 + 3(0) + 2(0) = 0$. $\qquad\square$

The Pivot Operation

The simplex method for standard maximum problems consists of (1) forming the initial tableau and (2) forming new tableaus by *pivoting*. We have just described the process of forming the initial tableau, and we turn now to the process of forming new tableaus by using the pivot operation.

The operation of pivoting in a tableau is essentially the operation used in gaussian elimination to replace an equation by another one in which certain variables have zero coefficients. In particular, it consists of row operations on the augmented matrix of the system of equations which convert one entry in the matrix to a 1 and all other entries in the column of that entry to 0s. The entry which is converted to a 1 is called the *pivot element*. If the entry which is made a 1 is in the ith row and the jth column, the new tableau has the variable of the jth column listed at the left of the ith row. For example, in the tableau of Tableau 7-2 suppose that the entry in row 2, column 1 is the pivot element. This entry is shown circled in Tableau 7-5.

The pivot element is located in the row of the matrix labeled by v and the column labeled x. In such a case the variable v is called the *departing variable* and x is called the *entering variable*. The new tableau is obtained by dividing row 2 by 10 (to make the pivot element a 1) and then adding a suitable multiple of row 2 to all other rows, including the row labeled p, to convert the other entries in the first column into 0s. For example, if after dividing row 2 by 10, -1 times this new row 2 is added to row 1 the entry in row 1 and column 1 is a zero. Similarly, if 5 times the new second row is added to the row labeled p, the entry in row 3 and column 1 is a zero. The tableau which results from this pivot operation is shown in Tableau 7-6.

TABLEAU 7-5

	x	y	z	u	v	Basic Solution
u	1	1	2	1	0	100
v	(10)	2	5	0	1	500
p	-5	-1	-3	0	0	0

TABLEAU 7-6

	x	y	z	u	v	Basic Solution
u	0	$\frac{4}{5}$	$\frac{3}{2}$	1	$-\frac{1}{10}$	50
x	1	$\frac{1}{5}$	$\frac{1}{2}$	0	$\frac{1}{10}$	50
p	0	0	$-\frac{1}{2}$	0	$\frac{1}{2}$	250

Note that in Tableau 7-6 the variables listed at the left of the tableau are u and x. As a result of the pivot operation the entering variable x has replaced the departing variable v. This is always the case with the simplex method: each pivot operation results in the replacement of one variable at the left by a new variable. The variables at the top of the tableau remain unchanged. The basic solution given by Tableau 7-6 is $u = 50$, $x = 50$, and $y = z = v = 0$. Moreover, for this basic solution the objective function $p = 5x + y + 3z$ has the value $5(50) + 0 + 3(0) = 250$, the value in the lower right-hand corner of the tableau. We note that this basic solution (corner point) is better than that of the initial tableau since the objective function now has value 250 whereas originally it had value 0.

As a second example of the pivot operation we return to Example 7-5 and Tableau 7-4. Suppose the element in the third row and second column of Tableau 7-4 is to be a pivot element. This element is circled in Tableau 7-7, and it indicates that the entering variable is y and the departing variable is w.

TABLEAU 7-7

	x	y	z	u	v	w	Basic Solution
u	3	0	9	1	0	0	54
v	1	3	6	0	1	0	36
w	0	⑤	1	0	0	1	15
p	-1	-3	-2	0	0	0	0

TABLEAU 7-8

	x	y	z	u	v	w	Basic Solution
u	3	0	9	1	0	0	54
v	1	0	$\frac{27}{5}$	0	1	$-\frac{3}{5}$	27
y	0	1	$\frac{1}{5}$	0	0	$\frac{1}{5}$	3
p	-1	0	$-\frac{7}{5}$	0	0	$\frac{3}{5}$	9

To carry out a pivot operation with the pivot element in the third row and second column we must multiply the third row by an appropriate number to convert the pivot element to a 1 (here the constant is $\frac{1}{5}$), and we must add suitable multiples of the third row to the other rows to convert all other entries in column 2 into 0s. The result is Tableau 7-8.

The basic solution given in Tableau 7-8 has $x = 0$, $y = 3$, $z = 0$, $u = 54$, and $v = 27$. The value of $p = x + 3y + 2z$ for this solution is $0 + 3(3) + 2(0) = 9$, as shown in the lower right-hand corner of the tableau. Since the problem under consideration is to maximize p, we see that the basic solution of Tableau 7-8 is better than the basic solution of the original tableau (Tableau 7-7).

The pivot operation is the essence of the simplex method. Pivots are carried out until either a maximum is obtained or it is established that no maximum exists. The criterion for distinguishing these two cases is given in the next section.

Exercises for Sec. 7-4

1. Form the initial tableau for the SMP.
   Find the maximum of $p = x + 3y$ for

$$x \geq 0 \qquad y \geq 0$$
$$2x + 3y \leq 12$$
$$6x + y \leq 9$$

2. Form the initial tableau for the SMP.
   Find the maximum of $p = 2x - y + z$ for

$$x \geq 0 \qquad y \geq 0 \qquad z \geq 0$$
$$5x - y + z \leq 20$$
$$2x + y + 2z \leq 30$$

3. Form the initial tableau for the SMP.
   Find the maximum of $p = 2x + 3y + z$ for

$$x \geq 0 \qquad y \geq 0 \qquad z \geq 0$$
$$5x \phantom{{}+ y} + z \leq 100$$
$$\phantom{5x + {}}y + 3z \leq 300$$
$$2x - y + z \leq 900$$

Carry out a pivot operation with the circled entry as the pivot element. What is the new basic solution?

4.

	x	y	u	v	Basic Solution
u	2	③	1	0	12
v	6	1	0	1	9
p	-1	-3	0	0	0

5.

	x	y	z	u	v	Basic Solution
u	1	0	5	1	0	50
v	0	⑩	1	0	1	30
p	-1	-5	-2	0	0	0

6.

	x	y	z	u	v	w	Basic Solution
u	5	0	1	1	0	0	100
v	0	①	3	0	1	0	300
w	2	-1	1	0	0	1	900
p	-2	-3	-1	0	0	0	0

Describe the SMP which corresponds to the initial tableau in:

7. Exercise 4   **8.** Exercise 5   **9.** Exercise 6

10. Carry out a pivot operation for the tableau shown below. The circled entry is to be the pivot element. Describe the basic solution given by the original tableau and the basic solution given by the new tableau (the tableau which results after the pivot operation is performed) and decide which is better in terms of the related SMP.

	x	y	z	u	v	Basic Solution
u	1	0	⑤	1	0	50
y	0	1	$\frac{1}{10}$	0	$\frac{1}{10}$	3
p	-1	0	$-\frac{3}{2}$	0	$\frac{1}{2}$	15

Proceed as in Exercise 10:

11.

	x	y	z	u	v	Basic Solution
z	$\left(\frac{1}{5}\right)$	0	1	$\frac{1}{5}$	0	10
y	$-\frac{1}{50}$	1	0	$-\frac{1}{50}$	$\frac{1}{10}$	2
p	$-\frac{7}{10}$	0	0	$\frac{3}{10}$	$\frac{1}{2}$	30

12.

	x	y	z	u	v	w	Basic Solution
u	3	0	9	1	0	0	54
v	1	0	$\left(\frac{27}{5}\right)$	0	1	$-\frac{3}{5}$	27
y	0	1	$\frac{1}{5}$	0	0	$\frac{1}{5}$	3
p	-1	0	$-\frac{7}{5}$	0	0	$\frac{3}{5}$	9

13.

	x	y	z	u	v	w	Basic Solution
u	$\left(\frac{4}{3}\right)$	0	0	1	$-\frac{5}{3}$	1	9
z	$\frac{5}{27}$	0	1	0	$\frac{5}{27}$	$-\frac{1}{9}$	5
y	$-\frac{1}{27}$	1	0	0	$-\frac{1}{27}$	$\frac{10}{45}$	2
p	$-\frac{20}{27}$	0	0	0	$\frac{7}{27}$	$\frac{20}{27}$	16

7-5 OPTIMAL VECTORS VIA THE SIMPLEX METHOD

In Sec. 7-4 we described the part of the simplex method which consists of obtaining an initial tableau and of pivoting to obtain new tableaus. To complete the description of the method we need to answer the questions:

A. Which element should be the pivot element?
B. When should the process of pivoting stop?

The answer to question B depends upon the answer to question A. Indeed, one should stop the process of pivoting as soon as one is no longer able to find a proper pivot element. The technique of finding the proper pivot element is as follows:

1. Examine the bottom row of the tableau and find the most negative entry (not including the entry under the column marked Basic Solution). The variable at the top of the column with the most negative entry is the entering variable. The situation in which the bottom row of the tableau contains no negative entries will be discussed below.
2. Divide each positive entry in the column of the entering variable into the corresponding entry (same row) in the column under Basic Solution. The variable in the row which gives the smallest quotient is the departing variable.

To illustrate rules 1 and 2 for finding a pivot element, we consider Tableau 7-9.

TABLEAU 7-9

	x	y	z	u	v	Basic Solution
u	1	1	2	1	0	100
$\rightarrow v$	(10)	2	5	0	1	500
p	-5	-1	-3	0	0	0

According to rule 1, we examine the bottom row of the tableau and find the most negative entry. This is the entry -5, and since it is under the column marked x, the variable x is the entering variable. Next, following rule 2, we divide each positive entry in the column under x into the corresponding entry under Basic Solution. The quotients obtained are 100 and 50 ($\frac{100}{1}$ and $\frac{500}{10}$). Since the smallest quotient occurs in the row of variable v, v is the departing variable. The pivot element is then the entry determined by the departing variable (a row) and the entering variable (a column). The entering and departing variables are shown by arrows, and the pivot element is circled in Tableau 7-9.

A pivot was carried out earlier in Tableau 7-9 with pivot element 10, and the result is reproduced in Tableau 7-10.

If we again apply rules 1 and 2 for obtaining a pivot element, we see that z is the entering variable ($-\frac{1}{2}$ is the only negative entry in the bottom row) and u is the departing variable ($50/\frac{3}{2} = 100/3$ is the smallest quotient). Thus the next pivot element is $\frac{3}{2}$, shown circled in Tableau 7-10. Pivoting at this element gives Tableau 7-11.

If we attempt to apply rules 1 and 2 to Tableau 7-11, we encounter a problem. *No entry* in the bottom row is negative. In such a case how should we interpret the rule "find the most negative entry"? The answer is that in such a situation no further pivoting is necessary:

> If the last row of a tableau contains no negative entries, then the basic solution given by the tableau is an optimal vector for the linear programming problem.

Thus in Tableau 7-11 the basic solution ($x = \frac{100}{3}$, $y = 0$, $z = \frac{100}{3}$) is an optimal vector for the original linear programming problem (Example 7-4). Moreover, as we noted earlier, the last entry under the column Basic Solution is the value of the quantity being maximized in the problem. Here the objective function is $p = 5x + y + 3z$, and the maximum value of p is $p = 5(\frac{100}{3}) + 3(\frac{100}{3}) = \frac{800}{3}$.

As a second illustration of the rules for finding the pivot element we carry out another example of the simplex method.

TABLEAU 7-10

	x	y	z	u	v	Basic Solution
$\to u$	0	$\frac{4}{5}$	$\boxed{\frac{3}{2}}$	1	$-\frac{1}{10}$	50
x	1	$\frac{1}{5}$	$\frac{1}{2}$	0	$\frac{1}{10}$	50
p	0	0	$-\frac{1}{2}$	0	$\frac{1}{2}$	250

TABLEAU 7-11

	x	y	z	u	v	Basic Solution
z	0	$\frac{8}{15}$	1	$\frac{2}{3}$	$-\frac{1}{15}$	$\frac{100}{3}$
x	1	$-\frac{1}{15}$	0	$-\frac{1}{3}$	$\frac{2}{15}$	$\frac{100}{3}$
p	0	$\frac{4}{15}$	0	$\frac{1}{3}$	$\frac{14}{30}$	$\frac{800}{3}$

EXAMPLE 7-6

Solve the following SMP by using the simplex method.
Find the maximum of $p = x + 3y + 2z$ for

$$x \geq 0 \qquad y \geq 0 \qquad z \geq 0$$
$$3x + 9z \leq 54$$
$$x + 3y + 6z \leq 36$$
$$5y + z \leq 15$$

Solution: In Example 7-5 we obtained the initial tableau for this problem. It is shown again in Tableau 7-12.

TABLEAU 7-12

	x	y	z	u	v	w	Basic Solution
u	3	0	9	1	0	0	54
v	1	3	6	0	1	0	36
→ w	0	⑤	1	0	0	1	15
p	−1	−3	−2	0	0	0	0

According to rule 1 for finding the pivot element we examine the last row of the tableau and find the most negative entry. This is the entry -3, and since it is under the column marked y, the variable y is the entering variable. Next, following rule 2, we divide each positive entry in the column under y into the corresponding entry under Basic Solution. The quotients obtained are 12, in the row corresponding to v ($\frac{36}{3}$), and 3, in the row corresponding to w ($\frac{15}{5}$). The smallest quotient is in the row corresponding to the variable w, and hence w is the departing variable. The pivot element is then the element at the intersection of the departing row variable (row w) and the entering column variable (column y). The entering and departing variables are shown by arrows, and the pivot element is circled in Tableau 7-12.

A pivot was carried out earlier with the element circled in Tableau 7-12, and the result is Tableau 7-13.

If we again apply rules 1 and 2 for obtaining a pivot element, we see that z is the entering variable ($-\frac{7}{5}$ is the most negative entry in the bottom row) and v is the departing variable ($27/\frac{27}{5} = 5$ is the smallest quotient). Thus the next pivot element is $\frac{27}{5}$, shown circled in Tableau 7-13. Pivoting at this element gives Tableau 7-14.

TABLEAU 7-13

	x	y	$\downarrow$ z	u	v	w	Basic Solution
u	3	0	9	1	0	0	54
$\rightarrow v$	1	0	$\left(\frac{27}{5}\right)$	0	1	$-\frac{3}{5}$	27
y	0	1	$\frac{1}{5}$	0	0	$\frac{1}{5}$	3
p	-1	0	$-\frac{7}{5}$	0	0	$\frac{3}{5}$	9

TABLEAU 7-14

	$\downarrow$ x	y	z	u	v	w	Basic Solution
$\rightarrow u$	$\left(\frac{4}{3}\right)$	0	0	1	$-\frac{5}{3}$	1	9
z	$\frac{5}{27}$	0	1	0	$\frac{5}{27}$	$-\frac{1}{9}$	5
y	$-\frac{1}{27}$	1	0	0	$-\frac{1}{27}$	$\frac{2}{9}$	2
p	$-\frac{20}{27}$	0	0	0	$\frac{7}{27}$	$\frac{4}{9}$	16

If we again apply the rules for finding a pivot element, we find that x is the entering variable, u is the departing variable, and $\frac{4}{3}$ is the pivot element. Pivoting here results in Tableau 7-15.

TABLEAU 7-15

	x	y	z	u	$\downarrow$ v	w	Basic Solution
x	1	0	0	$\frac{3}{4}$	$-\frac{5}{4}$	$\frac{3}{4}$	$\frac{27}{4}$
$\rightarrow z$	0	0	1	$-\frac{5}{36}$	$\left(\frac{5}{12}\right)$	$-\frac{1}{4}$	$\frac{15}{4}$
y	0	1	0	$\frac{1}{36}$	$-\frac{1}{12}$	$\frac{1}{4}$	$\frac{9}{4}$
p	0	0	0	$\frac{5}{9}$	$-\frac{2}{3}$	1	21

It is tempting to stop at this stage since we have now replaced each of the slack variables by one of the original variables; however, since there is still a negative entry in the last row of the tableau, the rules call for us to make another pivot. The entering variable is the slack variable v, and the departing variable is z. The pivot element is $\frac{5}{12}$, and carrying out a pivot at this point results in Tableau 7-16.

At this point the original linear programming problem is solved. There are no negative entries in the last row of Tableau 7-16, and hence the basic solution given by the tableau, $x = 18$, $y = 3$, $z = 0$, is optimal for the problem. The maximum value of p is 27.   □

TABLEAU 7-16

	x	y	z	u	v	w	Basic Solution
x	1	0	3	$\frac{1}{3}$	0	0	18
v	0	0	$\frac{12}{5}$	$-\frac{1}{3}$	1	$-\frac{3}{5}$	9
y	0	1	$\frac{1}{5}$	0	0	$\frac{1}{5}$	3
p	0	0	$\frac{8}{5}$	$\frac{1}{3}$	0	$\frac{3}{5}$	27

We have now provided a partial answer to question B posed above: When should pivoting be stopped? The partial answer is that pivoting should be stopped whenever the last row of the tableau has no negative entries. In such a case the basic solution given by the tableau is optimal. There is one other case where pivoting must also stop, namely, when the last row contains a negative entry (and hence an entering variable can be chosen) but the column of the entering variable *does not have a positive element* (and hence no departing variable can be chosen). In such a case *the linear programming problem does not have a solution*. We illustrate this possibility with an example.

EXAMPLE 7-7

Find the maximum of $p = 2y + z$ for

$$x \geq 0 \qquad y \geq 0 \qquad z \geq 0$$
$$x - y + z \leq 5$$
$$2x - z \leq 10$$

Solution: To form the initial tableau we first introduce slack variables u and v to form the system of equations

$$x - y + z + u = 5$$
$$2x - z + v = 10$$

The initial tableau is Tableau 7-17.

TABLEAU 7-17

		x	y↓	z	u	v	Basic Solution
?	u	1	−1	1	1	0	5
	v	2	0	−1	0	1	10
	p	0	−2	−1	0	0	0

If we apply the rules for finding a pivot element, we find that y is the entering variable, but since no entry in the column under y is positive, no departing variable can be selected. Thus, as we noted above, this linear programming problem has *no solution*. The quantity p can be made arbitrarily large and has no maximum value for these constraints. In examining the problem we note that the difficulty is that y can assume arbitrarily large values and still satisfy the constraints. Since y can be arbitrarily large, certainly $p = 2y + z$ can be arbitrarily large. □

There is one final item to consider in our discussion of the simplex method relating to the rules for finding pivot elements. We must decide how to proceed if the rules result in a tie, i.e., if either two entries are "most negative" in the bottom row or two quotients are "smallest." In such cases we have a choice in selecting the entering variable (if more than one most negative entry exists) and the departing variable (if more than one smallest quotient exists). Fortunately, such situations are not a serious problem, and any of the possible choices are acceptable. Thus if two variables are legitimate choices for the entering variable, either can be selected and the simplex method can be continued. A similar comment holds for the departing variable.

Summary of the Simplex Method

1. Check whether the problem is formulated as an SMP. If it is, continue to step 2. If it is not, stop. Application of the method introduced here to problems which are not SMPs may require techniques not considered in this chapter.
2. Introduce slack variables.
3. Form the initial tableau.
4. Carry out pivot operations until a new pivot element cannot be determined.
5. If a pivot element cannot be determined because a departing variable cannot be selected, the problem has no solution. If a pivot element cannot be determined because an entering variable cannot be selected, the basic solution in the final tableau gives an optimal vector and the entry in the last row and the last column gives the maximum value of the objective function.

Exercises for Sec. 7-5

For each of the following tableaus find the entering variable, the departing variable, and the pivot element:

1. a.

	x	y	u	v	Basic Solution
u	1	4	1	0	12
v	5	4	0	1	20
p	-1	-2	0	0	0

b.

	x	y	u	v	Basic Solution
y	$\frac{1}{4}$	1	$\frac{1}{4}$	0	3
v	4	0	-1	1	8
p	$-\frac{1}{2}$	0	$\frac{1}{2}$	0	6

c.

	x	y	u	v	Basic Solution
y	0	1	$\frac{5}{16}$	$-\frac{1}{16}$	$\frac{5}{2}$
x	1	0	$-\frac{1}{4}$	$\frac{1}{4}$	2
p	0	0	$\frac{3}{8}$	$\frac{1}{8}$	7

2. a.

	x	y	z	u	v	Basic Solution
u	2	3	-4	1	0	20
v	-1	1	1	0	1	5
p	-1	-3	1	0	0	0

b.

	x	y	z	u	v	Basic Solution
u	5	0	-7	1	-3	5
y	-1	1	1	0	1	5
p	-4	0	4	0	3	15

3. Find an SMP which has the tableau of part **a** Exercise 1 as the initial tableau.

4. Find an SMP which has the tableau of part **a** Exercise 2 as the initial tableau.

5. Solve the linear programming problem identified in Exercise 4 by using the simplex method.

6. Solve the following SMP by using the simplex method.
   Find the maximum of $p = x + y$ for

$$x \geq 0 \qquad y \geq 0$$
$$5x + 5y \leq 20$$
$$2x + 10y \leq 40$$

7. Solve the following SMP by using the simplex method.
   Find the maximum of $p = x$ for

$$x \geq 0 \qquad y \geq 0$$
$$y \leq 5$$

8. Find the maximum of $p = 5x - 4y + 3z$ for

$$x \geq 0 \qquad y \geq 0 \qquad z \geq 0$$
$$5x + 5z \leq 100$$
$$5y - 5z \leq 50$$
$$5x - 5y \leq 50$$

7-6 DUAL PROGRAMMING PROBLEMS

Linear programming problems may occur as either maximum problems or minimum problems. Examples of both types are solved by geometrical methods in Chap. 5. In this chapter we have restricted our attention to maximum problems (actually SMPs); however, our methods are in fact more general than they might at first appear. The reason they are more general is that linear programming problems do not occur in isolation but arise in natural pairs. If one member of the pair is a maximum problem, e.g., to maximize profit, then the second member of the pair is a minimum problem, e.g., to minimize costs. These two problems are intimately related by virtue of the fact that they are phrased in terms of the same data; and, indeed, a complete study of either programming problem requires consideration of both problems in the pair. In this section we study the second member of the pair when the first member is an SMP. In particular we show how the simplex method solves both problems with the same set of tableaus. Also we shall give an example with an interpretation of the minimizing problem to indicate why this problem is important economically as well as mathematically.

In discussing how linear programming problems occur in natural pairs, it is very convenient to have a shorthand notation for writing such problems. The matrix notation developed in Chap. 6 in connection with systems of linear equations is equally appropriate for systems of linear inequalities and for linear programming problems.

The column vector $\mathbf{x}$ with n coordinates was defined in Sec. 6-3 as

$$\mathbf{x} = \begin{bmatrix} x_1 \\ x_2 \\ \vdots \\ x_n \end{bmatrix}$$

If n is 2 or 3, then we prefer to use $\mathbf{x} = \begin{bmatrix} x \\ y \end{bmatrix}$ or $\mathbf{x} = \begin{bmatrix} x \\ y \\ z \end{bmatrix}$ to avoid subscripts.

In the discussion which follows we phrase the definitions in terms of general n, that is, with vectors $\mathbf{x}$ with coordinates $x_1, x_2, \ldots, x_n$. The simplifications for $n = 2$ or $n = 3$ are direct.

In Sec. 6-3 we defined equality of vectors in terms of equality of coordinates. Our first task here is to take a similar approach to defining inequality for vectors.

Definition   Let
$$\mathbf{x} = \begin{bmatrix} x_1 \\ x_2 \\ \vdots \\ x_n \end{bmatrix} \quad \text{and} \quad \mathbf{y} = \begin{bmatrix} y_1 \\ y_2 \\ \vdots \\ y_n \end{bmatrix}$$

be two column n-vectors. The *inequality* $\mathbf{x} \leq \mathbf{y}$ means that each coordinate of $\mathbf{x}$ is less than or equal to the corresponding coordinate of $\mathbf{y}$. Thus, $\mathbf{x} \leq \mathbf{y}$ is equivalent to the n inequalities $x_1 \leq y_1, x_2 \leq y_2, \ldots, x_n \leq y_n$.

With this definition and the matrix-vector product defined in Sec. 6-3 we can easily write systems of inequalities in matrix notation. To illustrate the notation, consider the system

$$x \le 200$$
$$2x + 3y \le 1200$$
$$25x + 15y \le 7500$$

If we define the matrix $\mathbb{A}$ and vectors $\mathbf{x}$ and $\mathbf{b}$ as

$$\mathbb{A} = \begin{bmatrix} 1 & 0 \\ 2 & 3 \\ 25 & 15 \end{bmatrix} \qquad \mathbf{x} = \begin{bmatrix} x \\ y \end{bmatrix} \qquad \mathbf{b} = \begin{bmatrix} 200 \\ 1200 \\ 7500 \end{bmatrix}$$

this system can be written as $\mathbb{A}\mathbf{x} \le \mathbf{b}$.

Conceptually a vector is an ordered set of numbers, and it does not matter whether we write these numbers in a vertical column or a horizontal row. However in multiplying matrices and vectors together it is important that the sizes be such that the products are defined. Although it is rare for confusion to arise, any potential for misunderstanding can be eliminated by distinguishing notationally between vectors which are written as a column and as a row. The symbol $\mathbf{x}$ represents a column vector, and we simply modify the symbol to denote a row vector

Definition

$$\text{If} \qquad \mathbf{x} = \begin{bmatrix} x_1 \\ x_2 \\ \vdots \\ x_n \end{bmatrix} \qquad \text{then} \qquad \mathbf{x}^T = \begin{bmatrix} x_1 & x_2 & \cdots & x_n \end{bmatrix}$$

The notation $\mathbf{x}^T$ is read "the vector $\mathbf{x}$ transpose." There is a more general notion of the transpose of an arbitrary matrix, but we shall not make use of it.

If $\mathbf{x}$ and $\mathbf{w}$ are two n vectors, then $\mathbf{x}$ can be viewed as a $n \times 1$ matrix and $\mathbf{w}^T$ can be viewed as a $1 \times n$ matrix. Therefore, the matrix product $\mathbf{w}^T\mathbf{x}$ is defined, and

$$\mathbf{w}^T\mathbf{x} = w_1 x_1 + w_2 x_2 + \cdots + w_n x_n$$

For example, if $\mathbf{x}$ is a 4-vector and $\mathbf{w}^T = \begin{bmatrix} 1 & 2 & 4 & -1 \end{bmatrix}$, then $\mathbf{w}^T\mathbf{x} = x_1 + 2x_2 + 4x_3 - x_4$.

We are now in a position to write linear programming problems in matrix form.

EXAMPLE 7-8

Write the following SMP in matrix form.

Find the maximum of $p = 3x + y + 2z$ for

$$x \geq 0 \qquad y \geq 0 \qquad z \geq 0$$
$$2x + y \leq 20$$
$$x + 3y + 5z \leq 40$$
$$3x + 5y + z \leq 60$$

Solution: We define

$$\mathbb{A} = \begin{bmatrix} 2 & 1 & 0 \\ 1 & 3 & 5 \\ 3 & 5 & 1 \end{bmatrix} \qquad \mathbf{b} = \begin{bmatrix} 20 \\ 40 \\ 60 \end{bmatrix} \qquad \mathbf{c} = \begin{bmatrix} 3 \\ 1 \\ 2 \end{bmatrix}$$

The constraints (other than $x \geq 0$, $y \geq 0$, $z \geq 0$) can be written as $\mathbb{A}\mathbf{x} \leq \mathbf{b}$. We write the nonnegativity constraints on the variables as $\mathbf{x} \geq \mathbf{0}$, where $\mathbf{0}$ denotes the vector all of whose coordinates are zero. The linear function p can be written as $\mathbf{c}^T\mathbf{x}$. Consequently the SMP can be written in matrix form as

Find the maximum of $\mathbf{c}^T\mathbf{x}$ for $\mathbf{x} \geq \mathbf{0}$ and $\mathbb{A}\mathbf{x} \leq \mathbf{b}$

In general, for any $m \times n$ matrix $\mathbb{A}$, $\mathbf{b}$ a nonnegative m-vector, and $\mathbf{c}$ an n-vector, the SMP defined by $\mathbb{A}$, $\mathbf{b}$ and $\mathbf{c}$ is

Find the maximum of $\mathbf{c}^T\mathbf{x}$ for $\mathbf{x} \geq \mathbf{0}$ and $\mathbb{A}\mathbf{x} \leq \mathbf{b}$

We abbreviate "SMP defined by $\mathbb{A}$, $\mathbf{b}$, and $\mathbf{c}$" by writing SMP $[\mathbb{A}, \mathbf{b}, \mathbf{c}]$. Of course, if one is given the SMP in matrix form, the process can be reversed and it can be written in terms of linear inequalities (see Exercises 4 and 5).   □

The Dual Problem of an SMP

Consider the SMP $[\mathbb{A}, \mathbf{b}, \mathbf{c}]$. This is the problem

Find the maximum of $\mathbf{c}^T\mathbf{x}$ for $\mathbf{x} \geq \mathbf{0}$ and $\mathbb{A}\mathbf{x} \leq \mathbf{b}$ (7-5)

If $\mathbb{A}$ is an $m \times n$ matrix, it is understood that the vector $\mathbf{x}$ in (7-5) is to be an n-vector.

Using the same matrix $\mathbb{A}$ and the same vectors $\mathbf{b}$ and $\mathbf{c}$, we can also state a minimum problem:

$$\text{Find the minimum of} \quad \mathbf{z}^T\mathbf{b} \quad \text{for} \quad \mathbf{z} \geq \mathbf{0} \quad \text{and} \quad \mathbf{z}^T\mathbb{A} \geq \mathbf{c} \quad (7\text{-}6)$$

With the same assumption about the size of $\mathbb{A}$ the vector $\mathbf{z}$ in (7-6) must be an m-vector.

Problems (7-5) and (7-6) are said to be *dual linear programming problems*, and each problem is said to be the *dual* of the other.

EXAMPLE 7-9

Let

$$\mathbb{A} = \begin{bmatrix} 1 & 3 \\ 6 & 2 \end{bmatrix} \quad \mathbf{b} = \begin{bmatrix} 6 \\ 12 \end{bmatrix} \quad \text{and} \quad \mathbf{c} = \begin{bmatrix} 18 \\ 18 \end{bmatrix}$$

Solve the SMP $[\mathbb{A}, \mathbf{b}, \mathbf{c}]$ and its dual.

Solution: The SMP $[\mathbb{A}, \mathbf{b}, \mathbf{c}]$ is the problem of finding the maximum of $p = 18x + 18y$ for $x \geq 0$, $y \geq 0$ and

$$x + 3y \leq 6$$
$$6x + 2y \leq 12$$

Using the simplex method, we have the initial tableau shown in Tableau 7-18: the tableaus obtained by pivoting are shown in Tableaus 7-19 and 7-20.

Since Tableau 7-20 has no negative entries in the last row, we see that a solution to the SMP $[\mathbb{A}, \mathbf{b}, \mathbf{c}]$ is given by $x = \frac{3}{2}$, $y = \frac{3}{2}$. The maximum value of $p = 18x + 18y$ is 54.

Next we seek to solve the dual problem of SMP $[\mathbb{A}, \mathbf{b}, \mathbf{c}]$. Following the

TABLEAU 7-18

	x	y	u	v	Basic Solution
u	1	3	1	0	6
$\rightarrow v$	(6)	2	0	1	12
p	-18	-18	0	0	0

TABLEAU 7-19

	x	y	u	v	Basic Solution
$\rightarrow u$	0	$(\frac{8}{3})$	1	$-\frac{1}{6}$	4
x	1	$\frac{1}{3}$	0	$\frac{1}{6}$	2
p	0	-12	0	3	36

TABLEAU 7-20

	x	y	u	v	Basic Solution
y	0	1	$\frac{3}{8}$	$-\frac{1}{16}$	$\frac{3}{2}$
x	1	0	$-\frac{1}{8}$	$\frac{9}{48}$	$\frac{3}{2}$
p	0	0	$\frac{9}{2}$	$\frac{9}{4}$	54

definition given above, for $\mathbf{z} = (r,s)$ the dual problem is to find the minimum of $6r + 12s$ for $r \geq 0$, $s \geq 0$ and

$$r + 6s \geq 18$$
$$3r + 2s \geq 18$$

These constraints and the objective function are obtained by considering the variable vector $\mathbf{z} = (r,s)$ and the constraints $\mathbf{z} \geq \mathbf{0}$, $\mathbf{z}^T \mathbb{A} \geq \mathbf{c}$. The objective function is $\mathbf{z}^T \mathbf{b}$.

At the moment our only technique to solve this minimum problem is the geometrical method of Chap. 5. Using this technique, we graph the feasible set determined by the constraints, and we evaluate the objective function at the corner points of that set. The graph is shown in Fig. 7-2. The corner points are $(0,9)$, $(18,0)$ and $(\frac{9}{2}, \frac{9}{4})$. Evaluating the objective function $6r + 12s$ at the corner points, we find that the minimum occurs at $(r,s) = (\frac{9}{2}, \frac{9}{4})$. This minimum value is 54, and since the objective function never assumes negative values, it follows (from the theorem of Sec. 5-7) that 54 is the smallest possible value of the objective function on the feasible set. Thus we have solved the dual of the SMP $[\mathbb{A}, \mathbf{b}, \mathbf{c}]$. $\square$

In Example 7-9 we have used a geometrical technique which is difficult to apply when there are more than two variables in the problem, and we need another

FIGURE 7-2

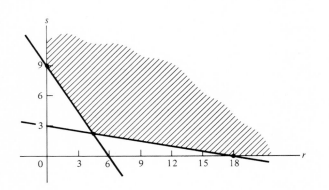

method to handle such problems. To this end we make the following observations about the solution of the SMP [$\mathbb{A}$,**b**,**c**] and the solution of the dual problem in Example 7-9.

1. The maximum value of the objective function in the original problem (SMP [$\mathbb{A}$,**b**,**c**]) is the same as the minimum value of the objective function in the dual problem. Both objective functions have an optimal value of 54.
2. The solution of the dual problem [both the optimal vector $(r, s) = (\frac{9}{2}, \frac{9}{4})$ and the optimal value 54] are displayed in the last row of the tableau which gives the solution of the original problem (SMP [$\mathbb{A}$,**b**,**c**]; see Tableau 7-20).

   Observations 1 and 2 indicate that we did not actually need to solve the dual problem by using the graphical approach from Chap. 5. At least for Example 7-9 the solution of the dual problem can be obtained by simply examining the final tableau obtained in using the simplex method to solve the original problem. Indeed, this example is not special in this regard, and dual problems always have the following two properties:

1. If one of the problems in a pair of dual linear programming problems has a solution, the other problem also has a solution. Moreover, the maximum value of the objective function in the maximum problem is equal to the minimum value of the objective function in the minimum problem.
2. In solving an SMP by the simplex method, if the method results in a tableau which yields an optimal vector for the SMP, this same tableau yields a solution for the dual problem. An optimal vector for the dual problem is displayed in the last row of the tableau under the entries which are the slack variables for the SMP. The value of the first variable in the dual problem is under the first slack variable, the value of the second variable in the dual problem is under the second slack variable, and so on for all the dual variables.

   To illustrate the significance of the dual problem in a practical setting, we consider again the SMP posed in Example 7-1.

EXAMPLE 7-10

As formulated in Example 7-1 the Scarumsilly Toy Company wishes to solve the following SMP.

Find the maximum of $p = x + 1.25y + 1.5z$ for $x \geq 0$, $y \geq 0$, $z \geq 0$ and

$$
\begin{aligned}
4x + 3y + 9z &\leq 160 \\
3x + 4y + z &\leq 50 \\
2x + 4y + 3z &\leq 50
\end{aligned}
\tag{7-7}
$$

TABLEAU 7-21

	x	y	$\downarrow$ z	u	v	w	**Basic Solution**
u	4	3	9	1	0	0	160
v	3	4	1	0	1	0	50
$\rightarrow w$	2	4	③	0	0	1	50
p	-1	$-\frac{5}{4}$	$-\frac{3}{2}$	0	0	0	0

We use the simplex method to solve this problem. The initial tableau is shown in Tableau 7-21, and the tableau after the first pivot is shown in Tableau 7-22.

TABLEAU 7-22

	x	y	z	u	v	w	**Basic Solution**
u	-2	-9	0	1	0	-3	10
v	$\frac{7}{3}$	$\frac{8}{3}$	0	0	1	$-\frac{1}{3}$	$\frac{100}{3}$
z	$\frac{2}{3}$	$\frac{4}{3}$	1	0	0	$\frac{1}{3}$	$\frac{50}{3}$
p	0	$\frac{3}{4}$	0	0	0	$\frac{1}{2}$	25

Tableau 7-22 has no negative entries in the last row; and hence, even though only one pivot operation has taken place, an optimal vector has been obtained. This vector has $x = 0$, $y = 0$, $z = \frac{50}{3}$, and the maximum value of p is 25. Recalling the definitions of x, y, and z, we see that in order to maximize profits the company should produce 0 Scary Harry dolls, 0 Horrible Harriet dolls, and $17\frac{1}{3}$ The Glob dolls per hour. In other words all resources should be used to produce The Glob dolls, and none should be used to produce other dolls. Moreover, the optimal vector had the values $u = 10$, $v = \frac{100}{3}$, and $w = 0$ for the slack variables. Thus, in producing only The Glob dolls there will be a surplus of plastic and a surplus of time for adding clothes but all possible time for adding special monster features will be used. In other words the company can increase production (and profits) by simply having more time available for adding special monster features. A reasonable question for the company to ask is: What is an acceptable price to pay for an increase in the time available for adding special features? In general to answer questions of this sort it is useful to consider the dual problem.

The dual problem here is the problem of finding the minimum of $160r + 50s + 50t$ for $r \geq 0$, $s \geq 0$, $t \geq 0$ and

$$4r + 3s + 2t \geq 1$$
$$3r + 4s + 4t \geq \tfrac{5}{4}$$
$$9r + s + 3t \geq \tfrac{3}{2}$$

The solution to this problem is given by Tableau 7-22. According to property 2, this solution is $r = 0$, $s = 0$, and $t = \frac{1}{2}$, and the minimum of $160r + 50s + 50t$ is 25. These numbers (and the dual problem itself) have an important interpretation in the context of the original SMP. We now turn to this topic. □

An Interpretation of the Dual Problem

Recall that in the SMP formulated in Example 7-1 the Scarumsilly Toy Company is interested in maximizing the profit obtained from their monster dolls. They manufactured these dolls in three steps, first molding plastic, then adding clothes, and finally adding special monster features. The profit (in dollars per hour) obtained from the dolls is

$$p = x + 1.25y + 1.5z$$

where x = number of Scary Harry dolls produced per hour

$\quad\quad y$ = number of Horrible Harriet dolls produced per hour

$\quad\quad z$ = number of The Glob dolls produced per hour

The constraints on these variables are given in the system (7-1), and the solution to the problem, given in Tableau 7-22, is $x = 0$, $y = 0$, and $z = \frac{50}{3}$. In this context, the dual problem with variables r, s, and t has the following interpretation and significance.

The optimal value of the variable r is the value which will be added to the company's product (the dolls) by the addition of 1 ounce more of plastic. Likewise the optimal values of the variables s and t represent the value added to the product per additional minute of time which is devoted to putting on clothes and adding special monster features, respectively. To indicate the importance of these values, suppose that 2 additional minutes are added to the time devoted to putting special monster features on the dolls. Then the constraints of system (7-7) become

$$
\begin{aligned}
4x + 3y + 9z &\le 160 \\
3x + 4y + z &\le 50 \\
2x + 4y + 3z &\le 52
\end{aligned}
\qquad (7\text{-}8)
$$

The final tableau which results when the new problem is solved by the simplex method is shown in Tableau 7-23. The optimal vector has coordinates $x = 0$, $y = 0$, $z = \frac{52}{3}$, $u = 4$, $v = \frac{98}{3}$, and $w = 0$. The maximum value of the profit function is $26 per hour. Thus, by increasing the amount of time available for adding

TABLEAU 7-23

	x	y	z	u	v	w	Basic Solution
u	-2	-9	0	1	0	-3	4
v	$\frac{7}{3}$	$\frac{8}{3}$	0	0	1	$-\frac{1}{3}$	$\frac{98}{3}$
z	$\frac{2}{3}$	$\frac{4}{3}$	1	0	0	$\frac{1}{3}$	$\frac{52}{3}$
p	0	$\frac{3}{4}$	0	0	0	$\frac{1}{2}$	26

special monster features from 50 to 52 minutes per hour, it is possible to increase profit from \$25 to \$26 per hour. In other words, if the company can arrange for additional minutes (perhaps by paying overtime rates), the return (or profit) is increased by 50¢ per minute added. If the cost of the additional time is more than 50¢ per minute, the company will not gain by the increase in production; if it is less than 50¢ per minute it will gain: and if the cost is exactly 50¢ per minute the company will come out even. Note that this critical value of 50¢ per minute is the *optimal value of t in the solution of the dual problem.*

Warning: It might appear from the discussion above that if the company could arrange for many more additional minutes of time for adding special features (perhaps by hiring a second person), they should do so whenever the cost is no more than 50¢ per minute. *This is not true. Additional time or resources should only be added at the rates determined by the dual problem if the solution to the dual problem does not change as a result of the addition.*

Note that the solution of the dual problem for the system (7-8) (as given in Tableau 7-23) is the same as the solution given in Tableau 7-22. Hence two additional minutes of "special feature" time can be added at any rate up to 50¢ per minute, and the company will gain (or at least not lose) by the addition. This is not true if large amounts of additional time are added at this rate. For example, if a second person is hired to work 24 minutes per hour adding special features at a cost of \$12 per hour, the company will lose money because profit is not increased by that amount (see Exercise 12).

Exercises for Sec. 7-6

1. Express the following SMP using matrix notation.
   Find the maximum of $x + y$ for

$$x \geq 0 \qquad y \geq 0$$
$$5x + 5y \leq 20$$
$$2x + 10y \leq 40$$

2. Express the following SMP using matrix notation.
   Find the maximum of $x + 3y - z$ for

$$x \geq 0 \qquad y \geq 0 \qquad z \geq 0$$
$$2x + 3y - 4z \leq 20$$
$$-x + y + z \leq 5$$

3. Express the following SMP using matrix notation.
   Find the maximum of $3x_1 + 4x_2 - x_3 + 4x_4$ for

$$x_1 \geq 0 \qquad x_2 \geq 0 \qquad x_3 \geq 0 \qquad x_4 \geq 0$$
$$x_1 + 2x_2 + 3x_3 + 4x_4 \leq 20$$
$$5x_1 + 6x_2 + 7x_3 + 8x_4 \leq 30$$
$$10x_1 - x_2 + x_3 + 10x_4 \leq 40$$
$$-x_1 + x_2 + 3x_3 + 7x_4 \leq 30$$

4. Using the given values of $\mathbb{A}$, $\mathbf{b}$, and $\mathbf{c}$, express each of the following SMP $[\mathbb{A}, \mathbf{b}, \mathbf{c}]$ in terms of linear inequalities:

   a. $\mathbb{A} = \begin{bmatrix} 2 & 2 \\ 3 & 6 \end{bmatrix} \qquad \mathbf{b} = \begin{bmatrix} 16 \\ 18 \end{bmatrix} \qquad \mathbf{c} = \begin{bmatrix} 3 \\ 8 \end{bmatrix}$

   b. $\mathbb{A} = \begin{bmatrix} 1 & -2 & 3 \\ -4 & 0 & 6 \end{bmatrix} \qquad \mathbf{b} = \begin{bmatrix} 20 \\ 30 \end{bmatrix} \qquad \mathbf{c} = \begin{bmatrix} 4 \\ -1 \\ 3 \end{bmatrix}$

5. Proceed as in Exercise 4 with the data

$$\mathbb{A} = \begin{bmatrix} 1 & 0 & -1 & 5 \\ -1 & 2 & 7 & 4 \\ 0 & 3 & -2 & 0 \\ 0 & 3 & 0 & 0 \\ 0 & -3 & 0 & 6 \end{bmatrix} \qquad \mathbf{b} = \begin{bmatrix} 25 \\ 50 \\ 20 \\ 24 \\ 30 \end{bmatrix} \qquad \mathbf{c} = \begin{bmatrix} 2 \\ -2 \\ 3 \\ 7 \end{bmatrix}$$

6. State the dual problem for the following SMP.
   Find the maximum of $p = 2x + y$ for $x \geq 0$, $y \geq 0$ and

$$x \leq 200$$
$$2x + y \leq 1200$$
$$25x + 15y \leq 7500$$

7. State the dual for the following problem.
   Find the minimum of $8x + 10y$ for $x \geq 0$, $y \geq 0$ and

$$x + y \geq 400$$
$$y \geq 200$$
$$2y - x \geq 0$$

8. State the dual problem for the following SMP.
Find the maximum of $x + y + z$ for $x \geq 0$, $y \geq 0$, $z \geq 0$ and

$$x + 4y + z \leq 1$$
$$4x + 7y + z \leq 1$$

9. Use the simplex method to solve the SMP of Exercise 6 and its dual problem.

10. Use the simplex method to solve the SMP of Exercise 7 and its dual problem.

11. Use the simplex method to solve the SMP of Exercise 8 and its dual problem.

12. Solve the SMP of Example 7-1 and its dual where the amount of time available for adding special monster features is increased from 50 to 74 minutes. Interpret the results of the dual problem.

Solve the SMP and its dual:

13. Exercise 1, Sec. 5-6   **14.** Exercise 9, Sec. 5-6   **15.** Exercise 3, Sec. 5-6

IMPORTANT TERMS

You should be able to describe, define, or give examples of each of the following:

Objective function
Feasible vector
Standard maximum problem (SMP)
Slack variable
Basic solution
Initial tableau
Pivot element

Entering variable
Departing variable
Pivot operation
Optimal vectors
Simplex method
The dual of an SMP

REVIEW EXERCISES

1. Using the definition of Sec. 7-2 decide whether the following problem is an SMP:
Find the maximum of $p = 4x - 2y + 3z$ for

$$x \geq 0 \qquad y \geq 0 \qquad z \geq 0$$
$$-x - y + 10z \leq 5$$
$$2x + 9y - z = -10$$

2. Formulate the following problem as a linear programming problem. The Trimetal Mining Company operates 2 mines from which they obtain gold, silver, and copper. Mine I costs $1000 per day to operate and it yields 30 ounces of gold, 1 ton of silver, and 4 tons of copper per day. Mine II costs $1200 per day to operate and it yields 40 ounces of gold, 2 tons of silver, and 3 tons of copper per day. The company has a contract to supply *at least* 240 ounces of gold, 10 tons of silver, and 12 tons of copper. How many days should each mine be operated so that the contract can be filled at minimum cost?

3. Formulate the following as a linear programming problem and find the initial tableau for the simplex method. The Fragrant Fertilizer Company carries three fertilizer brands: X, Y, and

Z. One sack (100 pounds) of each contains nitrogen, phosphorus, and potash in the amounts shown below. The profit per sack is shown at the right, and the available material is shown at the bottom. How many sacks of each brand should be produced to maximize profit?

Brand	Amount per Sack (pounds)			
	Nitrogen	Phosphorus	Potash	Profit
X	20	5	5	$10
Y	10	10	10	$ 5
Z	5	15	10	$15
Available	2000	1500	1500	

4. Find the pivot element in the following tableau and carry out the pivot.

	x	y	z	u	v	w	Basic Solution
u	50	100	25	1	0	0	100
v	100	25	−25	0	1	0	50
w	150	200	50	0	0	1	250
p	−50	25	−100	0	0	0	0

5. Use the simplex method to solve the following linear programming problem:

Find the maximum of $p = 5x - 2y$ for

$$x \geq 0 \qquad y \geq 0$$
$$x + y \leq 20$$
$$2x + 4y \leq 40$$
$$x \leq 15$$

6. Graph the feasible set for the problem of Exercise 5 and find all corner points. Show that the corner points are basic solutions for the linear programming problem.

7. Express the following linear programming in matrix notation:

Find the maximum of $x - y$ for

$$x \geq 0 \qquad y \geq 0$$
$$5x - 10y \leq 20$$
$$x + y \leq 5$$

8. Formulate the dual problem (in matrix notation) for the problem of Exercise 7.

9. Use the simplex method to solve the SMP of Exercise 7. What is the solution to the dual problem?

10. Formulate the dual of the problem in Exercise 3.

11. Solve the dual of the problem in Exercise 3.

PART THREE

APPLICATIONS

MARKOV
CHAINS

EIGHT

8-1   INTRODUCTION

Chapter 3 contains several examples of situations which can be modeled as a sequence of experiments, i.e., as a number of experiments which are performed one after the other. Frequently in such a sequence of experiments the probabilities of the various outcomes of one experiment depend upon the outcomes of preceding experiments. A production process in which items are produced and checked sequentially and an experiment in a psychology lab in which a mouse moves in a maze are situations which can be modeled as sequential experiments. Typically, the predictions based on such a model depend heavily on the detailed assumptions which underlie the model.

There are obviously many possibilities for the relationship between any two experiments in a sequence. The simplest assumption is that the outcomes of one experiment do not depend at all on the outcomes of the preceding experiments. This assumption results in a model based on an independent-trials process, and such models were used in Chap. 3 to make predictions for certain production processes. Chapter 3 also contained examples of situations in which the probabilities of the various outcomes of one experiment depend on the outcomes of all the previous experiments (Exercise 1 of Sec. 3-4). Of course there are many intermediate cases. One of special interest is that in which the outcome of one experiment depends on the outcome of the immediately preceding experiment but not on the outcomes of any other experiments. This case is of sufficient importance to have its own name. Processes in which the probabilities have this type of dependence are known as *Markov* * *chains*, and they and their applications have been studied extensively. The basic definitions and notation useful in studying Markov chains will be given in Sec. 8-2, and two important classes of Markov chains will be discussed in Secs. 8-3 and 8-4. We conclude this introduction with three examples to illustrate the nature of the processes to be studied. We shall return to each of these examples later in the chapter or in the exercises.

EXAMPLE 8-1

A year ago a survey of the employment status of the people employed in a small town showed that 30 percent of them were employed in industries, 50 percent were employed in small businesses, and 20 percent were self-employed. A similar survey this year shows that of those initially employed in industries, 70 percent of them still are, 20 percent are now employed in small businesses and 10 percent are self-employed. Also, of those initially employed in small businesses, 50 percent still are, 30 percent are now employed in industries, and 20 percent are self-employed. Finally, of those who were initially self-employed, 40 percent still are, and 30 percent are now employed in each of industries and small businesses. If it is assumed that the probabilities of shifts in employment are determined by current employment, what will be the distribution of individuals in these three types of employment 2 years after the initial survey and 10 years after the initial survey?   □

EXAMPLE 8-2

Many experiments concerned with learning or conditioning of animal subjects utilize the paradigm of a maze. We consider a very simple example of such a maze and the behavior of a subject, say a mouse in the maze. Suppose that the physical apparatus consists of a four-compartment maze, as shown in Fig. 8-1. The compartments have different levels of illumination, as indicated in the figure. The experiment is designed to measure the mouse's choice of compartments with various levels of illumination. It is initially released in one compartment, and its movements after that time are observed. On the basis of these observations hypotheses regarding the movements of the mouse can be formulated. Using these hypotheses, the future behavior of the mouse can be predicted. In particular, suppose that the behavior of the mouse depends only

---

* After the Russian mathematician A. A. Markov (1856–1922), who made many contributions to the development of probability theory and its applications.

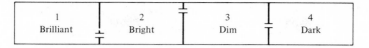

FIGURE 8-1

on the compartment it occupies at the time and not on its earlier movements. Suppose that the mouse is twice as likely to move toward a darker compartment as it is to move toward a lighter compartment. If the mouse is released in compartment 1, what is the probability that it is in compartment 3 after its seventh move? ◻

EXAMPLE 8-3

The service specialists of Metroburg Heating and Air-conditioning Company make their calls using a fleet of radio-dispatched service trucks. Established procedure is that when a call requesting service is received by the office, the dispatcher sends the next available truck to respond to the call. One of the service specialists feels that the dispatcher does not always follow this procedure, and, in particular, requires him to make an excessive number of crosstown trips. He decides to collect data, and if they support his feelings, to file a grievance with the union.

TABLE 8-1

District of Current Call	District of Next Call	Percent of Calls
East	East	30
	Central	20
	West	50
Central	East	20
	Central	50
	West	30
West	East	50
	Central	30
	West	20

The specialist begins by dividing the Metroburg area into three districts, East, Central, and West. Records of his calls for a 2-week period provide him with the information shown in Table 8-1. Also, the service specialist determines that the service calls received by the firm during the 2-week period originated equally from the three districts.

The specialist claims that the distribution of service calls is inconsistent with the frequency of his east-west trips across the city. Does he have a case?   □

8-2 BASIC CHARACTERISTICS OF MARKOV CHAINS

Situations of the sort which can be modeled using Markov chains (those described in Examples 8-1 to 8-3, for instance) can be characterized most conveniently with the aid of suitable notation and definitions. In order to have a vocabulary which can be used in many different situations, we introduce two of the basic components of Markov chains as undefined terms. In the applications these terms will have a specific meaning. We recall, as an example of a similar situation, that the term *set* introduced in Chap. 1 is never given a precise definition; i.e., it is used as a primitive or undefined concept.

This chapter will be concerned with systems which can be in any of N possible *states*. The systems are observed successively, and *transitions* between states are noted. The concepts of a state and of a system being in a state are taken as undefined terms. It is helpful to keep examples in mind. Recalling Example 8-3, we can view that system as having three states. It is in state 1 if the service specialist is in the east district of Metroburg, it is in state 2 if he is in the central district, and it is in state 3 if he is in the west district. Example 8-2 on the other hand, is naturally viewed as a system with four states. We say that the system is in state 1 if the mouse is in compartment 1, in state 2 if the mouse is in compartment 2, etc.

If the system is in state i at observation k and in state j at observation $k + 1$, we say that *the system has made a transition from state i to state j at the kth trial* (or stage or step) of the experiment. It may be that the system is in the same state, say state i, on two successive observations. In this case the transition is from state i to state i. For example, in Example 8-3, if the service specialist is in the east district of Metroburg at the first observation and in the central district on the second observation, we say that the system made a transition from state 1 to state 2 on the first trial of the experiment.

We suppose that the transitions of the system (a mouse in a maze, a serviceman in a city, or whatever) are unpredictable and the best we can do is to describe these transitions in probabilistic terms. Thus, we are interested in the probability that the system makes a transition from state i to state j on the kth trial of the experiment. We now make the fundamental assumption which distinguishes Markov chains from other sequential probabilistic processes:

The probability that the system makes a transition from state i to state j on the kth trial depends only on i and j and not on k, the trial of the experiment. This probability will be denoted by p_{ij}, $1 \leq i \leq N$, $1 \leq j \leq N$ (where N is the number of states).

One way of viewing this assumption in an intuitive way is to think of Markov chains as mathematical systems without memories. That is, the probability that the system makes a transition from one state (say state i) to another state (or even back into state i) depends only on the two states and not on the sequence of transitions through which the system arrived at state i. Thus, in Example 8-3, for instance, we assume that the likelihood of a service specialist in the east district being dispatched to the central district does not depend on how the service specialist reached the east district or on the time of day. It depends only on the fact that he is in the east district.

Another way of viewing this assumption is in terms of the tree diagrams used to describe sequential experiments in Chap. 3. Figure 8-2 provides a tree diagram for a sequential experiment in which the outcome at each stage of the experiment is either I or II. In this experiment we note that if the outcome is I on the first stage, the probability that the outcome is I on the second stage is $\frac{1}{4}$. And if the outcome is I on the second stage, the probability that the outcome is I on the third stage is $\frac{1}{5}$. We conclude that in this sequential experiment the probability of making the transition I to I on trial k varies with k. This probability is $\frac{1}{4}$ on the second trial ($k = 2$) and $\frac{1}{5}$ on the third trial ($k = 3$). Note that the same probability is $\frac{1}{3}$ on the first trial. We conclude that Fig. 8-2 is not the tree diagram of a Markov chain. On the other hand, Fig. 8-3 is the tree diagram of

FIGURE 8-2

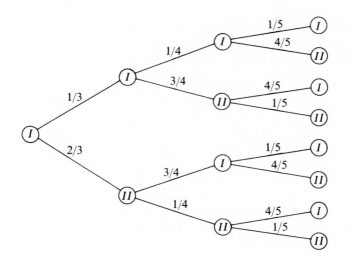

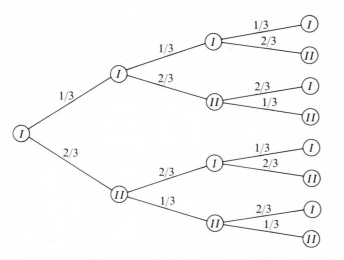

FIGURE 8-3

a Markov chain. The branch weights on the branches emanating from I are the same for any outcome I and the branch weights on the branches emanating from II are the same for any outcome II. If one views outcomes I and II as states, this means that the probability of going from one state (I or II) to any other state (I or II) depends only on the states involved, not on the stage of the process. Thus, for example, the probability of making a transition from state I to state II is always $\frac{2}{3}$ regardless of when the transition is made.

Since the transition probabilities in a Markov chain depend only on the state, the tree diagram for such an experiment can be simplified. We need only indicate the states and, on arrows connecting the states, the transition probabilities. Thus the transition probabilities shown in Fig. 8-3 can also be represented as shown in Fig. 8-4. Diagrams like that shown in Fig. 8-4 are known as *transition diagrams*.

FIGURE 8-4

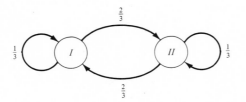

EXAMPLE 8-4 (CONTINUATION OF EXAMPLE 8-2)

Suppose that in Example 8-2 the mouse is observed every 2 minutes and every time it moves from one compartment to another. Construct a transition diagram to represent the assumption that the mouse is twice as likely to remain in the same compartment as to move and that all moves are equally likely.

Solution: We denote the states by 1, 2, 3, and 4. Since the mouse is observed every time it moves from one compartment to another, only transitions from a compartment to itself or to an immediately adjacent compartment are possible. Consider the process when it is in state 1. The mouse is twice as likely to remain in compartment 1 as it is to move, and if it moves, it must move to compartment 2. The weights associated with the two outcomes can be determined as in Chap. 2. Thus the probability of a transition from state 1 to state 1 is $\frac{2}{3}$ and from state 1 to state 2 is $\frac{1}{3}$. The remaining transition probabilities can be determined similarly. The transition diagram is shown in Fig. 8-5. □

FIGURE 8-5

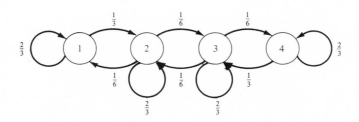

Although the transition diagram is a convenient means of representing transition probabilities, it is not especially suited to computation. For computational purposes it is preferable to organize the transition probabilities in a matrix. A Markov chain with N states has N^2 transition probabilities, which we denote by p_{ij}, $1 \le i \le N$, $1 \le j \le N$. The matrix $\mathbb{P}$, whose (i,j) element is p_{ij}, is the *transition matrix* for the Markov chain. Note that the (i,j) entry in $\mathbb{P}$ is the probability of a transition from state i to state j. We have

$$\mathbb{P} = \begin{bmatrix} p_{11} & p_{12} & \cdots & p_{1N} \\ p_{21} & p_{22} & \cdots & p_{2N} \\ \vdots & \vdots & & \vdots \\ p_{N1} & p_{N2} & \cdots & p_{NN} \end{bmatrix}$$

EXAMPLE 8-5

Find the transition matrix for the Markov chain whose transition diagram is given in Fig. 8-4.

Solution: There are two states (I and II), and therefore $N = 2$. We identify states I and II with 1 and 2, respectively. Then we have

$$\mathbb{P} = \begin{bmatrix} \frac{1}{3} & \frac{2}{3} \\ \frac{2}{3} & \frac{1}{3} \end{bmatrix}$$

□

In any particular situation, the transition probabilities p_{ij}, and consequently the transition matrix $\mathbb{P}$, depend upon what is assumed about the behavior of the system. In every case, however, the rows of the transition matrix must each have sum 1, because the entries in each row are the probabilities of all outcomes of a subexperiment. In terms of tree diagrams, each row of the matrix is a set of branch weights at one fork of the tree.

EXAMPLE 8-6 (CONTINUATION OF EXAMPLE 8-2)

Suppose that in Example 8-2 observations are taken every 2 minutes and whenever the mouse changes compartments. In addition, assume that the mouse is twice as likely to move to a darker compartment (if one exists) as it is to make any other move and other than that all options are equally attractive. Find the transition matrix for the Markov chain defined by these assumptions.

Solution: As in Example 8-4, the assumptions imply that there are positive probabilities only for transitions from a compartment to itself or to an adjacent compartment. The transition matrix $\mathbb{P}$ for the chain based on these assumptions is

$$\mathbb{P} = \begin{bmatrix} \frac{1}{3} & \frac{2}{3} & 0 & 0 \\ \frac{1}{4} & \frac{1}{4} & \frac{1}{2} & 0 \\ 0 & \frac{1}{4} & \frac{1}{4} & \frac{1}{2} \\ 0 & 0 & \frac{1}{2} & \frac{1}{2} \end{bmatrix}$$

□

The transition probabilities discussed thus far can be termed one-step transition probabilities since they are the probabilities of making a transition from one state to another in a single trial of the experiment. There are corresponding k-step transition probabilities which give the probability of making a transition from one state to another in k steps or trials of the experiment. Let us introduce the notation $p_{ij}(k)$ to denote the probability of making a transition from state i to state j in k steps or trials of the experiment. In order to illustrate the idea, consider the Markov chain with the transition diagram given in Fig. 8-4. The tree diagram for an experiment in which the system is initially in state I and makes three transitions is given in Fig. 8-3. From the

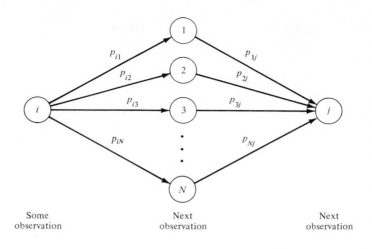

Some observation   Next observation   Next observation

FIGURE 8-6

tree diagram we can compute the probability that the system begins in state I and is in state I (state II) after three transitions. Computing the path weights by taking the products of branch weights, we have

$$p_{I,\,I}(3) = \tfrac{1}{3} \cdot \tfrac{1}{3} \cdot \tfrac{1}{3} + \tfrac{1}{3} \cdot \tfrac{2}{3} \cdot \tfrac{2}{3} + \tfrac{2}{3} \cdot \tfrac{2}{3} \cdot \tfrac{1}{3} + \tfrac{2}{3} \cdot \tfrac{1}{3} \cdot \tfrac{2}{3} = \tfrac{13}{27}$$

$$p_{I,\,II}(3) = \tfrac{1}{3} \cdot \tfrac{1}{3} \cdot \tfrac{2}{3} + \tfrac{1}{3} \cdot \tfrac{2}{3} \cdot \tfrac{1}{3} + \tfrac{2}{3} \cdot \tfrac{2}{3} \cdot \tfrac{2}{3} + \tfrac{2}{3} \cdot \tfrac{1}{3} \cdot \tfrac{1}{3} = \tfrac{14}{27}$$

We return now to the general case, and we consider the two-step transition probability $p_{ij}(2)$: *the probability that the system is in state i on one observation and in state j two observations later.* The system must be in some state on the intervening observation, and consequently we have the situation depicted in Fig. 8-6. The transition probabilities between the various states are shown on the lines connecting those states. The system can move from state i to state j in two steps by moving $i \to 1 \to j$. This happens with probability $p_{i1}p_{1j}$. Recall that the probability that the system makes a transition from state 1 to state j in one step is independent of the states it occupied before state 1. Likewise, the system can move from state i to state j through states 2, 3, ..., N. These events happen with probabilities $p_{i2}p_{2j}$, $p_{i3}p_{3j}$, ..., $p_{iN}p_{Nj}$, respectively. Since the system must move from state i to state j through exactly one intermediate state, we have

$$p_{ij}(2) = p_{i1}p_{1j} + p_{i2}p_{2j} + p_{i3}p_{3j} + \cdots + p_{iN}p_{Nj}$$

This expression for $p_{ij}(2)$ reminds us of the formula for the (i,j) element in the product of two matrices, and in fact, $p_{ij}(2)$ is exactly the (i,j) element in the matrix product $\mathbb{P}\mathbb{P}$ (see Sec. 6-4). Thus, if we introduce the notation $\mathbb{P}(2)$ to denote the matrix of two-step transition probabilities, we have the result

$$\mathbb{P}(2) = \mathbb{P}\mathbb{P}$$

This is a special case of the following more general result.

Theorem   Let $\mathbb{P}$ be the (one-step) transition matrix for a Markov chain. Then the matrix $\mathbb{P}(k)$ of k-step transition probabilities is given by

$$\mathbb{P}(k) = \underbrace{\mathbb{P}\mathbb{P}\cdots\mathbb{P}\mathbb{P}}_{k\ \textbf{Factors}} = \mathbb{P}^k \tag{8-1}$$

The theorem can be proved by using the special case $\mathbb{P}(2) = \mathbb{P}^2$ (which was verified above) and mathematical induction. The induction step is completed by an argument very similar to that used to establish the special case.

EXAMPLE 8-7

Find the matrix of two-step transition probabilities for the Markov chain defined in Example 8-6.

Solution:

$$\mathbb{P}(2) = \mathbb{P}^2 = \begin{bmatrix} \frac{1}{3} & \frac{2}{3} & 0 & 0 \\ \frac{1}{4} & \frac{1}{4} & \frac{1}{2} & 0 \\ 0 & \frac{1}{4} & \frac{1}{4} & \frac{1}{2} \\ 0 & 0 & \frac{1}{2} & \frac{1}{2} \end{bmatrix} \begin{bmatrix} \frac{1}{3} & \frac{2}{3} & 0 & 0 \\ \frac{1}{4} & \frac{1}{4} & \frac{1}{2} & 0 \\ 0 & \frac{1}{4} & \frac{1}{4} & \frac{1}{2} \\ 0 & 0 & \frac{1}{2} & \frac{1}{2} \end{bmatrix}$$

$$= \begin{bmatrix} \frac{10}{36} & \frac{14}{36} & \frac{12}{36} & 0 \\ \frac{7}{48} & \frac{17}{48} & \frac{12}{48} & \frac{12}{48} \\ \frac{1}{16} & \frac{2}{16} & \frac{7}{16} & \frac{6}{16} \\ 0 & \frac{1}{8} & \frac{3}{8} & \frac{4}{8} \end{bmatrix}$$

Thus, for example, the probability that the mouse moves from compartment 3 to compartment 2 in two steps is $\frac{2}{16}$, the (3,2) element in $\mathbb{P}(2)$. □

EXAMPLE 8-8

Find the matrix of three-step transition probabilities for the Markov chain whose transition diagram is given in Fig. 8-4.

Solution: From Example 8-5 we have

$$\mathbb{P} = \begin{bmatrix} \frac{1}{3} & \frac{2}{3} \\ \frac{2}{3} & \frac{1}{3} \end{bmatrix}$$

Therefore

$$\mathbb{P}(3) = \mathbb{P}^3 = \mathbb{P}\mathbb{P}\mathbb{P}$$

$$= \begin{bmatrix} \frac{1}{3} & \frac{2}{3} \\ \frac{2}{3} & \frac{1}{3} \end{bmatrix} \left(\begin{bmatrix} \frac{1}{3} & \frac{2}{3} \\ \frac{2}{3} & \frac{1}{3} \end{bmatrix} \begin{bmatrix} \frac{1}{3} & \frac{2}{3} \\ \frac{2}{3} & \frac{1}{3} \end{bmatrix} \right)$$

$$= \begin{bmatrix} \frac{1}{3} & \frac{2}{3} \\ \frac{2}{3} & \frac{1}{3} \end{bmatrix} \begin{bmatrix} \frac{5}{9} & \frac{4}{9} \\ \frac{4}{9} & \frac{5}{9} \end{bmatrix}$$

$$= \begin{bmatrix} \frac{13}{27} & \frac{14}{27} \\ \frac{14}{27} & \frac{13}{27} \end{bmatrix}$$

Notice that the (1,1) and (1,2) entries of $\mathbb{P}(3)$ are exactly the values of $p_{I, I}(3)$ and $p_{I, II}(3)$ computed earlier. $\qquad\square$

Since the entries $p_{i1}, p_{i2}, p_{i3}, \ldots, p_{iN}$ in the ith row of the transition matrix $\mathbb{P}$ are the probabilities that the system moves from state i to states 1, 2, 3, ..., N, respectively, and since in each trial the system must move from state i to some state (perhaps i itself), it follows that the ith row of the matrix $\mathbb{P}$ is a probability vector, i.e., consists of nonnegative numbers whose sum is 1. Since this is true for every i, we conclude that

Each row of a transition matrix $\mathbb{P}$ of a Markov chain is a probability vector.

Likewise, for each k the rows of the matrix $\mathbb{P}(k)$ are probability vectors. Indeed, the entries of the ith row of $\mathbb{P}(k)$ are the probabilities $p_{i1}(k)$, $p_{i2}(k)$, $p_{i3}(k)$, ..., $p_{iN}(k)$, and the system must move from state i to some state in k moves (perhaps i itself). It is easily verified that each of the rows of the matrices $\mathbb{P}$ of Examples 8-5 and 8-6 is a probability vector, that each of the rows of the matrix $\mathbb{P}(2)$ of Example 8-7 is a probability vector, and that each of the rows of the matrix $\mathbb{P}(3)$ of Example 8-8 is also a probability vector.

Exercises for Sec. 8-2

1. Find the transition diagram for the Markov chain defined in Example 8-6.

2. Find the transition matrix for the Markov chain defined in Example 8-4.

3. Let $\mathbb{P}$ (given below) be the transition matrix for a Markov chain and suppose that the system is initially in state 1.
   a. Draw the tree diagram for a sequential experiment consisting of three observations (including the first).
   b. Determine the path weights and the tree measure for this experiment.
   c. Determine the matrix $\mathbb{P}(2)$ using Eq. (8-1).
   d. Determine row one of the matrix $\mathbb{P}(2)$ by using the tree diagram and tree measure of parts **a** and **b**.

$$\mathbb{P} = \begin{bmatrix} \frac{1}{2} & \frac{1}{2} & 0 \\ \frac{1}{3} & 0 & \frac{2}{3} \\ \frac{1}{4} & \frac{1}{2} & \frac{1}{4} \end{bmatrix}$$

4. Determine $\mathbb{P}(2)$ and $\mathbb{P}(3)$ for the Markov chain with the transition matrix

$$\mathbb{P} = \begin{bmatrix} \frac{1}{4} & \frac{1}{4} & \frac{1}{2} \\ \frac{1}{4} & \frac{1}{2} & \frac{1}{4} \\ \frac{1}{2} & 0 & \frac{1}{2} \end{bmatrix}$$

Verify that the rows of $\mathbb{P}(2)$ and $\mathbb{P}(3)$ are probability vectors.

5. The situation described in Example 8-1 can be formulated as a Markov chain as follows. Consider an individual and say that the system is in state *I*, *SB*, or *SE* if the individual is employed in an industry, in a small business, or self-employed, respectively.
   a. Find the transition diagram for this Markov chain.
   b. Number the states 1, 2, and 3 (in any way you choose) and find the transition matrix $\mathbb{P}$ for the Markov chain.
   c. What is the probability that an individual who was employed in an industry at the time of the initial survey is employed in an industry 2 years later?

6. Formulate a Markov-chain model for the situation described in Example 8-3. Use the notion of state introduced at the beginning of Sec. 8-2 and the data of Example 8-3 to determine the transition matrix for the chain.

7. Suppose that in Example 8-4 every possible option for the mouse at any time in the experiment is assigned equal probability.
   a. What is the transition diagram for this Markov chain?
   b. What is the transition matrix for this Markov chain?
   c. What is the probability that the mouse moves from state 2 to state 3 in three steps?

8. A consumer purchases an American-built automobile every 2 years. She is twice as likely to purchase one from the same manufacturer as any other manufacturer; otherwise she is twice as likely to purchase from a major manufacturer (GM, Chrysler, Ford) as from AMC.
   a. Formulate a Markov-chain model for this situation and find the transition matrix.
   b. If the consumer bought a GM automobile 4 years ago, how likely is she to buy a Ford product this year?

8-3 REGULAR MARKOV CHAINS

In the preceding section we identified the basic property that distinguishes Markov chains from other stochastic processes. However, for many purposes the class of all Markov chains is too large, and within the class there are processes which behave in quite different ways. In this and the next section we shall examine briefly two subclasses of Markov chains. These subclasses illustrate in a certain sense the extremes of behavior of Markov chains.

It is helpful to introduce the concept of a state vector.

Definition

Consider a Markov chain with N states. A *state vector* for the Markov chain is a probability N-vector, that is a vector $\mathbf{x} = (x_1, x_2, ..., x_N)$ in which each x_i is non-negative and $x_1 + x_2 + \cdots + x_N = 1$. The ith coordinate x_i of the state vector $\mathbf{x}$ is to be interpreted as the probability that the system is in state i. By convention we also write a state vector as a row vector, that is, as a $1 \times N$ matrix. In particular, matrix notation is always used when the vector is displayed in a vector-matrix product.

EXAMPLE 8-9

The Markov chain defined in Example 8-6 has four states. The state vector $\mathbf{x} = (\frac{1}{3}, \frac{1}{2}, 0, \frac{1}{6})$ is to be interpreted as follows. With probability $\frac{1}{3}$ the mouse is in compartment 1, with probability $\frac{1}{2}$ it is in compartment 2, it is not in compartment 3, and with probability $\frac{1}{6}$ it is in compartment 4. $\qquad \square$

If the system is known to be in a specific state, the state vector has a particularly simple form. If the system is in the ith state, the ith coordinate of the state vector is 1 and all other coordinates are 0. For example, if in the Markov chain defined in Example 8-4 the system is known to be in state 1 (the mouse is in the first compartment), the state vector is $(1,0,0,0)$.

The behavior of a Markov chain can be described with a sequence of state vectors. The initial state of the system can be described with a state vector which we denote by $\mathbf{x}_0$. After one transition, the system can again be described with a state vector which we call $\mathbf{x}_1$. After two transitions it can be described by another state vector which we call $\mathbf{x}_2$. In general, after k transitions the system can be described by a state vector which we call $\mathbf{x}_k$. The relation between these state vectors can be represented with vector-matrix multiplication. We summarize the facts in a theorem.

Theorem

If $\mathbf{x}_k$ and $\mathbf{x}_{k+1}$ denote the state vectors which describe a Markov chain after k and $k + 1$ transitions, respectively, then $\mathbf{x}_{k+1} = \mathbf{x}_k \mathbb{P}$, where $\mathbb{P}$ is the transition matrix of the chain. In particular, $\mathbf{x}_1 = \mathbf{x}_0\,\mathbb{P}$, $\mathbf{x}_2 = \mathbf{x}_1\mathbb{P} = \mathbf{x}_0\,\mathbb{P}^2$, ..., and in general $\mathbf{x}_k = \mathbf{x}_0\,\mathbb{P}^k$. That is, the state vector $\mathbf{x}_k$ which describes the system after k transitions is the product of the initial state vector and the kth power of the transition matrix.

The theorem can be verified by examining each coordinate of the product $\mathbf{x}_k\,\mathbb{P}$ and using the meaning of $\mathbf{x}_k$ and $\mathbb{P}$. The conclusion is that the jth coordinate of $\mathbf{x}_k\,\mathbb{P}$ is exactly the probability that the system is in state j after $k + 1$ transitions, i.e., the jth coordinate of $\mathbf{x}_{k+1}$. Note that if the chain has N states, so that $\mathbf{x}_k$ is an N-vector and $\mathbb{P}$ is an $N \times N$ matrix, then $\mathbf{x}_k\,\mathbb{P}$ is also an N-vector.

EXAMPLE 8-10

If the initial state vector of the Markov chain defined in Example 8-6 is $\mathbf{x}_0 = (\frac{1}{3},\frac{1}{2},0,\frac{1}{6})$, find the state vector after one transition.

Solution: The transition matrix $\mathbb{P}$ for this Markov chain was determined in Example 8-6. Using the theorem, we find that the state vector after one transition is

$$\mathbf{x}_1 = \mathbf{x}_0\,\mathbb{P} = \begin{bmatrix} \frac{1}{3} & \frac{1}{2} & 0 & \frac{1}{6} \end{bmatrix}\begin{bmatrix} \frac{1}{3} & \frac{2}{3} & 0 & 0 \\ \frac{1}{4} & \frac{1}{4} & \frac{1}{2} & 0 \\ 0 & \frac{1}{4} & \frac{1}{4} & \frac{1}{2} \\ 0 & 0 & \frac{1}{2} & \frac{1}{2} \end{bmatrix}$$

$$= \begin{bmatrix} \frac{17}{72} & \frac{25}{72} & \frac{24}{72} & \frac{6}{72} \end{bmatrix} \qquad \square$$

We turn now to the primary concern of this section. There are several different ways of classifying Markov chains, and we choose one which distinguishes between chains on the basis of their long-run, or *asymptotic*, behavior. We interpret asymptotic behavior to mean the behavior of the k-step transition probabilities for large values of k. For example, if in Example 8-6 the mouse begins in compartment 1, what is the probability that it will be in compartment 1 after 1, 2, and 10 transitions? Those probabilities are $p_{11}(1) = p_{11}$, $p_{11}(2)$, and $p_{11}(10)$, respectively. The asymptotic behavior of the probabilities $p_{11}(k)$ gives the likelihood that if the mouse begins in compartment 1, it is in compartment 1 after many transitions.

In the general case of a Markov chain with N states the probability that the system is in the jth state after k trials depends upon the state in which it started. Thus $p_{1j}(k)$ is the probability that the system is in state j after k trials if it is initially in state 1. There are similar meanings attached to $p_{2j}(k)$, $p_{3j}(k)$, ..., and $p_{Nj}(k)$. There is no reason to have (or expect) equality among all these probabilities. However, for some Markov chains there is a positive probability

q_j associated with the jth state such that the k-step transition probabilities $p_{ij}(k)$ all become close to q_j for large k. That is, the likelihood that the system is in state j after k transitions is, for large k, nearly the same for all starting states. The precise definition is as follows:

Definition

A Markov chain is said to be *regular* if there is a positive probability vector $\mathbf{q} = (q_1, q_2, \ldots, q_N)$ such that for each state j the difference $|p_{ij}(k) - q_j|$ can be made as small as we choose by selecting k sufficiently large. The vector $\mathbf{q}$ is known as a *stable vector*, and its coordinates q_j are known as *stable probabilities* for the Markov chain.

It is important to note two aspects of this definition. First, the difference $|p_{ij}(k) - q_j|$ can be made as small as we choose *for all i* by selecting k sufficiently large; i.e., *the asymptotic behavior of $p_{ij}(k)$ is independent of the initial state.* Second, the probability vector $\mathbf{q}$ is to have strictly positive coordinates; that is, $q_j > 0$ for each j.

EXAMPLE 8-11

Consider a Markov chain whose transition matrix is

$$\mathbb{P} = \begin{bmatrix} \frac{1}{2} & \frac{1}{2} & 0 \\ 0 & \frac{1}{2} & \frac{1}{2} \\ \frac{1}{2} & \frac{1}{4} & \frac{1}{4} \end{bmatrix}$$

Then

$$\mathbb{P}(2) = \begin{bmatrix} \frac{4}{16} & \frac{8}{16} & \frac{4}{16} \\ \frac{4}{16} & \frac{6}{16} & \frac{6}{16} \\ \frac{6}{16} & \frac{7}{16} & \frac{3}{16} \end{bmatrix} \quad \text{and} \quad \mathbb{P}(3) = \begin{bmatrix} \frac{16}{64} & \frac{28}{64} & \frac{20}{64} \\ \frac{20}{64} & \frac{26}{64} & \frac{18}{64} \\ \frac{18}{64} & \frac{29}{64} & \frac{17}{64} \end{bmatrix}$$

Let us study the probabilities that the system is in state 3 after one, two, or three transitions. These probabilities form the third columns of the matrices $\mathbb{P}$, $\mathbb{P}(2)$, and $\mathbb{P}(3)$, respectively. They are

$$\begin{bmatrix} 0 \\ \frac{1}{2} \\ \frac{1}{4} \end{bmatrix} \quad \begin{bmatrix} \frac{4}{16} \\ \frac{6}{16} \\ \frac{3}{16} \end{bmatrix} \quad \begin{bmatrix} \frac{20}{64} \\ \frac{18}{64} \\ \frac{17}{64} \end{bmatrix}$$

The two-step transition probabilities are closer together than the one-step transition probabilities, and the three-step transition probabilities are still closer. Indeed, the probabilities that the system is in state 3 after three transitions from intial states 1, 2, and 3 differ from each other by at most $\frac{3}{64}$, whereas the differences after one transition can be as large as $\frac{1}{2}$. As we shall see in a moment, the probabilities $p_{13}(k)$, $p_{23}(k)$, and $p_{33}(k)$ all approach $\frac{2}{7}$ as k becomes large. □

The definition of a regular Markov chain has the following consequences. For each j and for k sufficiently large, each of the transition probabilities $p_{1j}(k)$, $p_{2j}(k)$, ..., $p_{Nj}(k)$ is close to q_j. That is, each of the entries in the jth column of the k-step transition matrix $\mathbb{P}(k)$ is close to q_j. Another way of saying this is that for large values of k the k-step transition matrix

$$\mathbb{P}(k) = \begin{bmatrix} p_{11}(k) & p_{12}(k) & \cdots & p_{1N}(k) \\ p_{21}(k) & p_{22}(k) & \cdots & p_{2N}(k) \\ \vdots & \vdots & & \vdots \\ p_{N1}(k) & p_{N2}(k) & \cdots & p_{NN}(k) \end{bmatrix}$$

is very close to the matrix $\mathbb{Q}$.

$$\mathbb{Q} = \begin{bmatrix} \mathbf{q} \\ \mathbf{q} \\ \vdots \\ \mathbf{q} \end{bmatrix} = \begin{bmatrix} q_1 & q_2 & \cdots & q_N \\ q_1 & q_2 & \cdots & q_N \\ \vdots & \vdots & & \vdots \\ q_1 & q_2 & \cdots & q_N \end{bmatrix}$$

Each of the rows of $\mathbb{Q}$ is the same, and each is the stable vector for the Markov chain.

EXAMPLE 8-12

Consider the transition matrix $\mathbb{P}$ of Example 8-6.

$$\mathbb{P} = \begin{bmatrix} \frac{1}{3} & \frac{2}{3} & 0 & 0 \\ \frac{1}{4} & \frac{1}{4} & \frac{1}{2} & 0 \\ 0 & \frac{1}{4} & \frac{1}{4} & \frac{1}{2} \\ 0 & 0 & \frac{1}{2} & \frac{1}{2} \end{bmatrix}$$

A straightforward computation of the k-step transition matrices (best carried out on a computer) gives

$$\mathbb{P}(4) = \begin{bmatrix} .155 & .287 & .336 & .222 \\ .108 & .245 & .340 & .307 \\ .063 & .170 & .384 & .383 \\ .042 & .154 & .383 & .422 \end{bmatrix} \qquad \mathbb{P}(8) = \begin{bmatrix} .085 & .206 & .364 & .345 \\ .077 & .200 & .368 & .359 \\ .068 & .184 & .373 & .375 \\ .065 & .180 & .375 & .381 \end{bmatrix}$$

and
$$\mathbb{P}(16) = \begin{bmatrix} .070 & .187 & .372 & .371 \\ .070 & .186 & .372 & .372 \\ .070 & .186 & .372 & .372 \\ .070 & .186 & .372 & .372 \end{bmatrix}$$

where the entries have been rounded off to the three decimal places shown. It is clear that the rows of $\mathbb{P}(16)$ are essentially equal. This illustrates the assertion that as k becomes large, the k-step transition matrix $\mathbb{P}(k)$ becomes close to a matrix $\mathbb{Q}$ all of whose rows are equal. $\qquad\square$

By definition, for a regular Markov chain the probabilities $p_{ij}(k)$ are for all large values of k nearly equal to stable probabilities q_j which are independent of the initial state i. The stable probabilities q_j can be obtained from the matrix $\mathbb{Q}$, which is closely approximated by $\mathbb{P}(k)$ for large values of k. However, obtaining $\mathbb{Q}$ from $\mathbb{P}(k)$ usually requires computing $\mathbb{P}^k$ for several values of k, an impractical method. Fortunately there is an alternative method of obtaining the stable probabilities.

Theorem

Let $\mathbb{P}$ be the transition matrix of a regular Markov chain. Then there is a unique probability vector $\mathbf{q}$ which satisfies $\mathbf{q}\mathbb{P} = \mathbf{q}$. The coordinates of this vector are the stable probabilities for the Markov chain.

This theorem (whose proof we omit) provides a direct method of obtaining the stable probabilities. Indeed, we need only solve a system of linear equations. The systems of linear equations which arise in this way always have infinitely many solutions, and we take the (unique) solution which is a probability vector.

EXAMPLE 8-13

The matrix

$$\mathbb{P} = \begin{bmatrix} \frac{1}{4} & \frac{3}{4} \\ \frac{2}{3} & \frac{1}{3} \end{bmatrix}$$

is the transition matrix of a regular Markov chain. Determine the vector $\mathbf{q}$ of stable probabilities for this Markov chain.

Solution: We make use of the theorem, and we solve the system

$$\mathbf{q}\mathbb{P} = \mathbf{q} \qquad \text{or equivalently} \qquad \mathbf{q}(\mathbb{P} - \mathbb{I}) = \mathbf{0}$$

If $\mathbf{q} = (q_1, q_2)$, the system $\mathbf{q}(\mathbb{P} - \mathbb{I}) = \mathbf{0}$ is

$$[q_1 \quad q_2]\begin{bmatrix} -\frac{3}{4} & \frac{3}{4} \\ \frac{2}{3} & -\frac{2}{3} \end{bmatrix} = \begin{bmatrix} 0 \\ 0 \end{bmatrix} \tag{8-2}$$

or

$$-\tfrac{3}{4}q_1 + \tfrac{2}{3}q_2 = 0$$
$$\tfrac{3}{4}q_1 - \tfrac{2}{3}q_2 = 0$$

Since one of these equations is the negative of the other, there are (as asserted above) infinitely many solutions to this system. Setting $q_1 = t$, we have

$$-\tfrac{3}{4}t + \tfrac{2}{3}q_2 = 0$$

or

$$q_2 = \frac{9t}{8}$$

Thus the vector $(t, 9t/8)$ is a solution of (8-2) for every value of t. We select that value of t for which the vector $(t, 9t/8)$ is a probability vector. The condition is that $t + (9t/8) = 1$ or $t = \frac{8}{17}$. Therefore, $\mathbf{q} = (\frac{8}{17}, \frac{9}{17})$ is the vector of stable probabilities for the Markov chain whose transition matrix is $\mathbb{P}$. $\qquad\square$

EXAMPLE 8-14 (CONTINUATION OF EXAMPLE 8-11)

Find the stable probabilities for the Markov chain whose transition matrix is

$$\mathbb{P} = \begin{bmatrix} \frac{1}{2} & \frac{1}{2} & 0 \\ 0 & \frac{1}{2} & \frac{1}{2} \\ \frac{1}{2} & \frac{1}{4} & \frac{1}{4} \end{bmatrix}$$

Solution: Since $N = 3$, there are three states; $\mathbf{q}$ is a 3-vector, $\mathbf{q} = (q_1, q_2, q_3)$. The system of equations $\mathbf{q}\mathbb{P} = \mathbf{q}$ or $\mathbf{q}(\mathbb{P} - \mathbb{I}) = \mathbf{0}$ is

$$[q_1 \quad q_2 \quad q_3]\begin{bmatrix} -\frac{1}{2} & \frac{1}{2} & 0 \\ 0 & -\frac{1}{2} & \frac{1}{2} \\ \frac{1}{2} & \frac{1}{4} & -\frac{3}{4} \end{bmatrix} = \begin{bmatrix} 0 \\ 0 \\ 0 \end{bmatrix} \tag{8-3}$$

The system (8-3) is

$$-\tfrac{1}{2}q_1 \qquad\quad + \tfrac{1}{2}q_3 = 0$$
$$\tfrac{1}{2}q_1 - \tfrac{1}{2}q_2 + \tfrac{1}{4}q_3 = 0$$
$$\tfrac{1}{2}q_2 - \tfrac{3}{4}q_3 = 0$$

and this system can be solved using the techniques of Sec. 6-3. There are infinitely many solutions of the form $(t, \frac{3}{2}t, t)$, where t can be any number. The solution which is a probability vector can be obtained by finding t such that $t + \frac{3}{2}t + t = \frac{7}{2}t = 1$ or $t = \frac{2}{7}$. The vector of stable probabilities is $(\frac{2}{7}, \frac{3}{7}, \frac{2}{7})$. This confirms the asertion made in Example 8-11 that as k becomes large, $p_{i3}(k)$ approaches $\frac{2}{7}$ for each initial state i, $i = 1, 2, 3$.   □

EXAMPLE 8-15   (CONTINUATION OF EXAMPLES 8-6 AND 8-12)

The Markov chain defined in Example 8-6 is regular. Find the vector of stable probabilities for this Markov chain.

Solution: The transition matrix for this Markov chain was determined in Example 8-6. It is

$$\mathbb{P} = \begin{bmatrix} \frac{1}{3} & \frac{2}{3} & 0 & 0 \\ \frac{1}{4} & \frac{1}{4} & \frac{1}{2} & 0 \\ 0 & \frac{1}{4} & \frac{1}{4} & \frac{1}{2} \\ 0 & 0 & \frac{1}{2} & \frac{1}{2} \end{bmatrix}$$

In order to compute the stable probabilities we solve the system $\mathbf{q}\,\mathbb{P} = \mathbf{q}$ with $\mathbf{q} = (q_1, q_2, q_3, q_4)$. We have

$$\mathbf{q}(\mathbb{P} - \mathbb{I}) = \begin{bmatrix} -\frac{2}{3}q_1 + \frac{1}{4}q_2 \\ \frac{2}{3}q_1 - \frac{3}{4}q_2 + \frac{1}{4}q_3 \\ \frac{1}{2}q_2 - \frac{3}{4}q_3 + \frac{1}{2}q_4 \\ \frac{1}{2}q_3 - \frac{1}{2}q_4 \end{bmatrix} = \begin{bmatrix} 0 \\ 0 \\ 0 \\ 0 \end{bmatrix} \tag{8-4}$$

Using the techniques of Sec. 6-3, we find that the vector

$$\left[t \quad \frac{8t}{3} \quad \frac{16t}{3} \quad \frac{16t}{3} \right]$$

is a solution of (8-4) for every value of t. We select t so that this vector is a probability vector, i.e., so that

$$t + \frac{8t}{3} + \frac{16t}{3} + \frac{16t}{3} = 1 \quad \text{or} \quad t = \frac{3}{43}$$

Therefore, the vector $\mathbf{q} = (\frac{3}{43}, \frac{8}{43}, \frac{16}{43}, \frac{16}{43})$ is the vector of stable probabilities for this

Markov chain. Expressed as decimals correct to three places, $\mathbf{q} = (.070, .186, .372, .372)$, which confirms the conclusions drawn in Example 8-12 based on $\mathbb{P}(16)$. $\qquad\qquad\square$

We now have a means of computing the vector of stable probabilities for any regular Markov chain. However, we do not yet have a means of distinguishing those Markov chains which are regular from those which are not. Fortunately, there is an easy way to determine whether a Markov chain is regular.

Theorem

A Markov chain with transition matrix $\mathbb{P}$ is regular if and only if there is a positive integer r such that the r-step transition matrix $\mathbb{P}(r)$ contains only positive entries.

Note that since the rows of $\mathbb{P}(k)$ are probability vectors, the entries in $\mathbb{P}(k)$ are automatically nonnegative for every positive integer k. This theorem states that the Markov chain is regular if (and only if) there is a positive integer r such that all the entries of $\mathbb{P}(r)$ are *positive*.

EXAMPLE 8-16

The transition matrix $\mathbb{P}$ of Example 8-6 is the transition matrix of a regular Markov chain since $\mathbb{P}(4)$ has only positive entries (Example 8-12). $\qquad\square$

Notice that in order to show that a Markov chain is regular it is sufficient to show that there is an integer r such that the r-step transition matrix has positive entries. It is unnecessary to determine the values of the entries, only whether they are positive or zero. There is a method of answering this less precise question which is simpler than actually computing the k-step transition matrices. We illustrate the technique with the transition matrix $\mathbb{P}$ of Example 8-6. If we adopt the convention that x and 0 denote positive and zero entries, respectively, the positivity of the entries of $\mathbb{P}(2)$ and $\mathbb{P}(3)$ can easily be determined. Since, in terms of positive entries,

$$\mathbb{P} = \begin{bmatrix} x & x & 0 & 0 \\ x & x & x & 0 \\ 0 & x & x & x \\ 0 & 0 & x & x \end{bmatrix}$$

we have, again in terms of positivity,

$$\mathbb{P}(2) = \mathbb{P}^2 = \begin{bmatrix} x & x & x & 0 \\ x & x & x & x \\ x & x & x & x \\ 0 & x & x & x \end{bmatrix} \quad \text{and} \quad \mathbb{P}(3) = \mathbb{P}^3 = \begin{bmatrix} x & x & x & x \\ x & x & x & x \\ x & x & x & x \\ x & x & x & x \end{bmatrix}$$

We conclude that $\mathbb{P}(3)$ contains only positive entries, and therefore $\mathbb{P}$ is the transition matrix of a regular Markov chain.

How many powers $\mathbb{P}^k$ might you be required to test to determine whether a Markov chain is regular? This question is answered by the fact that if $\mathbb{P}$ is the transition matrix for a regular Markov chain with N states, then one of the first $(N-1)^2 + 1$ powers of $\mathbb{P}$ contains only positive entries. [In most regular chains one has to test fewer than $(N-1)^2 + 1$ powers to find one with only positive entries.] Thus, if each of the first $(N-1)^2 + 1$ powers of the transition matrix contains at least one zero entry, the Markov chain is not regular.

Exercises for Sec. 8-3

1. Let $\mathbb{P}$ be the transition matrix of Example 8-6.
   a. If $\mathbf{x} = (0,1,0,0)$ is the initial state vector, find $\mathbf{x}_1$ and $\mathbf{x}_2$. Interpret $\mathbf{x}_2$ in the setting of Example 8-2.
   b. If $\mathbf{x}_0 = (\frac{1}{8}, \frac{1}{2}, \frac{1}{8}, \frac{1}{4})$ is the initial state vector, find $\mathbf{x}_1$.

2. A Markov-chain model for the situation described in Example 8-1 was formulated in Exercise 5 of Sec. 8-2. Find a state vector which describes the system:
   a. At the time of the initial survey
   b. 2 years after the initial survey

3. Find the vector of stable probabilities for the Markov chain whose transition matrix is

$$\begin{bmatrix} \frac{1}{3} & \frac{2}{3} \\ \frac{2}{3} & \frac{1}{3} \end{bmatrix}$$

4. Decide whether the Markov chain with transition matrix

$$\begin{bmatrix} 0 & 1 \\ \frac{1}{3} & \frac{2}{3} \end{bmatrix}$$

is regular. If it is, determine the vector of stable probabilities.

Repeat Exercise 4 for the transition matrix:

5. $\begin{bmatrix} 0 & 1 & 0 \\ 0 & 0 & 1 \\ \frac{1}{3} & 0 & \frac{2}{3} \end{bmatrix}$
   6. $\begin{bmatrix} 1 & 0 & 0 \\ \frac{1}{2} & \frac{1}{2} & 0 \\ 0 & \frac{2}{3} & \frac{1}{3} \end{bmatrix}$
   7. $\begin{bmatrix} 0 & \frac{1}{3} & 0 & \frac{2}{3} \\ \frac{1}{3} & 0 & \frac{2}{3} & 0 \\ 0 & \frac{1}{2} & 0 & \frac{1}{2} \\ \frac{1}{2} & 0 & \frac{1}{2} & 0 \end{bmatrix}$

8. Consider a Markov chain whose transition matrix is

$$\begin{bmatrix} \frac{1}{3} & 0 & \frac{2}{3} & 0 \\ \frac{1}{4} & 0 & \frac{3}{4} & 0 \\ \frac{1}{3} & \frac{1}{3} & 0 & \frac{1}{3} \\ 0 & \frac{1}{2} & 0 & \frac{1}{2} \end{bmatrix}$$

In which state is the system most likely to be found in the long run?

9. A Markov chain model for the situation described in Example 8-3 was formulated in Exercise 6 of Sec. 8-2.
   a. Determine whether the Markov chain is regular.
   b. If it is regular, find the stable probabilities.
   c. Use what you know about the situation to answer the question raised in Example 8-3.

10. (Continuation of Exercise 2) Let $\mathbb{P}$ be the transition matrix of the Markov chain model of Example 8-1 formulated in Exercise 5 of Sec. 8-2. Suppose that $\mathbb{P}(10)$ can be adequately approximated by the matrix each of whose rows is the vector of stable probabilities. If at the time of the initial survey employed individuals are distributed as described in Example 8-1, what is the distribution 10 years later?

8-4 ABSORBING MARKOV CHAINS

The behavior of a Markov chain can differ from that of the regular chains discussed in the preceding section in several ways. For example, since the stable probabilities of a regular Markov chain are positive, it follows that the k-step transition probabilities must all be positive for sufficiently large k. Another way of stating this fact is to say that for large enough k there is a positive probability that the system will move from one state to any other state in k transitions. One way in which this can fail to hold (and perhaps the most direct to observe) is for there to be a state from which transitions *are impossible*. This is illustrated by the next example.

EXAMPLE 8-17

Consider the maze introduced in Example 8-2 and suppose that the door into compartment 4 is one-way; i.e., it is possible to enter compartment 4 but impossible to leave it. If the other transitions are governed by the assumptions of Example 8-6, find the transition matrix of the resulting Markov chain.

Solution: Since the transitions from states 1, 2, and 3 are unaffected by the change to a one-way door on compartment 4, the first three rows of the transition matrix $\mathbb{P}$ are as in Example 8.6. The fourth row results from the fact that the only possible transition from state 4 is back into state 4. We have

$$\mathbb{P} = \begin{bmatrix} \frac{1}{3} & \frac{2}{3} & 0 & 0 \\ \frac{1}{4} & \frac{1}{4} & \frac{1}{2} & 0 \\ 0 & \frac{1}{4} & \frac{1}{4} & \frac{1}{2} \\ 0 & 0 & 0 & 1 \end{bmatrix} \qquad \square$$

The situation illustrated in this example is typical of a type of chain called an *absorbing Markov chain*. To make this concept precise we first identify the idea which was exemplified in the assumptions concerning compartment 4 of Example 8-17.

Definition   The ith state of a Markov chain is said to be an *absorbing state* if $p_{ii} = 1$ and $p_{ij} = 0$ for $j \neq i$. That is, state i is absorbing if the ith row of the transition matrix is the ith unit vector (a probability vector with a 1 as the ith coordinate).

In the Markov chain whose transition matrix is shown below the second state is absorbing. Note especially that the fifth state is not absorbing. Even though the fifth row contains a single 1, it is not in the fifth column; i.e., the fifth row is not the fifth unit vector.

$$\begin{bmatrix} \frac{1}{3} & 0 & \frac{1}{3} & 0 & \frac{1}{3} \\ 0 & 1 & 0 & 0 & 0 \\ \frac{1}{2} & 0 & \frac{1}{2} & 0 & 0 \\ \frac{1}{10} & \frac{2}{10} & \frac{3}{10} & \frac{4}{10} & 0 \\ 0 & 0 & 0 & 1 & 0 \end{bmatrix} \tag{8-5}$$

Absorbing chains must have absorbing states, but that is not enough. It must also be possible to go from nonabsorbing states to absorbing states.

Definition   A Markov chain is said to be *absorbing* if:

A. There is at least one absorbing state and
B. For every nonabsorbing state i there is some absorbing state j and a positive integer k such that the probability of a transition from state i to state j in k steps is positive.

Notice that there are two parts to the definition of an absorbing chain. A chain is not absorbing if *either* of these fails to hold. For example, the transition matrix (8-5) is the transition matrix of an absorbing chain. Indeed, condition A is satisfied since the second state is absorbing, and condition B is satisfied since transitions can occur from each of the other states to state 2, that is,

$$p_{12}(3) > 0 \qquad p_{32}(4) > 0, \qquad p_{42} > 0 \qquad \text{and} \qquad p_{52}(2) > 0$$

To verify this it is convenient to use the technique introduced at the end of Sec. 8-3 or the transition diagram for the chain. On the other hand, the transition matrix shown below is not the transition matrix for an absorbing Markov chain since condition B fails to hold for states 3 and 4.

$$\begin{bmatrix} \frac{1}{2} & \frac{1}{2} & 0 & 0 \\ 0 & 1 & 0 & 0 \\ 0 & 0 & \frac{1}{3} & \frac{2}{3} \\ 0 & 0 & \frac{3}{5} & \frac{2}{5} \end{bmatrix}$$

The analysis of situations modeled by absorbing Markov chains is facilitated if the transition matrices are written in a particular form. Since the absorbing states play a special role, it is conventional to collect them together and to label them as states 1 through k (we suppose that there are k absorbing states). The remaining states are labeled $k + 1$ through N. In relabeling states it is important to note that both the rows and columns of the transition matrix are altered. A transition matrix with the states labeled in this way is said to be in *canonical form*. A technique for writing a transition matrix in canonical form is illustrated in the next example.

EXAMPLE 8-18

Find a canonical form for the transition matrix (8-5).

Solution: The matrix as given is

$$\begin{bmatrix} \frac{1}{3} & 0 & \frac{1}{3} & 0 & \frac{1}{3} \\ 0 & 1 & 0 & 0 & 0 \\ \frac{1}{2} & 0 & \frac{1}{2} & 0 & 0 \\ \frac{1}{10} & \frac{2}{10} & \frac{3}{10} & \frac{4}{10} & 0 \\ 0 & 0 & 0 & 1 & 0 \end{bmatrix}$$

This matrix can be written in canonical form by relabeling states in such a way that the absorbing states (there is only one in this case) are listed first. The only absorbing state is state 2. Thus we relabel the states so that the original state 2 is relabeled as 1. The labels assigned to the remaining states are not important. We relabel the states as follows:

Old	1	2	3	4	5
New	2	1	3	4	5

After the states are relabeled, the transition matrix changes to reflect the new labels. For example the old entry p_{23} becomes the new entry p_{13}; the old p_{22}

becomes the new p_{11}, and so on. The transition matrix written in canonical form is

$$
\begin{array}{c}
 & \begin{array}{ccccc} \mathbf{2} & \mathbf{1} & \mathbf{3} & \mathbf{4} & \mathbf{5} \end{array} \\
\begin{array}{c} \mathbf{2} \\ \mathbf{1} \\ \mathbf{3} \\ \mathbf{4} \\ \mathbf{5} \end{array} &
\left[\begin{array}{ccccc}
1 & 0 & 0 & 0 & 0 \\
0 & \frac{1}{3} & \frac{1}{3} & 0 & \frac{1}{3} \\
0 & \frac{1}{2} & \frac{1}{2} & 0 & 0 \\
\frac{2}{10} & \frac{1}{10} & \frac{3}{10} & \frac{4}{10} & 0 \\
0 & 0 & 0 & 1 & 0
\end{array}\right]
\end{array}
$$

The labels beside the rows and above the columns are the original state labels. Thus the first row corresponds to relabeled state 1 which is state 2 in terms of the original transition matrix. ☐

EXAMPLE 8-19

Show that a transition matrix may have more than one canonical form by finding two canonical forms for the matrix of Example 8-17 by using different relabelings.

Solution: The transition matrix of Example 8-17 is

$$
\left[\begin{array}{cccc}
\frac{1}{3} & \frac{2}{3} & 0 & 0 \\
\frac{1}{4} & \frac{1}{4} & \frac{1}{2} & 0 \\
0 & \frac{1}{4} & \frac{1}{4} & \frac{1}{2} \\
0 & 0 & 0 & 1
\end{array}\right]
$$

The relabeling

Old	1	2	3	4
New	4	2	3	1

leads to the transition matrix in canonical form:

$$
\begin{array}{c}
 & \begin{array}{cccc} \mathbf{4} & \mathbf{2} & \mathbf{3} & \mathbf{1} \end{array} \\
\begin{array}{c} \mathbf{4} \\ \mathbf{2} \\ \mathbf{3} \\ \mathbf{1} \end{array} &
\left[\begin{array}{cccc}
1 & 0 & 0 & 0 \\
0 & \frac{1}{4} & \frac{1}{2} & \frac{1}{4} \\
\frac{1}{2} & \frac{1}{4} & \frac{1}{4} & 0 \\
0 & \frac{2}{3} & 0 & \frac{1}{3}
\end{array}\right]
\end{array}
$$

The relabeling

Old	1	2	3	4
New	2	3	4	1

can also be used. In this case the new transition matrix in canonical form is

$$
\begin{array}{c}
\quad\;\; \begin{matrix} \mathbf{4} & \mathbf{1} & \mathbf{2} & \mathbf{3} \end{matrix} \\
\begin{matrix} \mathbf{4} \\ \mathbf{1} \\ \mathbf{2} \\ \mathbf{3} \end{matrix}
\begin{bmatrix}
1 & 0 & 0 & 0 \\
0 & \frac{1}{3} & \frac{2}{3} & 0 \\
0 & \frac{1}{4} & \frac{1}{4} & \frac{1}{2} \\
\frac{1}{2} & 0 & \frac{1}{4} & \frac{1}{4}
\end{bmatrix}
\end{array}
\qquad (8\text{-}6)
$$

□

We conclude from Example 8-19 that the entries in a canonical form are not necessarily uniquely defined. However, the form of the matrix is unique. In particular, when written in canonical form, the transition matrix of an absorbing chain has a $k \times k$ identity matrix in the upper left-hand corner. Also the submatrix in the upper right-hand corner consists of all zeros. It is conventional to identify the submatrices in $\mathbb{P}$ as follows:

$$
\mathbb{P} = \begin{bmatrix} \mathbb{I} & \mathbb{O} \\ \mathbb{R} & \mathbb{Q} \end{bmatrix}
\qquad (8\text{-}7)
$$

Here $\mathbb{I}$ is a $k \times k$ identity matrix, $\mathbb{O}$ is a matrix with all zeros, and $\mathbb{R}$ and $\mathbb{Q}$ consist of transition probabilities which correspond to transitions which lead to absorption, $\mathbb{R}$, and transitions which do not lead to absorption, $\mathbb{Q}$.

In Example 8-19 the matrix (8-6) is in the form (8-7) with $\mathbb{I}$ a 1×1 identity matrix,

$$
\mathbb{R} = \begin{bmatrix} 0 \\ 0 \\ \frac{1}{2} \end{bmatrix} \quad \text{and} \quad \mathbb{Q} = \begin{bmatrix} \frac{1}{3} & \frac{2}{3} & 0 \\ \frac{1}{4} & \frac{1}{4} & \frac{1}{2} \\ 0 & \frac{1}{4} & \frac{1}{4} \end{bmatrix}
$$

In this example $N - k = 4 - 1 = 3$.

The matrix $\mathbb{Q}$ is always square, and for an identity matrix of the same size $\mathbb{I} - \mathbb{Q}$ is invertible.

Definition   Consider an absorbing Markov chain whose transition matrix is written in canonical form as (8-7). The matrix $\mathbb{N} = (\mathbb{I} - \mathbb{Q})^{-1}$ is called the *fundamental matrix* for the absorbing Markov chain. Since $\mathbb{Q}$ is $(N - k) \times (N - k)$, the identity matrix in the expression $\mathbb{I} - \mathbb{Q}$ is the same size, and so is $\mathbb{N}$.

The computation of $\mathbb{N}$ is illustrated in the next example.

EXAMPLE 8-20

Find the fundamental matrix for the absorbing Markov chain defined in Example 8-17.

Solution: The transition matrix for Example 8-17 is written in canonical form in (8-6). The matrix $\mathbb{Q}$ was determined above to be

$$\mathbb{Q} = \begin{bmatrix} \frac{1}{3} & \frac{2}{3} & 0 \\ \frac{1}{4} & \frac{1}{4} & \frac{1}{2} \\ 0 & \frac{1}{4} & \frac{1}{4} \end{bmatrix}$$

Therefore

$$\mathbb{I} - \mathbb{Q} = \begin{bmatrix} 1 & 0 & 0 \\ 0 & 1 & 0 \\ 0 & 0 & 1 \end{bmatrix} - \begin{bmatrix} \frac{1}{3} & \frac{2}{3} & 0 \\ \frac{1}{4} & \frac{1}{4} & \frac{1}{2} \\ 0 & \frac{1}{4} & \frac{1}{4} \end{bmatrix} = \begin{bmatrix} \frac{2}{3} & -\frac{2}{3} & 0 \\ -\frac{1}{4} & \frac{3}{4} & -\frac{1}{2} \\ 0 & -\frac{1}{4} & \frac{3}{4} \end{bmatrix}$$

The techniques introduced in Chap. 6 can be used to determine $\mathbb{N} = (\mathbb{I} - \mathbb{Q})^{-1}$. We find

$$\mathbb{N} = \begin{bmatrix} \frac{21}{8} & 3 & 2 \\ \frac{9}{8} & 3 & 2 \\ \frac{3}{8} & 1 & 2 \end{bmatrix}$$

The significance of the entries in $\mathbb{N}$ will be discussed after the next example. □

EXAMPLE 8-21

Find the fundamental matrix for the absorbing Markov chain whose transition matrix is

$$\begin{bmatrix} 1 & 0 & 0 & 0 \\ 0 & 1 & 0 & 0 \\ \frac{1}{3} & 0 & \frac{2}{3} & 0 \\ 0 & \frac{1}{5} & \frac{2}{5} & \frac{2}{5} \end{bmatrix} \tag{8-8}$$

Solution: The transition matrix is already written in canonical form. The matrix $\mathbb{Q}$ is $\begin{bmatrix} \frac{2}{3} & 0 \\ \frac{2}{5} & \frac{2}{5} \end{bmatrix}$, the matrix $\mathbb{I} - \mathbb{Q}$ is $\begin{bmatrix} \frac{1}{3} & 0 \\ -\frac{2}{5} & \frac{3}{5} \end{bmatrix}$, and the fundamental matrix $\mathbb{N}$ is $\begin{bmatrix} 3 & 0 \\ 2 & \frac{5}{3} \end{bmatrix}$. Again, $\mathbb{N}$ is determined by using the techniques of Chap. 6. □

Let us now consider the long-run behavior of an absorbing Markov chain. Since there is a positive probability that the system moves from each nonabsorbing state to an absorbing state in some number of transitions, we expect that sooner or later it reaches an absorbing state. Of course, once the system reaches an absorbing state, it does not leave it. However, this is a stochastic process, and we do not know exactly when the system will reach an absorbing state. As usual, in such circumstances we turn to the notion of expected value to help us describe the behavior of the process. It turns out that the fundamental matrix is very useful in this respect. In fact, it fully deserves its name.

In what follows we consider an absorbing Markov chain with transition matrix $\mathbb{P}$ written in canonical form (8-7) with absorbing states numbered 1 through k and nonabsorbing states numbered $k + 1$ through N. We refer to state $k + 1$ as the first nonabsorbing state, state $k + 2$ as the second nonabsorbing state, etc. The rows and columns of $\mathbb{Q}$ are identified with the nonabsorbing states in order. This identification carries over to the fundamental matrix $\mathbb{N}$.

Our results are contained in the following theorem, stated here without proof.

Theorem   The (i,j) entry in the fundamental matrix $\mathbb{N}$ gives the expected number of times that a system which begins in the ith nonabsorbing state will be in the jth nonabsorbing state before it reaches an absorbing state. The sum of the entries in the ith row of $\mathbb{N}$ gives the expected number of transitions of a system which begins in the ith nonabsorbing state and continues until it reaches an absorbing state.

EXAMPLE 8-22 (CONTINUATION OF EXAMPLE 8-21)

Suppose that the absorbing Markov chain whose transition matrix is (8-8) is initially in state 3, the first nonabsorbing state. Find the expected number of transitions before the system reaches an absorbing state.

Solution: The fundamental matrix $\mathbb{N}$ is $\begin{bmatrix} 3 & 0 \\ 2 & \frac{5}{3} \end{bmatrix}$. The expected number of times the system is in state 3 (the first nonabsorbing state) before absorption is 3, and the expected number of times it is in state 4 (the second nonabsorbing state) is 0. Therefore, the expected number of transitions before the system reaches an absorbing state is $3 + 0 = 3$.   □

EXAMPLE 8-23 (CONTINUATION OF EXAMPLE 8-17)

In the situation described in Example 8-17 if the system is initially in state 1, what is the expected number of transitions before it reaches an absorbing state?

Solution: If the states are relabeled as

Old	1	2	3	4
New	2	3	4	1

then the fundamental matrix $\mathbb{N}$ (as computed in Example 8-20) is

$$\mathbb{N} = \begin{bmatrix} \frac{21}{8} & 3 & 2 \\ \frac{9}{8} & 3 & 2 \\ \frac{3}{8} & 1 & 2 \end{bmatrix}$$

With the specified relabeling original state 1 becomes new state 2, the first nonabsorbing state. If the system begins in this state, the expected number of times it occupies the first nonabsorbing state before reaching an absorbing state is $\frac{21}{8}$; the expected number of times it occupies the second and third nonabsorbing states are 3 and 2, respectively. Therefore, the expected total number of transitions before it reaches an absorbing state is

$$\frac{21}{8} + 3 + 2 = \frac{61}{8} \qquad \square$$

It is useful to analyze Example 8-23 in terms of the original situation (the maze with the one-way door described in Example 8-17). With the relabeling introduced in Example 8-23 the first nonabsorbing state corresponds to compartment 1. Indeed, the first nonabsorbing state in the matrix (8-6) is state 2 of the system as described with this transition matrix, and this corresponds to the first state before relabeling. This state is associated with the first compartment. Thus, in terms of the original situation, this means that if the mouse begins in compartment 1, the expected number of transitions before it reaches compartment 4, the one with a one-way door, is $\frac{61}{8}$. Naturally, in any specific case the number of transitions may be more or less than $\frac{61}{8}$. This number is simply an average over many trials.

Notice that this method can be used to answer the question: If the mouse begins in compartment 1 of the original maze (no one-way doors), what is the expected number of transitions before *it first reaches compartment 4*? We simply imagine the original maze to be altered to one with a one-way door on compartment 4. The resulting system can be modeled as an absorbing Markov chain, and the expected number of transitions before absorption (given that the system begins in the nonabsorbing state corresponding to the first compartment) provides an answer to the question.

The fundamental matrix can also be used to obtain other types of information about the system. For example, if the system has a single absorbing state, the system will eventually reach that state, but if there is more than one absorbing state, it may in general be absorbed in any one of them. Given the state in which the system begins, the likelihood that it will be absorbed in various absorbing states can be computed using the fundamental matrix. A method for carrying out the computation is described in Exercise 10.

Exercises for Sec. 8-4

1. A transition matrix of a Markov chain is given as

$$\begin{bmatrix} 1 & 0 & 0 & 0 \\ 0 & \frac{1}{2} & \frac{1}{4} & \frac{1}{4} \\ 0 & 0 & 1 & 0 \\ 0 & \frac{1}{3} & 0 & \frac{2}{3} \end{bmatrix}$$

   a. Determine whether it is the transition matrix of an absorbing Markov chain.
   b. If not, which of the conditions of the definition of an absorbing chain is violated?
   c. If so, determine for each nonabsorbing state the minimum number of transitions necessary to reach some absorbing state.

Repeat Exercise 1 with the transition matrix:

2. $$\begin{bmatrix} 1 & 0 & 0 & 0 \\ \frac{1}{3} & 0 & \frac{2}{3} & 0 \\ 0 & 1 & 0 & 0 \\ 0 & 0 & 1 & 0 \end{bmatrix}$$ 3. $$\begin{bmatrix} 1 & 0 & 0 & 0 \\ 0 & 1 & 0 & 0 \\ 0 & 0 & \frac{1}{3} & \frac{2}{3} \\ 0 & 0 & 1 & 0 \end{bmatrix}$$

4. Write the transition matrix $\begin{bmatrix} 0 & \frac{1}{2} & \frac{1}{2} \\ 0 & 1 & 0 \\ \frac{1}{3} & \frac{1}{3} & \frac{1}{3} \end{bmatrix}$ in canonical form and find the associated fundamental matrix.

Repeat Exercise 4 for the transition matrix:

5. $$\begin{bmatrix} \frac{1}{3} & \frac{2}{3} & 0 & 0 \\ 0 & \frac{1}{2} & \frac{1}{2} & 0 \\ 0 & 0 & 1 & 0 \\ 1 & 0 & 0 & 0 \end{bmatrix}$$ 6. $$\begin{bmatrix} 1 & 0 & 0 \\ 0 & 1 & 0 \\ \frac{1}{3} & 0 & \frac{2}{3} \end{bmatrix}$$

7. Suppose that the absorbing Markov chain whose transition matrix is (8-8) is initially in state 4. Find the expected number of transitions before it reaches an absorbing state. (*Hint:* Use Example 8-22.)

8. In the situation described in Example 8-17, suppose the mouse begins in compartment 2. What is the expected number of times it is in compartment 1 before it reaches compartment 4? What is the expected total number of transitions before it reaches compartment 4 for the first time?

9. Consider the maze introduced in Example 8-2 and suppose that each possible option of the mouse at any point in the experiment is selected with equal probability. If the mouse begins in compartment 1, what is the expected number of transitions before it first reaches compartment 4?

10. The matrix $\mathbb{R}$ in the canonical form $\begin{bmatrix} \mathbb{I} & \mathbb{O} \\ \mathbb{R} & \mathbb{Q} \end{bmatrix}$ is an $(N - k) \times k$ matrix. Thus, the matrix product $\mathbb{NR}$ is also $(N - k) \times k$. Each of the rows of $\mathbb{NR}$ is associated with a nonabsorbing state, and each of the columns is associated with an absorbing state. Each of the rows of $\mathbb{NR}$ is a probability vector. The (i,j) entry of $\mathbb{NR}$ is the probability that if the system begins in the ith nonabsorbing state, it will be absorbed in the jth absorbing state.

   a. Suppose that the Markov chain with transition matrix (8-8) begins in the first nonabsorbing state. What is the probability that it is absorbed in the first absorbing state?

   b. Suppose that this Markov chain begins in the second nonabsorbing state. What is the probability that it is absorbed in the first absorbing state?

IMPORTANT TERMS

You should be able to describe, define, or give examples of each of the following:

State
Transition
Transition diagram
Transition matrix
Markov chain
State vector
Regular Markov chain

Asymptotic behavior
Stable probabilities
Absorbing state
Absorbing Markov chain
Canonical form
Fundamental matrix

REVIEW EXERCISES

1. A Markov chain has the transition matrix

$$\begin{bmatrix} \frac{1}{3} & \frac{1}{6} & \frac{1}{2} \\ \frac{1}{4} & \frac{3}{8} & \frac{3}{8} \\ \frac{1}{2} & 0 & \frac{1}{2} \end{bmatrix}$$

   a. Find the transition diagram for this Markov chain.
   b. Find the matrix of two-step transition probabilities.
   c. If the initial state vector is $(\frac{3}{5}, \frac{1}{5}, \frac{1}{5})$, what is the state vector after two transitions?

2. A Markov chain which has the transition matrix

$$\begin{bmatrix} .3 & .2 & .5 \\ .1 & .6 & .3 \\ .4 & .3 & .3 \end{bmatrix}$$

is initially in state 1. In what state is it most likely to be after three transitions?

3. A Markov chain has the transition matrix

$$\begin{bmatrix} \frac{1}{4} & 0 & 0 & \frac{3}{4} \\ \frac{1}{2} & \frac{1}{2} & 0 & 0 \\ 0 & \frac{1}{4} & \frac{1}{4} & \frac{1}{4} \\ 0 & 0 & \frac{1}{2} & \frac{1}{2} \end{bmatrix}$$

   a. Determine whether this Markov chain is regular.
   b. If it is regular, find the vector of stable probabilities.

4. Find an example of a regular Markov chain with three states for which the transition matrices $\mathbb{P}$, $\mathbb{P}(2)$, $\mathbb{P}(3)$, and $\mathbb{P}(4)$ all contain at least one zero entry.

5. The transition matrix of a Markov chain is

$$\begin{bmatrix} 1 & 0 & 0 & 0 \\ \frac{1}{2} & \frac{1}{2} & 0 & 0 \\ 0 & 0 & 1 & 0 \\ 0 & 0 & \frac{2}{3} & \frac{1}{3} \end{bmatrix}$$

a. Verify that it is the transition matrix of an absorbing Markov chain.

b. Determine the expected number of transitions to absorption for each initial state.

6. A Markov chain has the transition matrix

$$\begin{bmatrix} \frac{1}{4} & 0 & \frac{1}{4} & \frac{1}{2} \\ 0 & 1 & 0 & 0 \\ \frac{1}{3} & \frac{1}{3} & \frac{1}{3} & 0 \\ 0 & 0 & 0 & 1 \end{bmatrix}$$

If the system is initially in state 3, determine the expected number of transitions before it reaches an absorbing state.

7. Suppose that the Markov chain with the transition matrix given in Exercise 3 is initially in state 3. What is the expected number of transitions before it reaches state 1?

8. Suppose that the Markov chain with the transition matrix given in Exercise 2 is initially in state 1. What is the expected number of times it will be in state 2 before it reaches state 3 for the first time?

9. Suppose that a mouse moves in a random way in the maze shown in Fig. 8-7. In particular, suppose that the mouse moves in such a way that

1. It is twice as likely to make a move as to say put.
2. If it makes a move, it is twice as likely to move clockwise as counterclockwise.

Observations are made every minute and each time the mouse changes compartments.

a. Model this as a Markov chain and find the transition matrix.

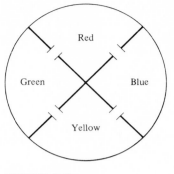

FIGURE 8-7

b. If the mouse begins in the red compartment, what is the expected number of transitions before it first reaches the yellow compartment?

10. The Highly Independent Consumers United Program (HICUP) surveyed 1000 people immediately after each had purchased a digestive aid. Four aids were considered, brands A, B, C, and X, and the probabilities of switching from one brand to another were determined to be

Probability of Switching	To			
	A	B	C	X
From A		.3	.2	.3
B	.1		.2	.3
C	.1	.1		.2
X	.3	.2	.2	

Assume that this situation can be modeled as a Markov chain and determine:

a. The probability that an A user will be a B user after two switches.

b. Which users (A, B, or C) will become brand X users in the minimum expected number of switches.

c. The anticipated long-run distribution of users.

TWO-PERSON ZERO-SUM GAMES

NINE

9-1 INTRODUCTION

Uncertainty is an inescapable part of the world in which we live and of the business and social communities whose operation influences our lives. We all make decisions in circumstances where uncertainty plays a significant role. The consequences of our decisions are determined not only by our own choices but also by chance and by the decisions made by others. In this chapter we introduce mathematical techniques which can be used to help in the analysis of situations involving chance, choice, and competition. Numerous everyday situations exhibit these properties, e.g., the selection of the best route across town in Friday afternoon rush-hour traffic. Also, these characteristics (chance, choice, and competition) are frequently present in such games as poker and bridge. In poker, for instance, a person's gain depends on chance (the cards dealt to him), on choice (his bets), and on competition (his opponents' bets and play). In view of the close association of the word *game* with situations involving chance, choice, and competition we shall use this word as a general term to describe any such situation. We hope to show by examples that our techniques are useful in a much broader setting than that of parlor games.

We begin our study with two simple examples which illustrate some of the characteristics of games. These examples will be analyzed later in the chapter. Next, we consider some of the general properties of games, and we continue with a development of certain mathematical techniques which are useful in solving games. Finally, we apply these ideas and techniques to the analysis of several examples.

9-2 TWO SIMPLE GAMES

We begin our discussion with two simple examples. The first has an especially simple structure which, as we shall see in the next section, permits a straight-forward analysis. The second game is slightly more complex, but because of this complexity it exhibits more of the features of general games. We use it to introduce many of the concepts which play an important role in game theory.

EXAMPLE 9-1 THE INVESTMENT GAME

Two investment-advisory services, Risehi and Callup, are competing for the right to manage a part of the assets of a large pension fund. The managers of the pension fund have decided that they will assign funds to the two advisory services in the following way. At the beginning of the competition each advisory service will make investment decisions for 50 percent of the assets of the pension fund. On the first day of each investment quarter, the two services must deliver their advice for the upcoming quarter. Their advice must be one of the following: "buy bonds" or "buy stocks". If in a given quarter both services give the same advice, both give equally correct or incorrect advice, or the prices of bonds and stocks stay essentially the same, no change will be made in the percentage of assets allocated to each service. However, if the services disagree and one is right (makes a good recommendation) and the other is wrong, the service which gives the better recommendation increases its share of the assets by 10 percent of the total and the other service loses the same amount.

Question: Suppose both services know that if a market change occurs in the present economic climate, stocks will go up and bond prices will go down with a probability .6 while the reverse situation will occur with probability .4. How should each service choose its advice? That is, how often should each say "buy stocks" and how often "buy bonds"? □

EXAMPLE 9-2 THE WORD-MATCHING GAME

Rolf is resting under a Chestnut tree when his friend Chester joins him and suggests that they match coins for fun and (perhaps) profit. Rolf expresses a dislike for strenuous exercise, and he suggests that they simultaneously state the words *head* or *tail* and make the payments accordingly. In Rolf's opinion the payoffs should be as follows: if they match at heads, Chester pays him 40¢; if they match at tails, Chester pays him 25¢; and if they fail to match, Chester pays him 10¢. Chester is not overly impressed with this proposal, but after some thought he agrees to play the game provided Rolf pays him 20¢ before the start of each play.

Question 1: Is this game fair to both players?

Question 2: How should the players play the game? That is, how should each player choose between heads and tails? □

We shall analyze these games and answer the questions in a specific sense later in the chapter. First, however, it is useful to discuss these examples with the objective of understanding the general attributes of games better. Also, we can begin to make precise our meaning of the term to *solve* a game.

In order to discuss the word-matching game between Rolf and Chester it is convenient to introduce some notation and terminology. We refer to the participants as *players* and the amounts which the players gain (or lose) as a result of playing the game one time as *payoffs*. In certain simple games involving only two players it is customary to represent the plays and payoffs in a payoff matrix. Such a matrix is shown in Fig. 9-1. The rows of this matrix correspond to the two options available to Rolf (the row player), and the columns correspond to the two choices available to Chester (the column player). The entries in the matrix are the payoffs between Rolf and Chester. These payoffs include the 20¢ that Rolf pays Chester before each play of the game.

Several features of this game should be noted since they can be used to characterize an important class of games. First, there are two players. Second, each player must choose one of two options at each play of the game, and he must choose without knowledge of the choice of his opponent. Also, after the

FIGURE 9-1

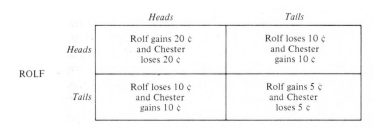

CHESTER

		Heads	Tails
ROLF	Heads	Rolf gains 20 ¢ and Chester loses 20 ¢	Rolf loses 10 ¢ and Chester gains 10 ¢
	Tails	Rolf loses 10 ¢ and Chester gains 10 ¢	Rolf gains 5 ¢ and Chester loses 5 ¢

players make their choices, there is an exchange between them to effect the payoff determined by their choices. Thus, the gain of one player is always equal to the loss of the other player. We shall see in the next section that these characteristics—the number of players, the method of play, and the method of payoffs—are used to classify games. We shall also see that the investment game and the game between Rolf and Chester are unusually simple ones, and it is possible to analyze them completely using only elementary mathematical ideas. On the other hand, if the game becomes more complicated, e.g., by increasing the number of players or by changing the method of payoff, it becomes increasingly difficult to solve the game in a meaningful way.

Since Rolf and Chester both know the payoffs resulting from different choices by each of them, it stands to reason that they will select one of their options (words) in such a way as to achieve their aims. But what are their aims? Here we must make an assumption about the objectives of Rolf and Chester in playing the game. *We assume that each is motivated by a desire to maximize his overall gain.* On a certain play of the game Rolf may win (show a net profit), and on another play of the game Chester may win. Thus, on any specific play of the game each player could show a profit or a loss. Of course, each player would like to win on every play, but in this game (assuming both players are reasonably intelligent) there is no method of selecting words that will ensure a win on every play to either player. In such circumstances the natural quantity for each player to consider and to attempt to maximize is his *average net gain per play of the game.* Thus, although each may sometimes win and sometimes lose, he would like to play in such a way that over many plays of the game his net gain will be as large as possible.

An initial consideration of Fig. 9-1 might indicate that Rolf should always choose the word *heads*, i.e., say heads, since this gives him a chance to achieve his greatest net gain. This gain occurs when Chester plays (says) heads, and it amounts to 20¢ since Rolf receives 40¢ from Chester and he has already paid 20¢ to Chester. However, if Rolf always plays heads, Chester will always play tails and Rolf will consistently show a net loss of 10¢. On the other hand, if Rolf always plays tails, Chester will always play heads and once again Rolf will show a loss of 10¢. Of course, the answer to this dilemma for Rolf is obvious. He should mix up his choices, sometimes playing heads and sometimes playing tails. But exactly how should they be mixed? This is the central question for Rolf and also for Chester since the reasoning is exactly the same for him. To discuss this problem of finding the best method of mixing choices properly we need a more formal mathematical setting than that of the present discussion. We provide this setting in the next section.

Exercises for Sec. 9-2

1. In the Rolf-Chester game, suppose each player plays heads one half of the time and tails the other half of the time. What is the expected value of the gain to Rolf per game? (*Hint:* Decide how often each payoff will occur and compute the expected value as defined in Sec. 3-7.)

2. Form a payoff matrix to represent the investment game (Example 9-1). Use the expected percent of assets to be shifted from one firm to the other as the payoff.

3. In each of the following areas describe a situation which has the characteristics of a game. In each case list the options open to each player and note the payoffs to each player for the different outcomes. It is not necessary that the sum of the payoffs for a particular outcome be zero, as it is in the investment game and the Rolf-Chester word-matching game.
   a. Labor-management negotiations for a contract
   b. Political parties campaigning in an election

4. Suppose that in the Rolf-Chester word-matching game the payoffs are changed so that when the players match, Rolf pays Chester 10¢ and when they do not match, Chester pays Rolf 20¢. Also suppose that Chester pays Rolf 5¢ at the start of each play of the game. Form a matrix which is similar to that of Fig. 9-1 to represent this new version of the game.

5. (Continuation of Exercise 4) Suppose that in the game of Exercise 4 both Rolf and Chester play heads and tails at random and each with probability $\frac{1}{2}$. What is the average return per game to each player?

9-3 SADDLE POINTS AND DOMINANCE

Each of the examples in Sec. 9-2 involves two players; each player has a fixed finite number of options (two in these cases) available on each play; and on each play of the game the gain of one player is the loss of another player. Such a situation is an example of a *two-person zero-sum* game. In general, zero-sum is used to describe a game with the property that the sum of the payoffs to the players on each play of the game is zero. In the examples of Sec. 9-2 there are two payoffs, and since one is the negative of the other, the sum is always 0. A precise definition is the following:

Definition   A *two-person zero-sum game* consists of two (finite) sets, $R = \{r_1, \ldots, r_m\}$ and $C = \{c_1, \ldots, c_n\}$, together with a payoff function p which assigns a number, $p(r_i,c_j)$, to each pair (r_i,c_j). The elements of the set R are called the *pure strategies for the row player* and the elements of the set C are called the *pure strategies for the column player*. The number $p(r_i,c_j)$ is the *payoff to the row player* for the pair of strategies (r_i,c_j) and $-p(r_i,c_j)$ is the corresponding *payoff to the column player*.

The restriction to games in which the sets R and C are finite enables us to use a convenient matrix representation for two-person zero-sum games. We begin by ordering the elements in the sets R and C, and we suppose that this order is preserved throughout the discussion. We use subscripts both to identify and to order the elements of R and C. Next, an $m \times n$ matrix is formed by defining the entry in the ith row and jth column of the matrix to be $p(r_i, c_j)$, for $i = 1, 2, \ldots, m$ and $j = 1, 2, \ldots, n$. The matrix $\mathbb{P}$, obtained in this way is called the *payoff matrix* for the game. In many cases the options corresponding to the rows and columns of the payoff matrix will not be labeled explicitly, and they are identified only by the order implied by the matrix.

The word-matching game played by Rolf and Chester is a convenient one to use to illustrate our definition of a two-person zero-sum game. The sets R and C consist of the options or pure strategies which are available to Rolf and Chester, respectively. In this case both players have the same two options [they must play heads (H) or tails (T) on each play of the game]. Thus the sets R and C are $R = C = \{H, T\}$. The entries in the matrix $\mathbb{P}$ can be determined from the description of the game given in Example 9-2, and they were noted in Fig. 9-1. Thus, the payoff matrix is

$$
\mathbb{P} = \begin{array}{c} \\ H \\ T \end{array} \begin{array}{c} H \qquad T \\ \left[\begin{array}{rr} 20 & -10 \\ -10 & 5 \end{array} \right] \end{array}
$$

In the matrix $\mathbb{P}$ the entry 20 is $p(H,H)$, the payoff to Rolf when he says heads and Chester says heads and we include the 20¢ that Rolf pays Chester at the start of each play. The other entries in the matrix have a similar interpretation. Note that $p(H,T)$ is negative since the net payoff is from Rolf to Chester.

The process of representing two-person zero-sum games by matrices is a reversible process. Any matrix with real numbers as entries can be viewed as the payoff matrix for a two-person zero-sum game. The entry in the ith row and the jth column of the matrix is the payoff to the row player when he uses his ith strategy and his opponent (the column player) uses his jth strategy. The negative of this entry is the return or the payoff to the column player. The exact nature of the strategies for each player is not known, nor is it important. It is possible to analyze, and in one sense solve, a two-person zero-sum game based only on the knowledge of the payoff matrix.

We have used the word *solve* in regard to games a number of times, and since we are in the process of making our discussion more precise, we should be more careful about what we mean by *solve a game* and a *solution of a game*. As we noted in our discussion of the word-matching game, it is natural for both Rolf and Chester to play by mixing their choice of words and not by saying the same word every time. This situation often occurs in two-person zero-sum games, but it does not always occur. In some games the "best" strategy is to select the same option or pure strategy every time. As an indication of this type of game and what it means to call a strategy best, we consider the following simple example.

EXAMPLE 9-3

Let the matrix below be the payoff matrix for a two-person zero-sum game. Decide how the row and column players should play the game.

$$\mathbb{P} = \begin{bmatrix} 1 & -1 & 2 \\ 0 & 1 & 1 \\ 0 & -2 & 1 \\ 2 & 1 & 3 \end{bmatrix}$$

Solution: Recall that the entries in this matrix represent payoffs *to* the row player *from* the column player. Note that there are four options or pure strategies for the row player and three pure strategies for the column player. We proceed by inspecting the payoff matrix with the goal of determining which pure strategies are most attractive to each player. Since we are concerned only with pure strategies in this example—indeed we have not yet defined any other type of strategy—let us use the term *strategy* here to refer to a pure strategy. We note that if the row player uses strategy (row) 4, then for any choice of strategy by the column player the payoff to the row player is at least as high as if he used any other strategy, and if the column player uses strategy (column) 3, the payoff to the row player is actually greater with strategy 4 than with any other strategy. Thus the row player can be assured of a maximum return against any column strategy by simply playing r_4 every time. On the other hand, the column player can also see that the row player should always play r_4 and thus the column player can minimize losses by always playing c_2. Hence, in this simple (and somewhat dull) game each player should always use the same strategy and the payoff will always be $+1$ to the row player. In a sense we have solved this game by determining a method of play for each player which in the long run is as good or better than any other method of play. Also we know the average gain (or loss) for each player if both players use these best methods of play. ☐

It is interesting that in Example 9-3 a knowledge of one's opponent's intentions cannot be used to reduce his payoff. Thus, if the row player announces in advance that he will use strategy r_4, the column player cannot use this information to reduce the payoff to the row player. Similarly, if the column player announces in advance that he plans to use strategy c_2, the row player cannot use this information to increase his own payoff, i.e., reduce the payoff to the column player. We shall contrast this state of affairs with a very different one in the next section.

The situation which occurs in Example 9-3 is an example of a game which is strictly determined. It has a saddle point. To be specific:

Definition   A two-person zero-sum game has a *saddle point* at row i and column j if the (i,j) entry in the payoff matrix is both the minimum of the ith row and the maximum of the jth column. Any game with a saddle point is *strictly determined.*

Let us reconsider Example 9-3 from a somewhat different point of view. In addition to providing a perspective for our work, this viewpoint will give a method of identifying saddle points when they exist. We begin by asking: What is the worst payoff that the row player can receive when he plays strategy 1? An inspection of the payoff matrix shows that the minimum payoff the row player can receive when he plays strategy 1 is -1. Likewise, 0 is the minimum payoff the row player can receive if he plays strategy 2; and -2 and 1 are the minimum payoffs for strategies 3 and 4, respectively. Write the row minimums to the right of the payoff matrix as shown below.

$$\begin{bmatrix} 1 & -1 & 2 \\ 0 & 1 & 1 \\ 0 & -2 & 1 \\ 2 & 1 & 3 \end{bmatrix} \quad \begin{array}{c} \textbf{Row} \\ \textbf{Minimums} \\ \mathbf{-1} \\ \mathbf{0} \\ \mathbf{-2} \\ \mathbf{1} \end{array}$$

In a similar manner we can analyze the options available to the column player, remembering that higher payoffs are undesirable from the column player's point of view (since they represent what he must pay out). The column maximums, the worst that can happen to the column player, are listed below the payoff matrix.

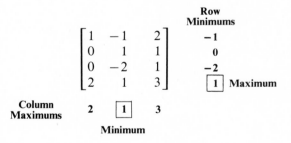

Next, the row player observes that if he selects strategy r_4, he is guaranteed a payoff of at least 1 unit and this is the best he can guarantee himself. Also, the column player observes that if he plays strategy c_2, he is guaranteed a loss of not more than 1 unit, and this is the best he can guarantee himself. In this example these two numbers are equal, and the (4,2) entry in the matrix is by definition a saddle point. The game is strictly determined since each player should always play the same (pure) strategy. There is no choice to be made at each play of the game. Thus if the row player seeks to maximize his average winning per game, then his best strategy is to play r_4 every time. Likewise, the column player's best strategy is to play c_2 every time. It is in this sense that the game is strictly determined.

To Find a Saddle Point

The procedure illustrated above is a general one which can always be used to find saddle points. The method is as follows:

> **Identify the row minimums (list them to the right of the payoff matrix for easy reference) and the column maximums (list them below the payoff matrix). Check whether the maximum of the row minimums is equal to the minimum of the column maximums. If so, there is at least one saddle point. If not, there are no saddle points. If there is a saddle point, there is one in a row corresponding to a maximum row minimum and in a column corresponding to a minimum column maximum.**

This method is illustrated in the next example.

EXAMPLE 9-4

Payoff matrices $\mathbb{P}_1$ and $\mathbb{P}_2$ for two games are shown below. Determine whether each game is strictly determined; i.e., decide whether there is a saddle point for each payoff matrix.

$$\mathbb{P}_1 = \begin{bmatrix} 1 & -1 & 2 \\ 1 & 1 & 3 \\ -1 & 1 & -2 \\ 3 & -2 & 3 \end{bmatrix} \qquad \mathbb{P}_2 = \begin{bmatrix} 1 & -1 & 2 \\ 1 & 1 & 3 \\ -1 & 1 & -2 \\ 3 & 2 & -3 \end{bmatrix}$$

Solution: The row minimums and column maximums for $\mathbb{P}_1$ are shown below:

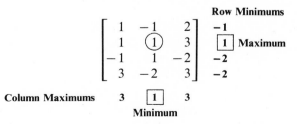

The maximum of the row minimums and the minimum of the column maximums, shown in boxes, are both 1. Thus there is a saddle point, the circled entry in row 2 and column 2. In this matrix there is only one saddle point. Note that other entries have the value 1, but in this matrix they are not saddle points. It is possible, however, to have more than one saddle point in a payoff matrix (see Exercise 1). The game with payoff matrix $\mathbb{P}_1$ is strictly determined. The best strategy for the row player is to always play r_2, and the best strategy for the column player is to always play c_2.

The row minimums, column maximums, the maximum row minimum, and the minimum column maximum for the matrix $\mathbb{P}_2$ are shown below:

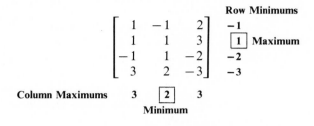

In this matrix the maximum of the row minimums (1) is not equal to the minimum of the column maximum (2), and there is no saddle point.   □

Even though the matrix $\mathbb{P}_2$ of Example 9-4 does not have a saddle point, it is possible to deduce some information about the best way to play the game by simply inspecting the matrix. For example, from the row player's point of view, pure strategies r_1 and r_3 should never be played because pure strategy r_2 is always at least as good as r_1 and r_3 and sometimes it is better (in terms of the payoff to the row player). Thus in deciding how to play this game, the row player can ignore the rows r_1 and r_3 and simply consider the matrix

$$\begin{matrix} & c_1 & c_2 & c_3 \\ r_2 & \begin{bmatrix} 1 & 1 & 3 \\ r_4 & 3 & 2 & -3 \end{bmatrix} \end{matrix}$$

At this stage, however, the column player can also simplify the game (we assume the players are equally intelligent, so that the column player knows the row player is not going to play r_1 or r_3). The column player observes that pure strategy c_2 is always a better play than c_1 since it never results in a larger payoff to the row player and it sometimes results in a smaller payoff. Thus the column player should never use c_1, and it can be eliminated from further consideration. Taking these observations into account, the matrix to be considered has the form

$$\begin{matrix} & c_2 & c_3 \\ r_2 & \begin{bmatrix} 1 & 3 \\ r_4 & 2 & -3 \end{bmatrix} \end{matrix} \tag{9-1}$$

At this stage no further simplifications are possible, and the game is similar to the word-matching game in that the best results cannot be achieved by using pure strategies. It is necessary to mix pure strategies in an appropriate way. The problem of determining the proper mix will be discussed in the next section.

The simplifications which were used to convert matrix $\mathbb{P}_2$ of Example 9-4 to matrix (9-1) are called *simplifications due to dominance*. In making this notion precise, it is convenient to use vector notation. Each row and each column of a payoff matrix can be considered to be a vector. In fact, the rows are row vectors, and the columns are column vectors, but the distinction between row and column vectors is unimportant here. We denote the vector which is the ith row of the payoff matrix by $\mathbf{r}_i$ and the vector which is the jth column of the matrix by $\mathbf{c}_j$. Vector inequalities, introduced in Sec. 7-6, provide a useful means of defining dominance.

Definition   Row i of a payoff matrix *dominates* row j if $\mathbf{r}_i \geq \mathbf{r}_j$. Column i of a payoff matrix dominates column j if $\mathbf{c}_i \leq \mathbf{c}_j$.

Recall that our convention is that the entries in the payoff matrix represent amounts paid by the column player to the row player.

In matrix $\mathbb{P}_2$ of Example 9-4 row 2 ($\mathbf{r}_2$) dominates both rows 1 and 3 ($\mathbf{r}_1$ and $\mathbf{r}_3$). Any row which is dominated by another row of the payoff matrix need not be considered as an option for the row player since it is not in his best interest to use it. In effect that row (and hence that pure strategy) can be eliminated from the payoff matrix. Similarly, if column i dominates column j, then column j (pure strategy c_j) need not be considered by the column player. That column can effectively be eliminated from the payoff matrix.

As a final comment about reducing the size of a payoff matrix using dominance, we note that the reductions may be made sequentially. For example, first the row player may eliminate a row by dominance, then, in the new reduced matrix, the column player may eliminate a column by dominance, then, in the twice-reduced matrix, the row player may eliminate another row by dominance, and so on until the matrix contains no further dominance. A matrix which is reduced by dominance to size 1×1 has a saddle point, the single remaining entry.

EXAMPLE 9-5

Reduce the following payoff matrix using dominance.

$$\begin{bmatrix} 2 & 3 & 1 & 0 & 1 & 1 & 2 \\ 2 & 3 & 1 & -1 & -2 & 1 & 2 \\ 2 & 2 & 1 & -1 & -1 & 0 & 1 \\ 1 & 1 & -1 & 1 & 0 & 1 & -1 \\ 2 & 2 & -1 & -1 & 2 & -1 & 0 \\ 1 & 4 & -1 & -1 & 2 & -2 & -2 \end{bmatrix}$$

Solution: Rows 2 and 3 are dominated by row 1 and can be eliminated from consideration. Also columns 1 and 2 are dominated by column 7, and they can be eliminated. The payoff matrix after this reduction is 4×5, and it has the form

$$
\begin{array}{c}
\\ r_1 \\ r_4 \\ r_5 \\ r_6
\end{array}
\begin{array}{cccccc}
c_3 & c_4 & c_5 & c_6 & c_7 \\
\left[\begin{array}{ccccc}
1 & 0 & 1 & 1 & 2 \\
-1 & 1 & 0 & 1 & -1 \\
-1 & -1 & 2 & -1 & 0 \\
-1 & -1 & 2 & -2 & -2
\end{array}\right]
\end{array}
$$

In this matrix $\mathbf{r}_5$ dominates $\mathbf{r}_6$ (note that this is not true in the original matrix). Also, in this new matrix $\mathbf{c}_3$ dominates $\mathbf{c}_5$. Removing $\mathbf{r}_6$ and $\mathbf{c}_5$, we have the payoff matrix

$$
\begin{array}{c}
\\ r_1 \\ r_4 \\ r_5
\end{array}
\begin{array}{ccccc}
c_3 & c_4 & c_6 & c_7 \\
\left[\begin{array}{cccc}
1 & 0 & 1 & 2 \\
-1 & 1 & 1 & -1 \\
-1 & -1 & -1 & 0
\end{array}\right]
\end{array}
$$

We continue the process. Now $\mathbf{r}_1$ dominates $\mathbf{r}_5$ and $\mathbf{c}_3$ dominates $\mathbf{c}_6$ and $\mathbf{c}_7$. Therefore, we have

$$
\begin{array}{c}
\\ r_1 \\ r_4
\end{array}
\begin{array}{ccc}
c_3 & c_4 \\
\left[\begin{array}{cc}
1 & 0 \\
-1 & 1
\end{array}\right]
\end{array}
$$

At this stage we must stop since there is no further dominance. However, a substantial reduction of the problem has been achieved. The original payoff matrix was 6×7, and by dominance it was reduced to 2×2. As we shall see in Sec. 9-4, a game with a matrix of this size is fairly easy to solve.   □

We conclude this section by once again considering the investment game of Sec. 9-2. In this game the payoffs depend upon chance (the behavior of the stock and bond markets). To simplify the setting it is assumed that stocks go up and bonds go down with probability .6 and the reverse occurs with probability .4. To form a payoff matrix we use the expected value of the change in the percent of assets controlled by each advisory service. The statement of the game specified that if one firm is correct and the other incorrect, the correct firm gains an additional 10 percent of the assets and the incorrect firm loses this 10 percent. Suppose that the firms are denoted by R and C. If R plays stocks and C plays bonds, then with probability .6 firm R will gain 10 percent and firm C will lose the same amount. Also, with probability .4 firm R will lose 10 percent and firm C will gain the same amount. Hence the expected payoff to R is $10(\frac{6}{10}) - 10(\frac{4}{10}) = 2$. Using this reasoning, we can determine all entries of the payoff matrix. We obtain the following matrix, where S means "buy stocks" and B means "buy bonds."

$$
\begin{array}{c}
 & & C & \\
 & & S \quad\quad B & \\
R & \begin{array}{c} S \\ B \end{array} & \begin{bmatrix} 0 & 2 \\ -2 & 0 \end{bmatrix}
\end{array}
$$

By inspection this game has a saddle point in the upper left-hand corner, and hence the best strategy for both firms is to play the strategy S, that is, to recommend to the pension fund buy stocks every quarter.

Exercises for Sec. 9-3

Decide whether each of the following payoff matrices has a saddle point. Find all such saddle points.

1. a. $\begin{bmatrix} 1 & 2 \\ 2 & 1 \end{bmatrix}$

b. $\begin{bmatrix} 1 & 2 & 3 \\ 2 & 2 & 3 \\ 2 & 2 & 2 \end{bmatrix}$

2. a. $\begin{bmatrix} 1 & -2 & 3 \\ 0 & 2 & -1 \\ 1 & -1 & 3 \end{bmatrix}$

b. $\begin{bmatrix} 1 & -2 & 3 \\ 0 & -1 & -1 \\ 1 & -1 & 3 \end{bmatrix}$

3. a. $\begin{bmatrix} 0 & -1 & -3 & 0 \\ 1 & 0 & 0 & 3 \\ 1 & 0 & 0 & 1 \\ 0 & -2 & -1 & 1 \end{bmatrix}$

b. $\begin{bmatrix} 1 & -1 & 2 & 3 \\ 0 & 3 & -3 & 1 \\ 1 & 0 & 2 & 4 \\ -1 & 2 & -4 & 0 \end{bmatrix}$

4. a. $\begin{bmatrix} 0 & 5 & 2 \\ 6 & 0 & 1 \\ 3 & 4 & 3 \end{bmatrix}$

b. $\begin{bmatrix} 5 & -1 & 1 & 5 \\ -5 & 1 & 1 & 9 \\ 5 & 1 & -3 & 1 \\ 9 & 1 & 1 & 7 \end{bmatrix}$

5. For each of the payoff matrices of Exercises 3 and 4 for which no saddle point exists use dominance to reduce the game as much as possible.

6. Give an example of a two-person zero-sum game which is strictly determined and which has exactly four saddle points.

7. In the investment game, Example 9-1, suppose the probability that stocks go up and bonds go down is .3 and the probability that stocks go down and bonds go up is .7. Find the payoff matrix for this game and decide whether this game has a saddle point.

9-4   SOLVING 2 × n AND m × 2 GAMES

If a game has a saddle point, there are pure strategies r_i for the row player and c_j for the column player such that the row player does best by always playing r_i and the column player does best by always playing c_j. On the other hand, if the game does not have a saddle point, it is not advantageous for either player always to use the same pure strategy. The goal of maximizing the average return per game is achieved only by mixing pure strategies in some way, i.e., by playing one pure strategy part of the time and other pure strategies at other times. The fraction of the time that various strategies are to be played must be determined for each game. To avoid being taken advantage of by an opponent, a person's method of play should not follow a pattern. Thus, in Example 9-2, if Rolf decides to play heads and tails each one-half of the time, he should not simply alternate between heads and tails. Instead he should use some chance device (such as flipping a coin) and let this chance device determine his play. We can represent Rolf's mixed strategy of playing each pure strategy one-half of the time by using the vector $(\frac{1}{2}, \frac{1}{2})$. Other mixed strategies for Rolf are $(\frac{1}{3}, \frac{2}{3})$, $(\frac{5}{6}, \frac{1}{6})$, and in general $(r, 1 - r)$ for any number r such that $0 \leq r \leq 1$. In this vector notation the first coordinate, r, is the probability that Rolf plays row 1 on any specific play of the game, and the second coordinate, $1 - r$, is the probability that he plays row 2. The problem for Rolf is to decide which choice of r gives him the "best" mixed strategy.

Although the concept of a best mixed strategy can be made precise in a formal mathematical setting, we shall not give a formal development. Instead we proceed somewhat informally by agreeing that a mixed strategy for the row player will be considered to be *optimal* (best) if its use guarantees the row player an average return (expected payoff) which is as large or larger than the average return guaranteed by the use of any other mixed strategy. Such optimal strategies always exist, and it is the goal of this and the next section to show how to find them. A similar comment can be made with regard to the column player. A mixed strategy for the column player is optimal if its guarantees an average return to the column player which is as large or larger than the average return guaranteed by any other mixed strategy.

A game is said to be solved when (1) optimal strategies are found for both players, and (2) the average guaranteed return to both players is determined. *Every two-person zero-sum game can be solved in this sense.* If one of the two players has only two pure strategies (or if by dominance the game can be reduced to one in which one of the players has two pure strategies), a solution for the game can be found by a relatively simple technique which involves the graphs of lines. We initially describe this method for a game where the row player has only two pure strategies. Later we shall indicate how the method can be adapted to a game in which the column player has two strategies. If a game has a payoff matrix with m rows and n columns, we say the game is $m \times n$. Thus, in this section we solve $2 \times n$ and $m \times 2$ games.

A Technique for Solving 2 × *n* Games

We illustrate the technique with an example.

EXAMPLE 9-6

Find a solution for the two-person zero-sum game with the following payoff matrix:

$$\mathbb{P} = \begin{bmatrix} 1 & -1 & 2 \\ -1 & 2 & -2 \end{bmatrix}$$

Solution: We first note that this game does not have a saddle point and cannot be reduced by dominance. Next we let $s(r) = (r, 1 - r)$ be an arbitrary mixed strategy for the row player. Thus, $0 \leq r \leq 1$. We consider the average payoff to the row player when he uses $s(r)$ and the column player uses pure strategy c_i (column i) on every play of the game. This payoff depends on r, and we denote it by $p_i(r)$. Using the definition of expected value (Sec. 3-7), we see that the expected return to the row player when he uses $s(r)$ and the column player uses c_1 is

$$p_1(r) = (1)(r) + (-1)(1 - r) = 2r - 1$$

This formula for $p_1(r)$ is obtained by noting that the row player plays row 1 with probability r and hence receives payoff 1 with probability r (see column 1 of the payoff matrix $\mathbb{P}$). Similarly, the row player plays row 2 with probability $1 - r$ and thus obtains payoff -1 with this probability.

In the same way that the formula for $p_1(r)$ was obtained, the formulas for $p_2(r)$ and $p_3(r)$ are found to be

$$p_2(r) = (-1)(r) + (2)(1 - r) = -3r + 2$$

$$p_3(r) = (2)(r) + (-2)(1 - r) = 4r - 2$$

$p_2(r)$ and $p_3(r)$ represent, of course, the expected payoff to the row player when he uses mixed strategy $s(r)$ and the column player uses pure strategies c_2 and c_3, respectively.

The functions p_1, p_2, and p_3 are linear in r and their graphs are shown in Fig. 9-2. The horizontal axis is the r axis, and the vertical axis represents the payoffs to the row player. If we take the point of view of the row player, then for each choice of r, the values $p_1(r)$, $p_2(r)$, and $p_3(r)$ represent, respectively, the expected payoffs when the strategy $s(r) = (r, 1 - r)$ is played against each of the three pure strategies of the column player. For example, if $r = \frac{1}{4}$, then $p_1(r) = -\frac{1}{2}$,

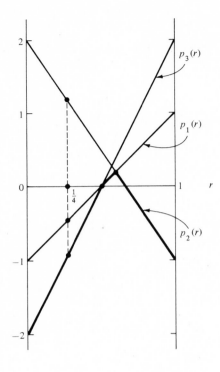

FIGURE 9-2

$p_2(r) = \frac{5}{4}$, and $p_3(r) = -1$. Thus, if the row player uses strategy $\mathbf{s}(\frac{1}{4}) = (\frac{1}{4}, \frac{3}{4})$, by using strategy c_3 every time the column player can hold the expected payoff to the row player to -1. Of course the row player would like to make this expected payoff higher, and hence the row player will choose r in such a way as to achieve the maximum possible expected payoff, considering at all times that the row player has no control over the column player's choice.

From Fig. 9-2 it is clear that in one sense the best choice of r is the value of r such that $p_1(r) = p_2(r)$. To be more precise, for this choice of r the minimum expected payoff to the row player is as large as possible. We claim that this choice of r is such that $\mathbf{s}(r) = (r, 1 - r)$ is an optimal mixed strategy for the row player. That is, using it, the row player maximizes his expected gain not only against the pure strategies of the column player but also against any mixed strategy of the column player. This method of finding an optimal strategy is called *max-min* reasoning since the row player chooses r in such a way that the worst thing that can happen to him, i.e., the minimum expected payoff he can receive, is the best possible, i.e., a maximum. The row player *max*imizes his *min*imum expected payoff. The graph of the function which gives the minimum expected payoff for each value of r is shown by the heavy line in Fig. 9-2. It is clear that in this example the maximum value of this minimum payoff is attained when r is selected so that $p_1(r) = p_2(r)$, that is, with $r = \frac{3}{5}$. Thus we claim that $\mathbf{s}(\frac{3}{5}) = (\frac{3}{5}, \frac{2}{5})$ is an optimal strategy for the row player.

To find an optimal strategy for the column player we consider only the pure strategies c_1 and c_2. Indeed, if the row player is using the strategy $\mathbf{s}(\frac{3}{5}) = (\frac{3}{5}, \frac{2}{5})$, the payoff to the row player will be smaller if the column player uses c_1 and c_2 than if the column player uses c_3 [in Fig. 9-2, $p_3(\frac{3}{5})$ is greater than $p_1(\frac{3}{5}) = p_2(\frac{3}{5})$]. Thus, the reduced game has the payoff matrix

$$\begin{array}{c} \\ r_1 \\ r_2 \end{array} \begin{array}{cc} c_1 & c_2 \\ \left[\begin{array}{cc} 1 & -1 \\ -1 & 2 \end{array} \right] \end{array}$$

To obtain the optimal mixed strategy for the column player we consider an arbitrary strategy $\mathbf{t}(c) = (c, 1 - c), 0 \leq c \leq 1$, and we consider the expected payoff when this strategy is played against each row strategy. As before, this payoff is a function of c. We denote the expected payoff when strategy $\mathbf{t}(c)$ is played against pure strategy r_i by $q_i(c)$. These payoffs are

$$q_1(c) = (1)(c) + (-1)(1 - c) = 2c - 1$$

$$q_2(c) = (-1)(c) + (2)(1 - c) = -3c + 2$$

Finally, to obtain the value of c which gives an optimal mixed strategy for the column player we solve for that value of c for which $q_1(c) = q_2(c)$. The desired value of c is $c = \frac{3}{5}$. Thus an optimal strategy for the column player is $(\frac{3}{5}, \frac{2}{5}, 0)$. The 0 in the third coordinate reflects the fact that pure strategy c_3 is never used.

The reasoning behind the use of the value of c determined by setting $q_1(c) = q_2(c)$ is analogous to the reasoning behind the selection of the value of r for an optimal strategy for the row player. Namely, one considers the graphs of the functions q_1 and q_2 (lines) as shown in Fig. 9-3. In this figure the interval $0 \leq c \leq 1$ on the horizontal axis represents the possible values of c which correspond to the mixed strategies available to the column player, and the vertical axis represents the payoffs to the row player. Thus for $c = \frac{1}{4}$, if the row player uses row 1 every time, the payoff to the row player is $-\frac{1}{2} = q_1(\frac{1}{4})$. On the other hand if the row player uses row 2 every time, the payoff to the row player is $\frac{5}{4} = q_2(\frac{1}{4})$. Since the column player makes these payoffs to the row player (the game is zero sum) and since the column player does not know which strategy the row player will use, the column player does not consider $c = \frac{1}{4}$ to be the best possible choice for minimizing the average gain to the row player. Indeed to hold the average gain of the row player to a minimum and thus to maximize the long-term gain to the column player, the column player must choose c such that $q_1(c) = q_2(c)$. For any other choice of c, the column player runs the risk of being exploited by the row player and having to pay more than would be the case with $c = \frac{3}{5}$. □

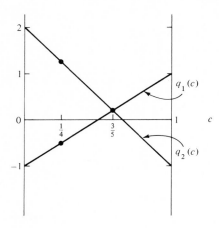

FIGURE 9-3

At this stage in our discussion we have supported the claim that optimal strategies for the row and column player are $\mathbf{s}^* = \left(\frac{3}{5}, \frac{2}{5}\right)$ and $\mathbf{t}^* = \left(\frac{3}{5}, \frac{2}{5}, 0\right)$, respectively. A rigorous proof that these are optimal can be given.

The average payoff to the row player in this game, called the *value* of the game, is the number $v = p_1\left(\frac{3}{5}\right) = p_2\left(\frac{3}{5}\right) = q_1\left(\frac{3}{5}\right) = q_2\left(\frac{3}{5}\right) = \frac{1}{5}$.

The method described in Example 9-6 works for any two-person zero-sum game in which one player has only two pure strategies. Using Example 9-6 as a guide, we can outline a method of solving games with payoff matrices which are $2 \times n$.

Method for Solving Games Using Graphs

1. Check whether the payoff matrix has a saddle point. If it does, solve the game using the technique of Sec. 9-2. Otherwise proceed to step 2.
2. Denote an arbitrary strategy for the row player by $\mathbf{s}(r) = (r, 1 - r)$ and denote the expected payoff when $\mathbf{s}(r)$ is played against column i by $p_i(r)$, $i = 1, 2, \ldots, n$.
3. Graph the functions $p_i(r)$, $i = 1, 2, \ldots, n$, for $0 \le r \le 1$.
4. Find the maximum point on the polygonal line which bounds the graphs of $p_1, p_2, \ldots, p_n$ from below. Let k and l be defined by the property that the graphs of $p_k(r)$ and $p_l(r)$ lie on this polygon and intersect at this highest point, $k < l$. Since the payoff matrix does not have a saddle point, k and l are certain to exist.
5. Solve for r^* such that $p_k(r^*) = p_l(r^*)$.
6. An optimal strategy for the row player is given by $\mathbf{s}^* = (r^*, 1 - r^*)$.
7. Let $\mathbf{t}(c)$ be the mixed strategy for the column player which plays column k with probability c and column l with probability $1 - c$, $0 \le c \le 1$. Also let $q_i(c)$ be the payoff when $\mathbf{t}(c)$ is played against row i, $i = 1, 2$.

8. Solve for the value of c^* such that $q_1(c^*) = q_2(c^*)$. Since the payoff matrix does not have a saddle point, c^* will lie in the interval $0 \leq c \leq 1$.
9. An optimal strategy for the column player is

$$\mathbf{t}^* = (0, \quad 0, \quad \cdots, \quad 0, \underbrace{c^*,}_{k\text{th Coord.}} 0, \quad \cdots, \quad 0, \underbrace{1 - c^*,}_{l\text{th Coord.}} 0, \quad \cdots, \quad 0)$$

10. The value of the game (the expected return to the row player) is
$v = p_k(r^*) = p_l(r^*) = q_1(c^*) = q_2(c^*).$

A Technique for Solving $m \times 2$ Games

The method outlined above is applicable to a game with a payoff matrix which is $2 \times n$. To solve a game which has an $m \times 2$ payoff matrix one simply interchanges the rows and columns of the payoff matrix, multiplies each entry by -1, and solves the new game by using the method given above (the new game is $2 \times m$). If $\tilde{\mathbf{s}}$ and $\tilde{\mathbf{t}}$ are optimal for the new game, and if the value is $\tilde{v}$, then $\mathbf{t}^* = \tilde{\mathbf{s}}$ and $\mathbf{s}^* = \tilde{\mathbf{t}}$ and $v = -\tilde{v}$ give a solution of the original game. The next example illustrates this method.

EXAMPLE 9-7

Find a solution for the two-person zero-sum game with the payoff matrix

$$\mathbb{P} = \begin{bmatrix} -1 & 2 \\ 3 & -2 \\ 2 & -1 \end{bmatrix}$$

Solution: We begin by multiplying each entry of the payoff matrix by -1 and interchanging rows and columns. This gives the payoff matrix $\tilde{\mathbb{P}}$ for a new game. The new game is 2×3 and

$$\tilde{\mathbb{P}} = \begin{bmatrix} 1 & -3 & -2 \\ -2 & 2 & 1 \end{bmatrix}$$

Using the method outlined above, we have $p_1(r) = 3r - 2$, $p_2(r) = -5r + 2$, and $p_3(r) = -3r + 1$. The graphs of these functions are shown in Fig. 9-4.

The lines in Fig. 9-4 are coincident at the point $(\frac{1}{2}, -\frac{1}{2})$ for which $r = \frac{1}{2}$. Hence an optimal strategy for the row player (in this new game) is $\tilde{\mathbf{s}} = (\frac{1}{2}, \frac{1}{2})$. The lines which bound the graph in Fig. 9-4 from below correspond to strategies 1 and 2 for the column player. When columns 1 and 2 and the method outlined above are used, an optimal strategy for the column player is given by $\tilde{\mathbf{t}} = (\frac{5}{8}, \frac{3}{8}, 0)$. The value of the new game is $\tilde{v} = -\frac{1}{2}$. Thus, a solution of

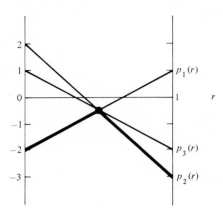

FIGURE 9-4

the original 3×2 game is given by $\mathbf{s}^* = (\frac{5}{8}, \frac{3}{8}, 0)$, $\mathbf{t}^* = (\frac{1}{2}, \frac{1}{2})$, $v = \frac{1}{2}$. Other solutions can also be obtained (see Exercise 7 for examples).   ☐

Although optimal strategies need not be unique (there are several optimal strategies for the column player in Example 9-7, for instance), the value of a game is unique. Thus the payoff for all pairs of optimal strategies is the same number v.

Exercises for Sec. 9-4

Using the graphing method outlined in this section, solve the game whose payoff matrix is

1. $\begin{bmatrix} 1 & -4 & 1 \\ -2 & 3 & -1 \end{bmatrix}$

2. $\begin{bmatrix} 0 & -1 & 4 & 4 \\ 5 & 5 & -2 & -1 \end{bmatrix}$

3. $\begin{bmatrix} -1 & 3 \\ 2 & -1 \\ 7 & -10 \end{bmatrix}$

4. $\begin{bmatrix} 2 & 4 & 1 & 8 \\ 4 & 1 & 8 & 0 \end{bmatrix}$

5. $\begin{bmatrix} -3 & 2 \\ -1 & -1 \\ 3 & -4 \\ 0 & -2 \end{bmatrix}$

6. If the graphs of the two functions $p_l(r)$ and $p_m(r)$ do not intersect at any point for $0 \le r \le 1$, what does this imply about the lth and mth columns of the payoff matrix?

7. In Example 9-7 find an optimal strategy for the column player which uses only rows 1 and 3.

Solve the two-person zero-sum game with the payoff matrix

8. $\begin{bmatrix} -2 & 2 & -3 \\ 6 & -4 & 5 \\ 6 & -5 & 4 \end{bmatrix}$

9. $\begin{bmatrix} 0 & 1 & -1 & -2 \\ 2 & 1 & -1 & 2 \\ 3 & -3 & -2 & 3 \\ -4 & 4 & -3 & 5 \end{bmatrix}$

9-5 SOLVING $m \times n$ GAMES USING THE SIMPLEX METHOD

Every two-person zero-sum game has a solution in the sense described in Sec. 9-3. We have discussed how a solution can be obtained in the special case of a game in which at least one of the players has only two pure strategies. In this section we shall describe a method which can be used to solve an $m \times n$ game. The basic idea is simple. First, a two-person zero-sum game is converted into a linear programming problem (an SMP in the terminology of Chap. 7), and then the SMP and its dual are solved using the simplex method (again, see Chap. 7). The solution of the SMP and its dual provides the optimal strategies for the column and row players, respectively. The method also yields the value of the game. We begin by illustrating the method with an example. Following the example we provide a summary of the steps for use in general situations.

EXAMPLE 9-8

Solve the two-person zero-sum game with the payoff matrix

$$\mathbb{P} = \begin{bmatrix} 1 & 0 & 0 \\ -1 & 1 & -1 \\ 0 & -1 & 1 \end{bmatrix}$$

Solution: We first observe that the payoff matrix of this game does not have a saddle point and it cannot be reduced in size by dominance. Thus the technique of Sec. 9-4 cannot be used.

To solve this game by the simplex method we first convert the payoff matrix to a matrix with all positive entries by adding the same suitable constant to every entry. In this case if we add 2 to every entry in $\mathbb{P}$, we obtain a new matrix $\mathbb{P}_2$ which has all positive entries:

$$\mathbb{P}_2 = \begin{bmatrix} 3 & 2 & 2 \\ 1 & 3 & 1 \\ 2 & 1 & 3 \end{bmatrix}$$

The second step in solving this game by the simplex method is to consider the SMP $[\mathbb{P}_2, \mathbf{b}, \mathbf{c}]$, where $\mathbf{b}$ and $\mathbf{c}$ are vectors with all coordinates equal to 1. The number of coordinates for $\mathbf{b}$ is the same as the number of rows of $\mathbb{P}_2$, and the number of coordinates for $\mathbf{c}$ is the same as the number of columns for $\mathbb{P}_2$. Since $\mathbb{P}_2$ is a 3×3 matrix, we have

$$\mathbf{b} = \begin{bmatrix} 1 \\ 1 \\ 1 \end{bmatrix} \quad \text{and} \quad \mathbf{c} = \begin{bmatrix} 1 \\ 1 \\ 1 \end{bmatrix}$$

Next, we solve the SMP $[\mathbb{P}_2, \mathbf{b}, \mathbf{c}]$ using the simplex method. The tableaus used in the solution are shown in Tableaus 9-1 to 9-4.

TABLEAU 9-1

	x	y	z	u	v	w	**Basic Solution**
→u	③	2	2	1	0	0	1
v	1	3	1	0	1	0	1
w	2	1	3	0	0	1	1
p	-1	-1	-1	0	0	0	0

TABLEAU 9-2

	x	y	z	u	v	w	**Basic Solution**
x	1	$\frac{2}{3}$	$\frac{2}{3}$	$\frac{1}{3}$	0	0	$\frac{1}{3}$
→v	0	⑦⁄₃	$\frac{1}{3}$	$-\frac{1}{3}$	1	0	$\frac{2}{3}$
w	0	$-\frac{1}{3}$	$\frac{5}{3}$	$-\frac{2}{3}$	0	1	$\frac{1}{3}$
p	0	$-\frac{1}{3}$	$-\frac{1}{3}$	$\frac{1}{3}$	0	0	$\frac{1}{3}$

TABLEAU 9-3

	x	y	z ↓	u	v	w	Basic Solution
x	1	0	$\frac{4}{7}$	$\frac{9}{21}$	$-\frac{2}{7}$	0	$\frac{1}{7}$
y	0	1	$\frac{1}{7}$	$-\frac{1}{7}$	$\frac{3}{7}$	0	$\frac{2}{7}$
→w	0	0	$\left(\frac{12}{7}\right)$	$-\frac{5}{7}$	$\frac{1}{7}$	1	$\frac{3}{7}$
p	0	0	$-\frac{2}{7}$	$\frac{2}{7}$	$\frac{1}{7}$	0	$\frac{3}{7}$

TABLEAU 9-4

	x	y	z	u	v	w	Basic Solution
x	1	0	0	$\frac{8}{7}$	$-\frac{1}{3}$	$-\frac{1}{3}$	0
y	0	1	0	$-\frac{1}{12}$	$\frac{5}{12}$	$-\frac{1}{12}$	$\frac{1}{4}$
z	0	0	1	$-\frac{5}{12}$	$\frac{1}{12}$	$\frac{7}{12}$	$\frac{1}{4}$
p	0	0	0	$\frac{1}{6}$	$\frac{1}{6}$	$\frac{1}{6}$	$\frac{1}{2}$

Since the last row of Tableau 9-4 has no negative entries, the SMP $[\mathbb{P}_2,\mathbf{b},\mathbf{c}]$ and its dual have been solved. These two problems are the following:

SMP $[\mathbb{P}_2,\mathbf{b},\mathbf{c}]$	Dual
Find the maximum of $x + y + z$, for $x \geq 0$, $y \geq 0$, $z \geq 0$, and $$\begin{aligned} 3x + 2y + 2z &\leq 1 \\ x + 3y + z &\leq 1 \\ 2x + y + 3z &\leq 1 \end{aligned}$$	Find the minimum of $r + s + t$, for $r \geq 0$, $s \geq 0$, $t \geq 0$, and $$\begin{aligned} 3r + s + 2t &\geq 1 \\ 2r + 3s + t &\geq 1 \\ 2r + s + 3t &\geq 1 \end{aligned}$$

According to Tableau 9-4, an optimal vector for the SMP $[\mathbb{P}_2,\mathbf{b},\mathbf{c}]$ is $(x,y,z) = (0,\frac{1}{4},\frac{1}{4})$, and an optimal vector for the dual problem is $(r,s,t) = (\frac{1}{6},\frac{1}{6},\frac{1}{6})$. The relationship between these optimal vectors and optimal strategies for the players in the game with the payoff matrix $\mathbb{P}$ is as follows:

1. Let M be the sum of the coordinates of the optimal vector for the SMP $[\mathbb{P}_2,\mathbf{b},\mathbf{c}]$, $M = 0 + \frac{1}{4} + \frac{1}{4} = \frac{1}{2}$. An optimal strategy for the column player is

$$\mathbf{t} = \left(\frac{0}{M}, \frac{\frac{1}{4}}{M}, \frac{\frac{1}{4}}{M}\right) = (0, \tfrac{1}{2}, \tfrac{1}{2})$$

and an optimal strategy for the row player is

$$\mathbf{s} = \left(\frac{\frac{1}{6}}{M}, \frac{\frac{1}{6}}{M}, \frac{\frac{1}{6}}{M} \right) = (\tfrac{1}{3}, \tfrac{1}{3}, \tfrac{1}{3})$$

2. The value of the game with the payoff matrix $\mathbb{P}$ is $v = 1/M - k$, where k is the amount added to the entries of $\mathbb{P}$ to make each entry positive. Here, $k = 2$ and $M = \tfrac{1}{2}$, and so $v = 2 - 2 = 0$.

   Thus, according to statements 1 and 2, the game with the payoff matrix $\mathbb{P}$ has the following solution: the optimal strategy for the row player is the mixed strategy $(\tfrac{1}{3},\tfrac{1}{3},\tfrac{1}{3})$, the optimal strategy for the column player is the mixed strategy $(0,\tfrac{1}{2},\tfrac{1}{2})$, and the value of the game is 0.   □

The method used to solve the game in Example 9-8 can be used to solve any two-person zero-sum game. The complete method can be summarized as follows.

Method for Solving Games Using the Simplex Method

1. Convert the payoff matrix $\mathbb{P}$ for the game to a payoff matrix with all positive entries by adding the same positive constant k to each entry. Denote the new payoff matrix by $\mathbb{P}_k$.
2. Use the simplex method to solve the SMP $[\mathbb{P}_k,\mathbf{b},\mathbf{c}]$, where $\mathbf{b}$ and $\mathbf{c}$ are vectors with each coordinate having value 1. The number of coordinates of $\mathbf{b}$ is the same as the number of rows of $\mathbb{P}_k$, and the number of coordinates of $\mathbf{c}$ is the same as the number of columns of $\mathbb{P}_k$.
3. Let the optimal vector obtained for the SMP $[\mathbb{P}_k,\mathbf{b},\mathbf{c}]$ be $\mathbf{x} = (x_1,x_2,\ldots,x_n)$, and let the optimal vector for the dual be $\mathbf{z} = (z_1,z_2,\ldots,z_m)$. Also let $M = x_1 + x_2 + \cdots + x_n$. Since the SMP $[\mathbb{P}_k,\mathbf{b},\mathbf{c}]$ maximizes $x_1 + \cdots + x_n$ and the dual problem minimizes $z_1 + z_2 + \cdots + z_m$, we also have $M = z_1 + z_2 + \cdots + z_m$ (see Sec. 7-6).
4. An optimal strategy for the row player in the game with payoff matrix $\mathbb{P}$ is $(z_1/M, z_2/M, \ldots, z_m/M)$. An optimal strategy for the column player is $(x_1/M, x_2/M, \ldots, x_n/M)$. The value of the game is $v = 1/M - k$.

An application of these rules in a practical setting is given in the next example.

EXAMPLE 9-9

The investment advisor for REIT (Real Estate Investment Trust) is trying to decide how to allocate a sum of money recently obtained through the foreclosure and sale of the property in which the REIT previously invested. The possible areas of investments are in an apartment complex (AC), a shopping center (SC), and a single-family-dwelling residential center (SF). The REIT has had problems with the local zoning board in the past, and the return on each type of investment will depend on who controls the board after the next election. If the three parties running in the election are designated by the labels Infinite Growth, Special Growth, and Neutral, REIT estimates the returns on investments in each area to be as shown in Table 9-1.

If the REIT assumes the electorate will vote to minimize the return of the REIT, what percent of the total amount to be invested should be put into each type of investment?

TABLE 9-1

	Percent Return on Investment	Zoning Board		
		Infinite	Special	Neutral
Investment	AC	40	-20	20
	SC	20	10	20
	SF	10	30	0

Note: The investment decision must be made now, and the election will be held in the near future before the electorate knows how the investments were made. Thus neither side (REIT nor electorate) knows the action of the other when the decisions are made.

Solution: The REIT views its decision as one of finding the optimal strategy for the row player in the two-person zero-sum game with the payoff matrix $\mathbb{P}$:

$$\mathbb{P} = \begin{bmatrix} 40 & -20 & 20 \\ 20 & 10 & 20 \\ 10 & 30 & 0 \end{bmatrix}$$

In order to solve this game using the simplex method we first add 30 to each entry (step 1). We construct the vectors **b** and **c** and solve the problem by using Tableaus 9-5 to 9-7 (step 2).

TABLEAU 9-5

	↓ x	y	z	u	v	w	Basic Solution
→u	(70)	10	50	1	0	0	1
v	50	40	50	0	1	0	1
w	40	60	30	0	0	1	1
p	−1	−1	−1	0	0	0	0

TABLEAU 9-6

	x	↓ y	z	u	v	w	Basic Solution
x	1	$\frac{1}{7}$	$\frac{5}{7}$	$\frac{1}{70}$	0	0	$\frac{1}{70}$
v	0	$\frac{230}{7}$	$\frac{100}{7}$	$-\frac{5}{7}$	1	0	$\frac{2}{7}$
→w	0	$\left(\frac{380}{7}\right)$	$\frac{10}{7}$	$-\frac{4}{7}$	0	1	$\frac{3}{7}$
p	0	$-\frac{6}{7}$	$\frac{2}{7}$	$\frac{1}{70}$	0	0	$\frac{1}{70}$

TABLEAU 9-7

	x	y	z	u	v	w	Basic Solution
x	1	0	$\frac{27}{38}$	$\frac{39}{25}(38)$	0	$-\frac{1}{380}$	$\frac{5}{380}$
v	0	0	$\frac{510}{38}$	$-\frac{14}{38}$	1	$-\frac{23}{38}$	$\frac{1}{38}$
y	0	1	$\frac{1}{38}$	$\frac{4}{380}$	0	$\frac{7}{380}$	$\frac{3}{380}$
p	0	0	$\frac{32}{38}(7)$	$\frac{2}{380}$	0	$\frac{6}{380}$	$\frac{8}{380}$

The last row of Tableau 9-7 has no negative entries, and hence the SMP $[\mathbb{P}_{30},\mathbf{b},\mathbf{c}]$ is solved. According to steps 3 and 4, optimal strategies for the row and column players of the game are, respectively,

$$\mathbf{s} = \left(\frac{\frac{2}{380}}{\frac{8}{380}}, 0, \frac{\frac{6}{380}}{\frac{8}{380}}\right) = \left(\tfrac{1}{4}, 0, \tfrac{3}{4}\right)$$

$$\mathbf{t} = \left(\frac{\frac{5}{380}}{\frac{8}{380}}, \frac{\frac{3}{380}}{\frac{8}{380}}, \frac{0}{\frac{8}{380}}\right) = \left(\tfrac{5}{8}, \tfrac{3}{8}, 0\right)$$

The value of the game (in percent return on investment) is

$$v = \tfrac{380}{8} - 30 = 17.5\%$$

Thus, REIT should invest 25 percent of its funds in the apartment complex and 75 percent in the SF homes. No money should be invested in the shopping center.

Note: The computations involved in Tableaus 9-5 and 9-7 are unwieldly because of the fractions involved, i.e., quantities like $\frac{7}{380}$. The computations can be simplified by first dividing all entries to Table 9-1 by 10 and then solving as before. The optimal strategies of the new game are the same as those of the original game, and the values of the games differ by the factor of 10. See Exercises 4 and 5. ☐

Exercises for Sec. 9-5

1. Solve the two-person zero-sum game with the payoff matrix $\mathbb{P}$ shown below. Use both the simplex method and the graphical method.

$$\mathbb{P} = \begin{bmatrix} 2 & 3 & -5 \\ 1 & -4 & 4 \end{bmatrix}$$

2. Use the simplex method to solve the two-person zero-sum games with the following payoff matrices:

   a. $\mathbb{P} = \begin{bmatrix} -2 & 3 \\ 4 & -6 \end{bmatrix}$

   b. $\mathbb{P} = \begin{bmatrix} 1 & -1 & 2 \\ 2 & 1 & 1 \\ -1 & 0 & 1 \end{bmatrix}$

3. Solve the two-person zero-sum game with the payoff matrix

$$\mathbb{P} = \begin{bmatrix} 2 & 1 & 2 & -5 \\ -1 & 4 & -2 & -1 \\ -3 & -2 & 7 & 1 \end{bmatrix}$$

4. Solve the problem of REIT in Example 9-9 with the table of returns on investments, Table 9-1, changed to

$$\begin{bmatrix} 4 & -2 & 2 \\ 2 & 1 & 2 \\ 1 & 3 & 0 \end{bmatrix}$$

   How does the value of this game compare with the value of the game in Example 9-9, and how do the payoffs compare?

5. In each of parts **a** to **c** solve the two games with the given payoff matrices. Next, find a rule for the effect on a solution of a game when the payoff matrix is multiplied by a positive number.

   a. $\begin{bmatrix} 20 & -30 \\ -10 & 30 \end{bmatrix}$ and $\begin{bmatrix} 2 & -3 \\ -1 & 3 \end{bmatrix}$

b. $\begin{bmatrix} 16 & -4 \\ -8 & 20 \end{bmatrix}$ and $\begin{bmatrix} 4 & -1 \\ -2 & 5 \end{bmatrix}$

c. $\begin{bmatrix} \frac{1}{2} & -\frac{1}{4} & \frac{1}{8} \\ -\frac{1}{8} & \frac{3}{4} & -\frac{1}{4} \end{bmatrix}$ and $\begin{bmatrix} 4 & -2 & 1 \\ -1 & 6 & -2 \end{bmatrix}$

6. Solve the game of Exercise 7, Sec. 9-4, by the simplex method.

7. Solve the game of Exercise 8, Sec. 9-4, by the simplex method.

IMPORTANT TERMS

You should be able to describe, define, or give examples of each of the following:

Two-person zero-sum game
Pure strategy
Payoff
Mixed strategy
Payoff matrix
Saddle point
Strictly determined game

Dominance
Optimal mixed strategy
Value of a game
Solution of a game
Solutions by graphing
Solutions by the simplex method

REVIEW EXERCISES

Find all saddle points of the game with the payoff matrix

1. $\begin{bmatrix} 1 & 2 & 3 & 1 \\ -1 & 5 & -3 & 0 \\ 0 & -2 & 6 & -2 \\ 1 & 4 & 1 & 1 \end{bmatrix}$

2. $\begin{bmatrix} 7 & -1 & -2 \\ 4 & -3 & 6 \\ 2 & -1 & 3 \\ 6 & -3 & -4 \\ -5 & -1 & 3 \end{bmatrix}$

Use the technique based on graphing lines to solve the game with the payoff matrix

3. $\begin{bmatrix} -1 & 3 & -2 & 4 \\ 2 & -1 & 3 & -2 \end{bmatrix}$

4. $\begin{bmatrix} 2 & -2 \\ -3 & 3 \\ 4 & -4 \\ -5 & 5 \\ 6 & -3 \end{bmatrix}$

Use the simplex method to solve the game with the payoff matrix

5. $\begin{bmatrix} 2 & 1 & -4 \\ 3 & 2 & -4 \\ 5 & -2 & 5 \end{bmatrix}$

6. $\begin{bmatrix} 1 & -1 & 2 \\ -2 & 2 & -4 \\ 3 & -2 & 6 \end{bmatrix}$

Solve (by any method) the game with the following payoff matrix

7. $\begin{bmatrix} 3 & -2 & 1 \\ -1 & -1 & 0 \\ 3 & -1 & 1 \end{bmatrix}$

8. $\begin{bmatrix} 3 & -2 & 1 \\ -1 & 2 & 0 \\ 3 & -1 & 1 \end{bmatrix}$

DIGRAPHS
AND NETWORKS

CHAPTER
TEN

10-1   INTRODUCTION

In the last two or three decades many mathematical concepts have been developed into tools which help solve problems in the social and management sciences. Among the most versatile of these concepts are those of a digraph and a network. This chapter consists primarily of two examples showing how these concepts can be used. These examples were selected because they illustrate important uses of the concepts and because they are elementary. The subject is rich, varied, and still growing rapidly.

The model introduced in Sec. 6-6 to describe a chain of command is an example of a model which can be formulated in graph-theoretic terms. In this chapter we revisit this example and show how it can be phrased in terms of a digraph. We also show how this model can be analyzed to yield additional information about the people in the system and their relations with each other. A version of a question asking for such information, in a slightly more general situation, is described in Example 10-1.

EXAMPLE 10-1

An organization is interested in developing measures of "status" to use in determining the salary levels of its employees. For each position there is a job

367

description which identifies the direct supervisor of an employee in that position. Suppose that it is agreed that if *A* supervises *B* and *B* supervises *C* then *A* gains more status than if *A* supervised *B* and *C* directly. What are some appropriate measures of status in the organization, and how can the status associated with each position be calculated most efficiently? ☐

Situations like that described in Sec. 6-6 and Example 10-1 can be approached by using the concept of a digraph in one way. There are, however, problems of a very different type in which digraphs are also useful. One such problem is described in Example 10-2.

EXAMPLE 10-2

Metroelectric Co., the electric utility in Metroburg, needs to lay an additional electric line to provide electricity to a new trade building. The power is available at a substation several blocks away. The trade building and the substation are located on the grid of city streets, as shown in Fig. 10-1. The electric utility must provide the additional power by adding new lines in the existing conduits in the streets. Each number shown on a line segment (a street block) of the grid in Fig. 10-1 is the cost associated with adding a new line along that block. Assume that the power company prefers to string a direct line, i.e., one which goes only north and east from the power substation to the trade building. Along which blocks should the lines be laid to provide the desired power at minimum cost? ☐

In providing answers to practical questions of the sort raised in Example 10-2 we immediately confront the problems involved in carrying out computations for such digraphs. These problems generate the need for efficient algorithms. A typical algorithm is the topic of Sec. 10-3.

FIGURE 10-1

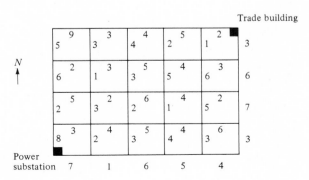

10-2 DEFINITIONS AND NOTATION

Because the subject of graphs uses a specialized vocabulary, it is necessary to introduce some terminology before proceeding. We use the example of Sec. 6-6 to illustrate the ideas. The situation considered in that example was the following. A business has six officers: a president P, a vice president for finance VPF, a vice president for production VPP, a treasurer T, a director of labor relations DLR, and a director of research DR. The organization chart which describes the relationships between these executives is shown in Fig. 10-2, where the arrows indicate the flow of directives.

Definition

A *digraph* (short for directed graph) G consists of a set V and a set E of ordered pairs (u,v), where u, $v \in V$ and $u \neq v$.

Remark: The concept of a *graph* is more general than that of a digraph, and it is also very useful in applications. A graph is similar to a digraph in that it has a set of vertices and a set of edges, but the edges are not directed. In a graph the edges are simply connections between the vertices. In this chapter we consider digraphs only and, in particular, digraphs with only a finite number of vertices.

Digraphs are frequently pictured as diagrams of points and lines. The elements of V are the points, or *vertices*, in the diagram, and the lines are the *edges*. Notice that since the elements of E are ordered pairs, each edge has a direction associated with it. We refer to (u,v) as the edge *from u to v*, and we denote it by a line or curve originating at the point u and terminating at the point v. The direction is usually denoted by an arrowhead on the edge. We write $G = [V,E]$ to denote the relation between a digraph G and its set of vertices V and its set of edges E.

FIGURE 10-2

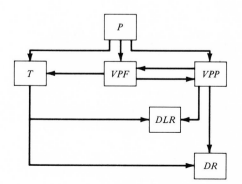

If the organization chart pictured in Fig. 10-2 is viewed as a digraph, we have

$$V = \{P, \, VPF, \, VPP, \, T, \, DLR, \, DR\}$$

and

$$E = \{(P,T), \, (P,VPF), \, (P,VPP), \, (VPF,T), \, (VPF,VPP), \, (VPP,VPF),$$
$$(VPP,DLR), \, (VPP,DR), \, (T,DLR), (T,DR)\}$$

The fact that both (VPP,VPF) and (VPF,VPP) are edges in the digraph is indicated by the pair of arrows between VPF and VPP in Fig. 10-2.

Several matrices associated with a digraph are useful in studying the situation modeled by the digraph. In order to describe these matrices it is useful to suppose that the vertices are labeled with the labels 1, 2, ..., n, where n is the number of vertices in V. That is, we suppose that $V = \{v_1, v_2, \ldots, v_n\}$.

Definition

The *adjacency matrix* of a digraph G with $V = \{v_1, v_2, \ldots, v_n\}$ is the $n \times n$ matrix $\mathbb{A} = (a_{ij})$ with

$$a_{ij} = \begin{cases} 1 & \text{if } (v_i, v_j) \in E \\ 0 & \text{if } (v_i, v_j) \notin E \end{cases} \qquad \text{for } i, j = 1, 2, \ldots, n$$

If the vertices of a digraph with n vertices are not labeled directly with the labels 1, 2, ..., n (for example, if the vertices are identified with the units of an organization which have different labels), the adjacency matrix of the diagraph can be determined by introducing an order into the set of vertices. In practice it is common to proceed as follows. Write the label of one of the vertices beside the first row and column of an $n \times n$ matrix (this vertex is treated as v_1); write the label of another vertex beside the second row and column of the matrix (this vertex is treated as v_2); and continue until the label of the last vertex is written beside the last row and column of the matrix (this vertex is treated as v_n). Then determine the adjacency matrix as in the definition.

The adjacency matrix for the situation pictured in Fig. 10-2 was computed in Chap. 6 (because of the setting of the example it was called a communication matrix in that discussion). The matrix is

$$
\mathbb{A} = \begin{array}{c} \\ P \\ VPF \\ VPP \\ T \\ DLR \\ DR \end{array}
\begin{array}{c}
\begin{array}{cccccc} P & VPF & VPP & T & DLR & DR \end{array} \\
\left[\begin{array}{cccccc}
0 & 1 & 1 & 1 & 0 & 0 \\
0 & 0 & 1 & 1 & 0 & 0 \\
0 & 1 & 0 & 0 & 1 & 1 \\
0 & 0 & 0 & 0 & 1 & 1 \\
0 & 0 & 0 & 0 & 0 & 0 \\
0 & 0 & 0 & 0 & 0 & 0
\end{array} \right]
\end{array}
\qquad (10\text{-}1)
$$

EXAMPLE 10-3

The adjacency matrix for the digraph shown in Fig. 10-3 is

$$
\begin{array}{c c}
 & \begin{array}{c c c c c} A & B & C & D & F \end{array} \\
\begin{array}{c} A \\ B \\ C \\ D \\ F \end{array} &
\left[\begin{array}{c c c c c}
0 & 1 & 1 & 0 & 1 \\
1 & 0 & 1 & 1 & 1 \\
1 & 1 & 0 & 1 & 1 \\
0 & 0 & 0 & 0 & 0 \\
0 & 1 & 0 & 1 & 0
\end{array}\right]
\end{array}
$$

The digraph shown in Fig. 10-3 might be proposed as a model for an industrial organization in which the vertices denote subunits, for example,

A: Secretarial pool
B: Executive offices
C: Food services
D: Production
F: Maintenance

and the edges denote direct service. Thus, the edge (A,F) denotes that the secretarial pool provides direct service for the maintenance department. It is clear from the figure that although the secretarial pool does not provide direct service to the production department, it provides indirect service since it services (directly) the executive office which provides (direct) service to the production department. This indicates that there may be useful connections in a digraph other than the direct ones (edges). ☐

FIGURE 10-3

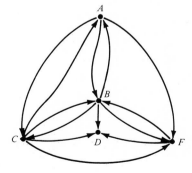

Let $G = [V,E]$ be a digraph. A *path* between vertices v_1 and v_m in G is a sequence,

$$v_1, e_1, v_2, e_2, \ldots, v_{m-1}, e_{m-1}, v_m \tag{10-2}$$

where each vertex v_i is in V, each $e_i = (v_i, v_{i+1})$ and each edge e_i is in E. This path can be identified by simply naming the vertices v_1, v_2, ..., v_m or $v_1 v_2 \cdots v_m$. When this notation is used, it is understood that $(v_i, v_{i+1}) \in E$ for each $i = 1, 2, ..., m - 1$.

For example, there is path $ABFD$ between A and D in the digraph of Fig. 10-3. $ACBD$ is another path between A and D. There is no path between D and A.

The path (10-2) is said to be of *length m* − 1. If there is a path between two vertices u and v in a digraph, then v is said to be *reachable* from u. If v is reachable from u, then the *distance* $d(u,v)$ between u and v is the length of the shortest path between u and v. If v is not reachable from u, we define $d(u,v)$ to be the symbol ∞ (infinity).

Another matrix which is useful in analyzing models formulated as digraphs is the distance matrix of a digraph G. Again we suppose that the vertex set V is $\{v_1, v_2, ..., v_n\}$.

Definition   The *distance matrix* $\mathbb{D} = (d_{ij})$ of a digraph G is defined by

$$d_{ij} = \begin{cases} 0 & \text{if } i = j \\ d(v_i, v_j) & \text{if } i \neq j \end{cases} \quad \text{for } i, j = 1, 2, ..., n$$

EXAMPLE 10-4

The distance matrix for the digraph shown in Fig. 10-3 is

$$
\mathbb{D} = \begin{array}{c} \\ A \\ B \\ C \\ D \\ F \end{array}
\begin{array}{c} \begin{array}{ccccc} A & B & C & D & F \end{array} \\
\left[\begin{array}{ccccc}
0 & 1 & 1 & 2 & 1 \\
1 & 0 & 1 & 1 & 1 \\
1 & 1 & 0 & 1 & 1 \\
\infty & \infty & \infty & 0 & \infty \\
2 & 1 & 2 & 1 & 0
\end{array} \right] \end{array}
\qquad \square
$$

The distance matrix, which contains some of the more useful information about the structure of the system modeled by a digraph, can usually be determined by inspection in a simple digraph, but in very complex digraphs it may be quite difficult to determine the shortest path between two vertices. Thus it is helpful to have a means of computing the distance matrix which does not depend upon determining shortest paths by ad hoc methods. The following theorem provides such a means.

Theorem The distance matrix $\mathbb{D}$ can be determined as follows:

If $i = j$, set $d_{ij} = 0$.
If $i \neq j$ and v_j is not reachable from v_i, set $d_{ij} = \infty$.
If $i \neq j$ and v_j is reachable from v_i, set d_{ij} equal to the smallest power n to which $\mathbb{A}$ must be raised so that $a_{ij}^{(n)} > 0$.

The use of this result will be illustrated by constructing the distance matrix for the digraph shown in Fig. 10-2. The adjacency matrix $\mathbb{A}$ and the powers $\mathbb{A}^2$ and $\mathbb{A}^3$ are (as determined in Sec. 6-6):

$$\mathbb{A} = \begin{bmatrix} 0 & 1 & 1 & 1 & 0 & 0 \\ 0 & 0 & 1 & 1 & 0 & 0 \\ 0 & 1 & 0 & 0 & 1 & 1 \\ 0 & 0 & 0 & 0 & 1 & 1 \\ 0 & 0 & 0 & 0 & 0 & 0 \\ 0 & 0 & 0 & 0 & 0 & 0 \end{bmatrix} \qquad \mathbb{A}^2 = \begin{bmatrix} 0 & 1 & 1 & 1 & 2 & 2 \\ 0 & 1 & 0 & 0 & 2 & 2 \\ 0 & 0 & 1 & 1 & 0 & 0 \\ 0 & 0 & 0 & 0 & 0 & 0 \\ 0 & 0 & 0 & 0 & 0 & 0 \\ 0 & 0 & 0 & 0 & 0 & 0 \end{bmatrix}$$

$$\mathbb{A}^3 = \begin{bmatrix} 0 & 1 & 1 & 1 & 2 & 2 \\ 0 & 0 & 1 & 1 & 0 & 0 \\ 0 & 1 & 0 & 0 & 2 & 2 \\ 0 & 0 & 0 & 0 & 0 & 0 \\ 0 & 0 & 0 & 0 & 0 & 0 \\ 0 & 0 & 0 & 0 & 0 & 0 \end{bmatrix}$$

where we have adopted the order of the vertices shown in (10-1). We begin to construct the distance matrix $\mathbb{D}$ by setting all diagonal entries equal 0. Next, enter 1 for d_{ij} whenever $a_{ij} \neq 0$, $i \neq j$. To continue, set $d_{ij} = 2$ for each (i, j) such that $a_{ij}^{(2)} \neq 0$ and $a_{ij} = 0$ [recall that $a_{ij}^{(2)}$ is the (i, j) entry in $\mathbb{A}^2$]. Finally, note that every nonzero entry in $\mathbb{A}^3$ occurs where there is already an entry in $\mathbb{D}$. Consequently, there are no paths between vertices which have not already been accounted for, and the remaining entries in $\mathbb{D}$ are set equal to ∞. We have

$$\mathbb{D} = \begin{bmatrix} 0 & 1 & 1 & 1 & 2 & 2 \\ \infty & 0 & 1 & 1 & 2 & 2 \\ \infty & 1 & 0 & 2 & 1 & 1 \\ \infty & \infty & \infty & 0 & 1 & 1 \\ \infty & \infty & \infty & \infty & 0 & \infty \\ \infty & \infty & \infty & \infty & \infty & 0 \end{bmatrix} \qquad (10\text{-}3)$$

$\square$

When the distance matrix is computed by the method used in Example 10-5, that is, by finding A, A^2, ..., it is necessary to know in advance when one is justified in stopping. An easy rule is the following. If for an integer k no new entries in D result from the computation of A^k, then no higher powers of A need be computed. The justification is straightforward. If there are vertices u and v such that v is reachable from u in k (>1) steps but in no fewer, there must be a vertex w which is reachable from u in $k - 1$ steps but no fewer. Thus, if there is an entry with value k in D, there must be at least one entry in the same row with value $k - 1$. We conclude that if there is an integer k with the property that no entry of D has value k, then no entry of D can have a value larger than k.

EXAMPLE 10-5

A fabricating firm merchandises its products through a set of distributors who place their orders directly with a factory. The factory is provided with information from the engineering division, distributors, and customers. The executive office provides information through engineering and marketing. Communication normally proceeds as shown in Fig. 10-4.

This situation can be modeled as a digraph with the vertices $u_1, \ldots, u_6$ as labeled. The adjacency matrix A for this digraph is

$$A = \begin{bmatrix} 0 & 1 & 0 & 0 & 0 & 0 \\ 1 & 0 & 1 & 0 & 0 & 0 \\ 0 & 0 & 0 & 0 & 1 & 0 \\ 0 & 1 & 0 & 0 & 0 & 0 \\ 0 & 1 & 0 & 0 & 0 & 0 \\ 0 & 0 & 1 & 1 & 0 & 0 \end{bmatrix}$$

FIGURE 10-4

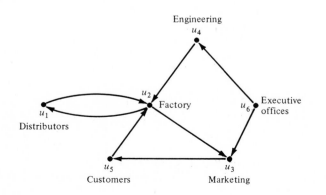

The next three powers of the adjacency matrix are

$$
A^2 = \begin{bmatrix} 1 & 0 & 1 & 0 & 0 & 0 \\ 0 & 1 & 0 & 0 & 1 & 0 \\ 0 & 1 & 0 & 0 & 0 & 0 \\ 1 & 0 & 1 & 0 & 0 & 0 \\ 1 & 0 & 1 & 0 & 0 & 0 \\ 0 & 1 & 0 & 0 & 1 & 0 \end{bmatrix} \qquad A^3 = \begin{bmatrix} 0 & 1 & 0 & 0 & 1 & 0 \\ 1 & 1 & 1 & 0 & 0 & 0 \\ 1 & 0 & 1 & 0 & 0 & 0 \\ 0 & 1 & 0 & 0 & 1 & 0 \\ 0 & 1 & 0 & 0 & 1 & 0 \\ 1 & 1 & 1 & 0 & 0 & 0 \end{bmatrix}
$$

$$
A^4 = \begin{bmatrix} 1 & 1 & 1 & 0 & 0 & 0 \\ 1 & 1 & 1 & 0 & 1 & 0 \\ 0 & 1 & 0 & 0 & 1 & 0 \\ 1 & 1 & 1 & 0 & 0 & 0 \\ 1 & 1 & 1 & 0 & 0 & 0 \\ 1 & 1 & 1 & 0 & 1 & 0 \end{bmatrix}
$$

Using the technique described above, the distance matrix D of this model is found to be

$$
D = \begin{bmatrix} 0 & 1 & 2 & \infty & 3 & \infty \\ 1 & 0 & 1 & \infty & 2 & \infty \\ 3 & 2 & 0 & \infty & 1 & \infty \\ 2 & 1 & 2 & 0 & 3 & \infty \\ 2 & 1 & 2 & \infty & 0 & \infty \\ 3 & 2 & 1 & 1 & 2 & 0 \end{bmatrix}
$$

This shows, for example, that the executive offices can reach all the units included in the model in three or fewer steps. $\square$

Indices

In situations which can be appropriately modeled using digraphs, one is frequently concerned with assigning a numerical measure of some attribute to each vertex or, more accurately, to whatever the vertex represents. For example, if the digraph represents consultations among the doctors in a clinic, one might wish to evaluate each doctor in terms of his usefulness to his colleagues. In order to compare individuals it is convenient to represent the evaluations in terms of numbers and to use the ordinary order of the numbers to make comparisons.

We introduce here one index which has been used in applications. Another is the topic of Exercises 11 and 12. We begin with a useful definition.

Definition   The *distance sum of a vertex* v_j in a digraph G is the sum of the finite entries in the jth row of the distance matrix of G.

In order to provide background for the introduction of an index of status consider a digraph which represents a business or government organization chart. Suppose that the vertices represent individuals (or positions) in the organization and the edge (u,v) denotes that u is the direct supervisor of v. How might the status of an individual in this organization be measured? There are, of course, many possibilities, and any specific hypothesis becomes acceptable only after confirmation by empirical work. We adopt a definition of status which depends on the supervisory relationships given in the organization chart.

What characteristics might reasonably be required of a measure of status based on supervisory relationships? A simple condition which seems entirely natural is that an individual who supervises two others should have higher status than an individual who supervises only one other individual. A more difficult question is how the statuses of u and v in Fig. 10-5 should be compared. It is reasonable to suppose that the status of u exceeds the status of v. Indeed, u and v each supervise (directly or indirectly) two others. Those supervised by v are supervised directly. On the other hand, u influences w directly but x only through w. It is common in such situations to view the impression of u on x as magnified by the presence of w as a sort of intermediary.

A straightforward definition of status which has the features just discussed is the following.

Definition   Let v be a vertex in a digraph G. The *status* $S(v)$ of v is the distance sum of v.

Why this definition provides a measure of status with the desired properties is made clear by the following equation for $S(v)$:

$$S(v) = 1(\text{number of vertices at distance 1 from } v)$$
$$+ 2(\text{number of vertices at distance 2 from } v)$$
$$+ 3(\text{number of vertices at distance 3 from } v)$$
$$+ \cdots \tag{10-4}$$

Since G has only a finite number of vertices, this sum contains only a finite number of terms.

FIGURE 10-5

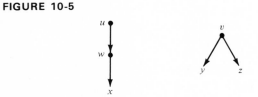

EXAMPLE 10-6

Find the status of each individual in the organization described in Fig. 10-2.

Solution: The distance matrix of the digraph of the organization pictured in Fig. 10-2 was determined above. It is given in Eq. (10-3). Using this distance matrix, we have

$$
\begin{array}{ll}
S(P) = 7 & S(T) = 2 \\
S(VPF) = 6 & S(DLR) = 0 \\
S(VPP) = 5 & S(DR) = 0
\end{array}
$$
$\square$

Exercises for Sec. 10-2

Find the adjacency matrix for the diagraph in:

1. Fig. 10-6*a*   **2.** Fig. 10-6*b*   **3.** Fig. 10-6*c*

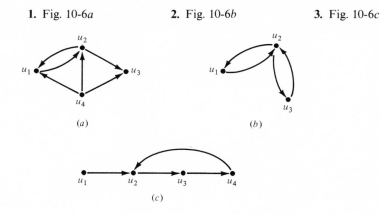

(*a*)   (*b*)

(*c*)

FIGURE 10-6

4. Let $G = [V,E]$ with $V = \{v_1, v_2, v_3, v_4\}$ and
   $E = \{(v_1,v_2), (v_2,v_3), (v_3,v_2), (v_3,v_4), (v_4,v_3)\}$.
   a. Draw a diagram of this digraph.
   b. Find the adjacency matrix.
   c. Find the distance matrix.

5. Let $G = [V,E]$ with $V = \{v_1, v_2, v_3, v_4, v_5, v_6\}$ and
   $E = \{(v_1,v_6), (v_6,v_1), (v_1,v_2), (v_6,v_2), (v_3,v_2), (v_5,v_3), (v_2,v_5), (v_4,v_3)\}$.
   a. Draw a diagram of this digraph.
   b. Find the adjacency matrix.
   c. Find the distance matrix.

Find the distance matrix for the digraph in:

6. Fig. 10-6*a*   **7.** Fig. 10-6*b*

8. Determine the status of the customers in the digraph shown in Fig. 10-4.

9. Which vertex in the digraph pictured in Fig. 10-3 has the highest status?

10. In the organization described in Sec. 10-1 and pictured in Fig. 10-2, suppose that in addition to all relationships shown, the treasurer could issue directives to the vice president for finance. Whose status changes? What is the new position of the treasurer in the corporate hierarchy?

The topic of the next two exercises is another index which has proved useful in applications. A definition is needed.

Definition

The *distance sum of a digraph G* is the sum of all finite entries in the distance matrix of G. Let v be a vertex in a digraph G, and suppose that the distance sum of v is not zero. The *index of relative centrality of v* is

$$\frac{\text{Distance sum of } G}{\text{Distance sum of } v}$$

The index of relative centrality is not defined for vertices for which the distance sum is zero.

11. Find the index of relative centrality of the vertex v of the digraph in Fig. 10-7.

FIGURE 10-7

12. Find the index of relative centrality of each vertex in each of the digraphs in Fig. 10-8. Do the results agree with your intuition?

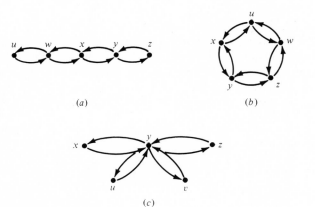

(a)

(b)

(c)

FIGURE 10-8

13. Let $G_1 = [V,E_1]$ and $G_2 = [V,E_2]$ be two digraphs with the same set of vertices. The *intersection* digraph is defined to be the digraph with vertex set V and edge set $E_1 \cap E_2$.
    a. Find the intersection digraph of the two digraphs in Fig. 10-6 *a* and *c*.
    b. Find the adjacency matrix of this intersection digraph.

14. Suppose that five members of a class are asked the following two questions:
    Which members of this group would you most like to have as your partners in a term-paper-writing project?
    With which members of this group would you most like to go to a concert?
    On the basis of the answers to these two questions two digraphs are constructed. In Fig. 10-9 suppose that G_1 is the digraph constructed on the basis of the answers to the first question and G_2 is the digraph constructed on the responses to the second question. An edge (u,v) means that u selects v.
    a. Find the adjacency matrices for these digraphs.
    b. Find the intersection graph and its adjacency matrix.
    c. Interpret the intersection graph in light of the original situation.

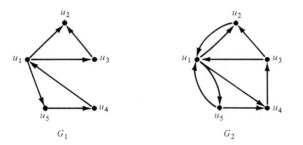

FIGURE 10-9
G_1 G_2

15. Determine the status of each of the vertices in Fig. 10-2 by using Eq. (10-4) instead of the definition (which involves the distance matrix).

10-3   NETWORKS

A digraph represents relations or connections between things (individuals, units in an organization, etc.). In many applications it is necessary to consider not only whether there is a relation or connection between two things but also the intensity or magnitude of the relation. For example, it may be germane that it is not only possible to fly from location A to location B but also that it takes 2 hours and costs \$150 to do so. The following definition, formulated in response to this need, introduces a special kind of digraph which includes additional information of a certain type.

Definition   A *network* is a digraph each of whose edges has a positive number associated with it.

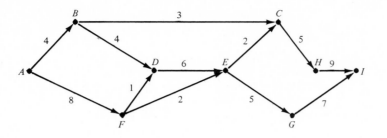

FIGURE 10-10

Depending upon the situation, the numbers associated with the edges of a network may be interpreted in different ways. For example, they might be viewed as costs, in time or money, or as capacities associated with the edges.

EXAMPLE 10-7

The N. T. Quaited Freight Line operates by shuttling trucks back and forth between regional warehouses. One Monday morning trucks are scheduled between warehouses A to I in the directions shown by the arrows on the digraph in Fig. 10-10. The numbers beside the edges show the unused capacity (in thousands of pounds) on each route. Thus, since there is an arrow from B to C with the number 3 attached to it, there are trucks running from warehouse B to warehouse C and these trucks have 3000 pounds of unused capacity. ☐

The *value* of a path in a network is the sum of the numbers associated with the edges which make up the path. If the numbers refer to costs (or capacities), it is common to refer to the cost (capacity) of a path.

EXAMPLE 10-8

Interstate highway distances between St. Louis and Dallas are shown on Fig. 10-11. This can be viewed as a network with vertices S, K, L, M, T, O and D, and edges (K,S), (S,K), (S,M), (M,S), etc. The value of the path $SKTOL$ is 945. ☐

Definition   Let u and v be two vertices in a network, and suppose that v is reachable from u. A path from u to v which has the smallest value of all such paths is said to be a *geodesic* from u to v.

The path $STOD$ is a geodesic from S to D in Fig. 10-11. Problems of determining geodesics arise in many forms.

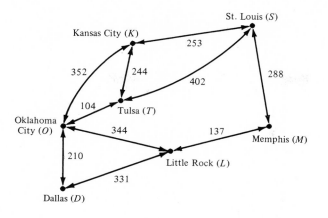

FIGURE 10-11

If the number of vertices in a network is small, it is easy to determine a geodesic between two vertices u and v simply by finding the values of all paths from u to v and selecting one of the paths with the smallest value. But if the number of vertices is even moderately large, this direct evaluation approach results in a very large number of computations. The computational task increases rapidly as the size of the digraph increases, and even with computer assistance it is unrealistic for digraphs of the size which commonly occur in applications.

To illustrate the computational difficulties of a direct approach, let us return to the situation in Example 10-2. The figure from that example is reproduced in Fig. 10-12, where vertices are labeled with letters and, since the utility company is interested in a direct route, the edges have been given directions as shown. How many paths are there from the power substation A to the trade building D'? Each path has to include five blocks on which one goes east and four blocks on which one goes north, a total of nine blocks; e.g., the path which alternates going east and north is the path $ABHIOPVWC'D'$. It follows that in order to count the number of direct paths from A to D' it is sufficient to count the number of ways that one can select five blocks (on which to head east) out of nine. There are

$$\binom{9}{5} = \frac{9!}{5!\,4!} = 126$$

ways to select 5 blocks from 9, and consequently there are 126 paths. Since each path includes 9 blocks, 9 additions are required to determine the value of each path. Therefore, we see that if the problem is attacked by brute force and the value of each path is determined, there are $126 \times 9 = 1134$ operations involved. In addition to performing the arithmetic operations it is necessary to keep track of which path is being used and to be certain that all paths are considered. This is still a relatively small problem, and certainly it can be handled by a computer,

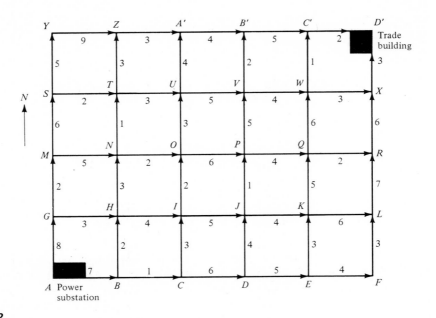

FIGURE 10-12

but similar problems which arise in industrial or business applications may have hundreds of vertices. Clearly a method which is more efficient than simple evaluation is needed.

The method, or *algorithm*, which we introduce now is in some respects representative of those developed to solve other problems in graph theory. The study of such algorithms is an active branch of mathematical research. We begin with an example which is similar to the electric utility situation of Example 10-2 but simpler.

EXAMPLE 10-9

Find a geodesic from A to G in the digraph pictured in Fig. 10-13.

FIGURE 10-13

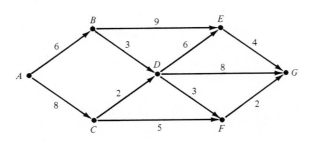

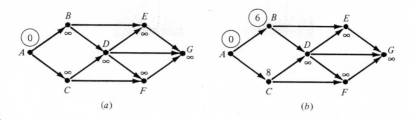

FIGURE 10-14

Solution: We begin by labeling vertex A with the permanent label 0, and we tentatively label all other vertices with ∞. The graph appears as in Fig. 10-14a. Next we compare one by one the labels on each of the vertices B through G with the sum of the label at vertex A and the direct distance from A to that vertex. (Remember that if there is no edge between two vertices, we assign ∞ to the distance between them. The symbol ∞ is to be considered larger than any number, and the sum of ∞ and any number is again ∞. Finally, the result of adding ∞ to ∞ is ∞.) The smaller of these two numbers is the new tentative label. The results of this step are shown in Fig. 10-14b. Since only two vertices (B and C) are directly connected to A, the labels at these vertices are the only non-∞ labels at this stage. Next we determine the smallest of these new temporary labels and make it permanent. We identify permanent labels by circling them. In this example the vertex B was given the permanent label 6. If at this step more than one vertex has the smallest label, we arbitrarily pick one of them to be assigned a permanent label.

To continue, at each vertex with a temporary label we compare the temporary label with the sum of the permanent label of vertex B and the direct distance from B to that vertex. The smaller of these two numbers becomes the new temporary label at that vertex. We then determine the smallest of the temporary labels and declare it permanent. The labeled network which results from this step is shown in Fig. 10-15. In that figure vertex C has permanent label 8.

The algorithm continues by repeating the step described just above, beginning with the vertex most recently given a permanent label. The next three applications

FIGURE 10-15

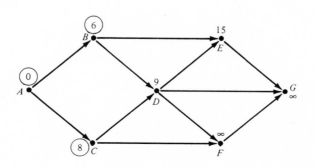

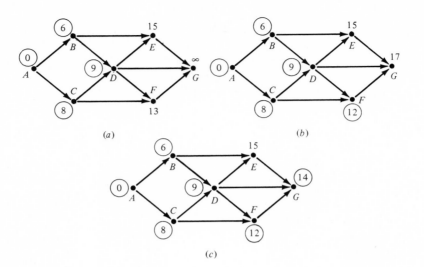

(a)

(b)

(c)

FIGURE 10-16

of the basic step of the algorithm result in the labeled networks shown in Fig. 10-16a, b, and c. In Fig. 10-16c the vertex G has been given a permanent label, and this signals that our problem is solved. The shortest path from A to G has value 14.

To determine the path, one simply works backward from G, picking out the labeled vertices which differ in value by exactly the value of the connecting edge. In this case the path is constructed by noting that the label of F differs from that of G by 2, the value of (F,G). [Notice that the label of D does not differ from that of G by the value of (D,G).] Next, the label of D differs from that of F by the value of (D,F), and the label of B differs from that of D by the value of (B,D). Finally, the label of B is the value of (A,B). Therefore, the shortest path from A to G is ABDFG, and its value is 14.   □

Notice that exactly one vertex is given a permanent label at each step. Also, one vertex is given a permanent label at the beginning of the problem. Consequently, if there are N vertices in the network, at most N-1 applications of the basic step will provide a geodesic between two specified vertices. In Example 10-9 there are seven vertices, and therefore at most six repetitions of the basic step will provide a geodesic between A and G. Actually only five such repetitions were necessary. The vertex E was never assigned a permanent label.

Summary

Let us now summarize the algorithm without reference to a specific network:

Suppose that our problem is to find a geodesic from u to v in a network which contains N vertices.

Initial step: Assign u the permanent label 0. Circle all permanent labels to distinguish them from temporary labels. Assign all vertices other than u the temporary label ∞.

Note: When the algorithm is implemented on a computer, the symbol ∞ must be replaced by a large number. For example, if M is the sum of the costs assigned to all edges, ∞ can be replaced by M and the algorithm proceeds without further change.

Basic step: (To be repeated.) Suppose that w is the vertex most recently assigned a permanent label. At each vertex which still has a temporary label compare the temporary label with the sum of the permanent label at vertex w and the cost of the edge from w to that vertex. The smaller of these two numbers is the new temporary label of the vertex. After all new temporary labels have been determined, select the smallest label (or select arbitrarily one of them if there are several) and identify it as permanent.

After at most $N - 1$ repetitions of the basic step the vertex v will be assigned a permanent label. This label is the cost of a geodesic from u to v. The geodesic itself can be obtained by working backward from v in the following manner: Find a vertex w' such that the sum of the permanent label at w' and the cost

FIGURE 10-17

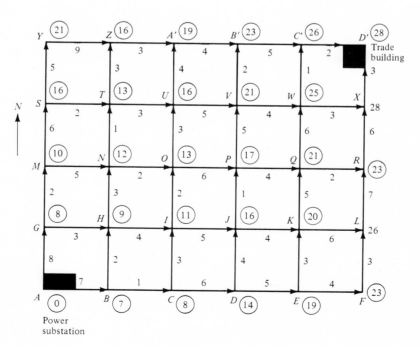

associated with (w',v) equals the permanent label at v. Next, find a vertex w'' such that the sum of the permanent label at w'' and the cost associated with (w'',w') equals the value of the permanent label at w'. Continuing at most $N-1$ steps, one reaches the vertex u. The path $u \cdots w''w'v$ is a cost geodesic from u to v. Although the cost of a geodesic from u to v is unique, in general the geodesic itself will not be.

EXAMPLE 10-10

Find a geodesic from vertex A to vertex D' in the network shown in Fig. 10-12.

Solution: The labeled network is shown in Fig. 10-17. The cost of the geodesic from A to D' is 28, and one such geodesic is $ABHNTUVWC'D'$.   □

Exercises for Sec. 10-3

1. Find a cost geodesic in the network shown in Fig. 10-13 with the cost of edge (C,F) equal to 1 (instead of 5).

2. Find a cost geodesic from A to G in the network shown in Fig. 10-18.

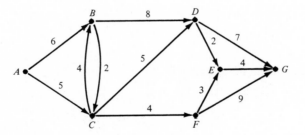

FIGURE 10-18

3. Find a cost geodesic from A to I in the network shown in Fig. 10-19.

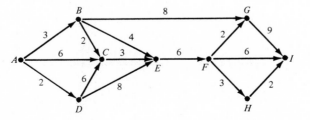

FIGURE 10-19

4. Find a cost geodesic from A to O in the network shown in Fig. 10-20.

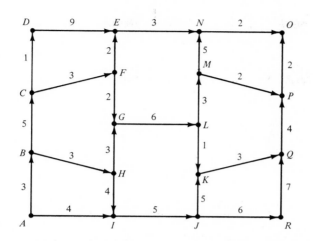

FIGURE 10-20

5. Suppose in Example 10-7 that trucks do not leave a warehouse until after all incoming trucks have arrived. Find the maximum amount which can be shipped from A to I by taking advantage of the unused capacity shown in Fig. 10-10. (*Note*: This is not a problem of finding a cost geodesic, and consequently the algorithm of this section cannot be applied. Use a trial-and-error technique to solve the problem.)

IMPORTANT TERMS

You should be able to describe, define, or give examples of each of the following:

Digraph
Vertex
Edge
Adjacency matrix
Path
Length of a path

Distance matrix
Status
Network
The value of a path in a network
Geodesic

REVIEW EXERCISES

1. Find the adjacency matrix for the digraph in Fig. 10-21.

2. Find the distance matrix for the digraph in Fig. 10-21.

3. Let $G = [V,E]$ with $V = \{v_1, v_2, v_3, v_4\}$ and $E = \{(v_1,v_4), (v_4,v_2), (v_4,v_3), (v_2,v_1), (v_3,v_2)\}$. Draw a diagram of this graph.

4. Find the adjacency matrix and the distance matrix for the digraph given in Exercise 3.

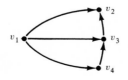

FIGURE 10-21

5. Find the distance matrix for the digraph represented by Fig. 10-22.

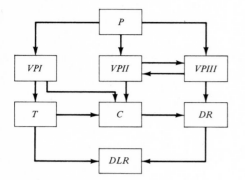

FIGURE 10-22

Find the status of each of the individuals in the organization represented by:

6. Fig. 10-21   **7.** Fig. 10-22

8. Find a cost geodesic from B to G in the network shown in Fig. 10-18.

9. Find a cost geodesic from B to P in the network shown in Fig. 10-20.

10. Find a cost geodesic from A to R in the network shown in Fig. 10-23.

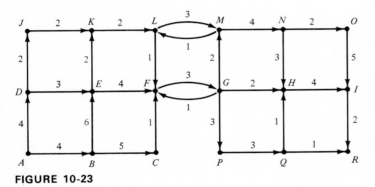

FIGURE 10-23

EVALUATING INVESTMENT OPTIONS

ELEVEN

11-1 INTRODUCTION

A common problem in allocating financial resources is: How can I use the money I have most efficiently? A simple example of a decision problem of this type is the following. Suppose that your work for your employer has been favorably reviewed and you are offered a $1000 bonus. You are given the following alternatives: (1) you may take the $1000 now, or (2) you may wait 6 months and receive $1100. Suppose also that your employer is trustworthy (you really will get the $1100 in 6 months) and you do not expect to need the $1000 during the coming 6 months. What should you do, take the money now or wait and receive more later?

To answer this question several factors must be examined. In particular you must consider the various ways of using and/or investing the $1000 if it is taken immediately; and to make an intelligent decision you must be able to compare the returns on these alternatives with the return obtained by waiting 6 months and receiving $1100. For example, you could deposit the $1000 in a savings account, or you could buy a bond, or you could buy stocks, or you could use it as a down payment on a new car. The goal of this chapter is to discuss some of the methods of determining the values of such investments so that they can be compared. We begin by considering the general definition of interest and, in particular, the return on a savings account. We proceed with a discussion of investments such as annuities, bonds, and stocks.

11-2   INTEREST

It costs money to borrow money, and, conversely, one can make money by lending money. *Interest* is the term used to describe the fee which is paid for the use of money. The *interest rate* is the percentage obtained by expressing the amount of interest as a percentage of the amount of money used. Since the money is used for a certain period of time, the interest rate is expressed in terms of a percentage per unit time. For example, if $1000 is used for 1 year and the interest is $50, the interest rate is $\frac{50}{1000} = .05 = 5$ *percent per year*. If $1000 is used for 6 months and the interest is $30, the interest rate is $\frac{30}{1000} = .03 = 3$ *percent per half year*. Because interest rates are so frequently compared, it is customary to express them in per year terms to facilitate comparisons. This is easily done by assuming that the same rate of payment is applied for 1 year as is applied to the given period. Thus, in the last example we assume that the fee for using the $1000 for the second 6 months of the year is also $30 (as it was for the first 6 months), and hence the fee for the year is $60 and the interest rate is 6 percent per year. As another example, suppose that $1000 is used for 2 months and the interest is $20, then the interest rate will be said to be 12 percent per year $[(\frac{20}{1000})(\frac{12}{2}) = .12]$ rather than 2 percent per 2-month period.

In any transaction involving borrowing money the amount of money borrowed is called the *principal* of the loan. The basic formula involving the principal P, the interest I, the interest rate r, and the time period t is

$$I = Prt \tag{11-1}$$

Thus, given any three of the four quantities I, P, r, t, the fourth can be obtained from Eq. (11-1).

EXAMPLE 11-1

Find the interest on a loan of $5000, given that the interest rate is 18 percent per year and the loan is for 6 months.

Solution: Using Eq. (11-1) with P and I measured in dollars, we have $P = 5000$, $r = .18$, and $t = \frac{1}{2}$. Thus the amount of interest is

$$I = (5000)(.18)(\tfrac{1}{2}) = 450 \qquad \square$$

EXAMPLE 11-2

Suppose that you can afford to pay $200 interest on a loan for 6 months. If the prevailing interest rate is 10 percent per year, how much can you afford to borrow?

Solution: We suppose that the entire $200 is used to pay interest. Then, using Eq. (11-1), the amount P which can be borrowed is $P = I/rt$. Therefore

$$P = \frac{\$200}{(.10)(\frac{1}{2})} = \$4000$$

$\square$

If a person deposits money in a savings account at a bank, the bank has, in effect, borrowed that person's money, and accordingly the bank should pay the person a fee (interest). The bank may compute and pay this fee in different ways with different rates, depending upon the type of account and the prevailing interest rates in the economy. One of the most common types of accounts is one in which the interest is *compounded* and automatically deposited in the account periodically. We now describe this method in detail.

EXAMPLE 11-3

The Fourth Federal Bank of Mulberry offers a savings account in which the interest is paid at a rate of 6 percent per year, computed every 3 months (quarterly) and automatically credited to the account. At the end of each quarter the interest is computed using as principal the minimum amount in the account during the quarter. Assuming no deposits or withdrawals, the amount on deposit at the beginning of each quarter consists of the original amount in the account plus all interest previously credited to the account.

Peni Saver opens such an account, in which she initially deposits $1000. How much will be in the account after 5 years, assuming no deposits or withdrawals other than the interest being credited to the account by the bank?

Solution: At the start of the first quarter (3-month period) there is $1000 in her account. At the end of this quarter the bank computes the interest by using Eq. (11-1) and obtains

$$I = Prt = (\$1000)(.06)(\tfrac{1}{4}) = \$15$$

This $15 is now added to the account, and we assume that it is left in the account for the second quarter. At the end of the second quarter the bank computes the interest on $1015 and adds this amount to the account. Again using Eq. (11-1), the interest is $I = Prt = (\$1015)(.06)(\tfrac{1}{4}) = \$15.23$. Repeating this argument a total of 20 times (20 quarters is 5 years) gives the final amount. This type of account is said to pay *compound interest* since the interest is always based on the original principal plus any interest already credited to the account. If the interest paid at each quarter is based only on the original principal, the account is said to pay *simple interest*. Table 11-1 gives a comparison of the

TABLE 11-1

	Simple Interest		Compound Interest	
Quarter	Quarterly Payment	Total in the Account at End of Quarter	Quarterly Payment	Total in the Account at End of Quarter
1	$15	$1015	$15.00	$1015.00
2	15	1030	15.22	1030.22
3	15	1045	15.45	1045.67
4	15	1060	15.69	1061.36
5	15	1075	15.92	1077.28
6	15	1090	16.16	1093.44
7	15	1105	16.40	1109.84
8	15	1120	16.65	1126.49
9	15	1135	16.90	1143.39
10	15	1150	17.15	1160.54
11	15	1165	17.41	1177.95
12	15	1180	17.67	1195.62
13	15	1195	17.93	1213.55
14	15	1210	18.20	1231.75
15	15	1225	18.47	1250.23
16	15	1240	18.75	1268.98
17	15	1255	19.03	1288.02
18	15	1270	19.32	1307.34
19	15	1285	19.61	1326.95
20	15	1300	19.91	1346.86

amounts in the account at various times when simple and compound interest are paid.[1]

The answer to Peni's question is contained in the last row of Table 11-1. However, it is unlikely that one will always have a table available (or be willing to construct one), and thus it is useful to be able to compute the last row of the table without the other rows.   □

Computing Compound Interest

The entries in Table 11-1 under the headings Quarterly Payment are all obtained from the basic interest formula $I = Prt$. In the case of simple interest P, r, and t are the same every quarter and hence I is always $15[= (\$1000)(.06)(\frac{1}{4})]$. In the case of compound interest the quantities r and t are always the same, but the principal P increases each quarter by the amount of interest paid the previous

---

[1] In actual practice a savings institution may obtain slightly different values for the entries in Table 11-1, depending on the number of digits used in their computations and on their policy of rounding off numbers.

quarter. If we denote the original principal by P, the principal at the end of quarter 1 by P_1, and the principal at the end of quarter i by P_i, the interest paid at the end of quarter $i + 1$ is I_{i+1}, where $I_{i+1} = P_i rt$. Thus $P_{i+1} = P_i + I_{i+1} = P_i + P_i rt$. There is a very useful formula for computing the amounts P_n directly without computing all P_i and I_i for i less than n. The formula can be deduced by considering a number of special cases from Example 11-3. Since $P = 1000$, we have

$$P_1 = P + Prt = 1000 + (1000)(.06)(\tfrac{1}{4}) = 1000(1 + .015)$$
$$P_2 = P_1 + P_1 rt = 1000(1 + .015) + 1000(1 + .015)(.06)(\tfrac{1}{4}) = 1000(1 + .015)^2$$
$$P_3 = P_2 + P_2 rt = 1000(1 + .015)^2 + [1000(1 + .015)^2](.015) = 1000(1 + .015)^3$$

Similarly, for any number n of periods

$$P_n = 1000(1 + .015)^n \qquad (11\text{-}2)$$

We note that the quantity .015 in (11-2) is the *interest rate per quarter*, $(.06)(\tfrac{1}{4}) = .015$, and n is the *number of quarters*. The general formula for compound interest has a similar form.

Formula for Compound Interest

If an amount P is invested for n units of time and interest is compounded and added at a rate of k per unit of time, then the total investment after the n units of time is

$$P_n = P(1 + k)^n \qquad (11\text{-}3)$$

Note: In Eq. (11-3) k is not necessarily the interest rate per year but is the interest rate per unit of time for the unit of time considered. If the unit of time is 1 year, then k is the interest rate per year.

EXAMPLE 11-4

If \$1000 is invested for 10 years and interest is compounded and added every 6 months at a rate of 10 percent per year, what is the total amount of interest added to the investment by the end of the tenth year?

Solution: We use Eq. (11-3) to compute the total of the original investment plus the interest added. Since the interest rate is 10 percent *per year*, it is 5 percent *per 6-month period*, that is, $k = .05$. Also, the number of 6-month periods is 20, and so the total investment after 10 years is $P_{20} = (1000)(1 + .05)^{20} = 1000(1.05)^{20} = 1000(2.653298) = \$2653.30.$*

Therefore, the amount of interest added is $\$2653.30 - \$1000.00 = \$1653.30.$

$\square$

Frequently, especially in their advertising, savings institutions extoll the virtues of their own savings account and in particular call attention to the frequency with which they compute and pay compound interest. Certainly the frequency of payment makes a difference when interest is compounded, and it is interesting (no pun intended) to consider and evaluate this difference. Suppose that four different institutions pay the same basic rate, say 6 percent per year, but one compounds and pays interest yearly, one every 6 months, one every 3 months, and one monthly. What is the difference in the amounts in the various accounts after 10 years, given the same initial investment of \$1000?

We can answer the question just posed by using Eq. (11-3) for each of the four accounts. The results are shown in Table 11-2.

TABLE 11-2

Frequency of Interest Payments	Initial Investment	Formula for P_n	Amount in Account after 10 years
Annually	\$1000	$1000(1 + .06)^{10}$	\$1790.85
Semiannually	1000	$1000(1 + .03)^{20}$	1806.11
Quarterly	1000	$1000(1 + .015)^{40}$	1814.02
Monthly	1000	$1000(1 + .005)^{120}$	1819.40

It is clear from Table 11-2 that a saver profits by having interest compounded and paid frequently. However, the gain in going from 2 periods per year to 4 periods (\$7.91) is not as great as the gain going from 1 period to 2 periods per year (\$15.26). Moreover, the gain in going from 4 periods per year to 12 is smaller yet (\$5.38). Thus, at least for this example, it appears that there would be relatively little gain in compounding and paying interest even more frequently, say weekly or daily. Indeed, this is the case in general, and the relative gain obtained by computing interest hourly or every minute is very small. To indicate how the amount of interest paid increases as it is compounded and paid more and more frequently it is necessary to examine the quantity P_n in Eq. (11-3) and to consider how it changes when the number of periods per year

---

* It is tedious and difficult to carry out the arithmetic to obtain P_{20} and similar terms by hand, but it is a simple matter to carry out these computations using a relatively inexpensive hand calculator. Such a calculator was used for many of the examples in this chapter and it will be needed for many of the exercises.

gets very large. Thus suppose that interest is being paid at the rate of r per year and that interest is compounded and paid M times per year. Thus the interest rate per period is r/M. Using Eq. (11-3), at the end of 1 year a principal P will have increased in value to an amount

$$P_M = P\left(1 + \frac{r}{M}\right)^M$$

In order to study how this quantity changes when interest is compounded more often, one must study the behavior of the quantity $(1 + r/M)^M$ as M increases. For $r = .06$, the size of this quantity for different choices of M is shown in Table 11-3.

Since M is the number of periods per year, we see from Table 11-3 that there is little change in the quantity $(1 + .06/M)^M$ as the number of periods per year gets very large. In fact, the relative change between $M = 24$ and $M = 8640$ is only of the order of 10^{-4}. The bottom two rows of the table correspond to computing interest daily ($M = 360$) and hourly ($M = 8640$). For all large values of M, say $M = 1,000,000$ or $M = 10,000,000$, the quantity $(1 + .06/M)^M$ has essentially the same value because $P_M = (1 + .06/M)^M$ has a *limit* as M gets large. To say that P_M has a limit as M gets large means that there is a number L such that P_M is *always close to L* when M is a large number, and, moreover, the value of P_M can be made *as close to the number L as one wishes* by simply taking M large enough. The limit of $(1 + r/M)^M$ as M gets large depends on r, and it is denoted by e^r, where e is the base for natural logarithms ($e \approx 2.71828$). Verification can be based on the fact that the number e is the limit of the quantity $(1 + 1/N)^N$ as N gets large. Indeed, the quantity P_M can be written

$$P_M = \left(1 + \frac{r}{M}\right)^M = \left[\left(1 + \frac{1}{M/r}\right)^{M/r}\right]^r$$

TABLE 11-3

M	$\left(1 + \dfrac{.06}{M}\right)^M$
1	1.06000
2	1.06090
3	1.06121
4	1.06136
6	1.06152
12	1.06168
24	1.06176
360	1.06183
8640	1.06184

The ratio M/r is large if M is large (r is a fixed number), and we see that P_M has the limit e^r as M gets large. The number M is the number of equally spaced times that interest is computed each year. Thus, taking larger and larger values of M corresponds to computing interest more and more frequently. In the limit we say we are computing interest *continuously*. The value at the end of 1 year when \$1 is invested and interest is computed continuously at a yearly rate r is e^r dollars. The value at the end of 1 year when P dollars are invested and interest is paid continuously at the rate of r per year is Pe^r dollars.

Similarly, if interest is compounded continuously for a period which is different from that of a year, say for t years, where t is any positive number, then the value of an initial investment of P dollars after t years is P_t, where

$$P_t = Pe^{rt} \tag{11-4}$$

We remark that in Eq. (11-4) r is the *yearly interest rate* and interest is *computed continuously* for t years.

Values of e^r for various choices of r can be obtained from tables or with the help of electronic calculators, and these values can be used to compute interest continuously. As Table 11-3 indicates, there is very little difference between the interest determined by computing interest monthly and continuously (the difference is about \$.16 on \$1000 left in an account 1 year). Thus the value Pe^r is a very good estimate of the amount in an account after 1 year when the initial amount is P, the interest rate is r, and interest is computed monthly, weekly, daily, or more frequently.

We conclude this section with a discussion of the term *effective rate of interest*. Whenever interest is compounded and paid more frequently than once a year, one can ask what rate of interest compounded and paid just once a year would give the same return. The once-a-year rate is called the *effective rate of interest* or the annual percentage rate or APR.

EXAMPLE 11-5

What is the effective rate of interest on an account which pays 6 percent per year and which compounds interest quarterly?

Solution: From Table 11-3 we see that the value of \$1 at the end of 1 year in an account in which the interest rate is 6 percent per year and in which interest is compounded 4 times is $(1 + .015)^4 = \$1.06136$. Since the original investment is \$1, the interest is $\$1.06136 - \$1 = \$.06136$. Thus, using the formula $I = Prt$ with $I = .06136$, $P = \$1$, and $t = 1$ year, we have an effective rate of interest of

$$\text{effr} = \frac{.06136}{1(1)} = 6.136 \text{ percent per year} \qquad \square$$

To find the effective rate of interest for an account which pays interest at a rate of r per year compounded M times per year we carry out a computation similar to that of Example 11-5. Thus we consider the value after 1 year of $1 invested at rate r per year compounded M times per year. This value is $(1 + r/M)^M$. Since the original investment is $1, the interest in 1 year is $(1 + r/M)^M - 1$. The effective rate of interest, denoted effr, is the rate which when paid once, at year end, yields this interest on an investment of $1 for 1 year. Equation (11-1) for the effective rate is $I = P(\text{effr})t$. Thus with $t = 1$, $P = 1$, and $I = (1 + r/M)^M - 1$, we have

$$\text{effr} = \left(1 + \frac{r}{M}\right)^M - 1 \tag{11-5}$$

EXAMPLE 11-6

Find the effective rate of interest on an account which pays 12 percent per year compounded monthly.

Solution: We use Eq. (11-5) with $r = .12$ and $M = 12$. We have

$$\text{effr} = \left(1 + \frac{.12}{12}\right)^{12} - 1 = (1.01)^{12} - 1 = 1.1268 - 1 = .1268 = 12.68\% \quad \square$$

Exercises for Sec. 11-2

1. Find the amount in an account at the end of the period indicated when interest is compounded at the specified rate and frequency.
   a. Principal = $10,000, term = 3 years, interest at 5 percent per year compounded twice a year.
   b. Principal = $5000, term = 2 years, interest at 10 percent per year compounded 4 times per year.
   c. Principal = $1000, term = 10 years, interest rate of 6 percent per year compounded continuously.

2. Find the effective rate of interest for each of the problems in Exercise 1.

3. What is the rate of interest offered to the individual in the introduction to this chapter? That is, if $1000 is left for 6 months and then it has value $1100, what is the effective rate of interest on this investment?

4. Suppose that interest on an investment is compounded continuously at a rate of 6 percent per year and after 2 years the investment is worth $563.26. What is the principal?

5. On January 1, 1976, Peni Saver deposited $2000 in an account which pays 6 percent per year compounded quarterly. On July 1, 1976, she withdrew $100, and on January 1, 1977, she withdrew another $100. How much was in the account on July 1, 1977?

6. If $10,000 is to be invested for 5 years, is it more profitable to invest it in an account which pays 6 percent per year compounded continuously or in an account which pays 6.25 percent per year compounded quarterly?

7. An amount P is invested in an account which pays 6 percent per year compounded quarterly. After 2 years the account contains $2500. What was the amount P of the original investment?

8. Nick L. Saver purchased a tract of unimproved land in 1960 for $80,000 and sold it in 1975. After all expenses were paid, he had a net profit (before taxes) of $60,000. What is the effective annual yield on his investment of $80,000? That is, at what interest rate r compounded yearly would $80,000 grow into $140,000 in 15 years?

9. The Safety First Savings and Loan Association increases the rate paid on savings accounts from 5 to $5\frac{1}{2}$ percent, each compounded quarterly. What are the old and new effective rates of interest?

10. If inflation is at an annual rate of 7.5 percent, what rate must the Safety First Savings and Loan Association offer its customers for their savings to grow at least as fast as inflation? Remember that Safety First compounds interest quarterly.

11-3 PRESENT VALUE AND ANNUITIES

How much should one pay for the right to receive a certain amount of money at a specified time in the future? To answer this question some additional information is needed. In particular one must know exactly when the money is to be paid and also what interest rate could be obtained on money invested between now and the time of payment. Moreover, in addition to knowing the rate of interest one should also know how often it will be compounded and paid.

The term *present value* is usually used to indicate the amount which one would pay to receive a certain payment in the future. Specifically:

Definition

The *present value* of an amount A which is to be paid after M periods of time with an interest rate of k per period is the principal P which when invested at this rate for this number of periods increases to the value A.

In Sec. 11-2 it was shown that a principal P invested for M periods of time with interest compounded at a rate of k per period will increase in value to $P(1 + k)^M$. In this situation we know A, k, and M, and we seek P such that $A = P(1 + k)^M$. Thus

$$P = \frac{A}{(1 + k)^M} \tag{11-6}$$

is the present value of A.

EXAMPLE 11-7

What is the present value of $1000 to be paid in 5 years if the interest rate is 10 percent and it is compounded semiannually?

Solution: The semiannual interest rate is 5 percent, and in 5 years there are 10 semiannual periods. Using Eq. (11-6), the present value P of the $1000 is

$$P = \frac{\$1000}{(1 + .05)^{10}} = \frac{\$1000}{1.6289} = \$613.91$$

□

EXAMPLE 11-8

The Slapumup construction company owes the Fourth Federal Bank $10,000 payable in 4 years and $20,000 payable in 6 years. The company asks the bank to consolidate the loans into one loan due in 5 years. If interest is 12 percent compounded semiannually, what should the amount due after 5 years be?

Solution 1: One method of solving this problem is to compute the combined present values of the two loans and then to consider what this amount will be in 5 years if it is invested at 12 percent compounded semiannually.

The present value of the $10,000 loan is

$$P_1 = \frac{\$10,000}{(1.06)^8} = \frac{\$10,000}{1.593848} = \$6274.12$$

and the present value of the $20,000 loan is

$$P_2 = \frac{\$20,000}{(1.06)^{12}} = \frac{\$20,000}{2.012196} = \$9939.39$$

Thus the present value of both loans is

$$P_3 = P_2 + P_1 = \$16,213.51$$

In 5 years a principal with value P_3 will increase to an amount

$$A_1 = P_3(1.06)^{10} = (\$16,213.51)(1.790848) = \$29,035.93$$

In other words, using this method, the company should replace the two loans with one loan of $29,035.93 payable in 5 years.

Solution 2: A second method of converting these two loans into a single 5-year loan is to consider the value of the $10,000 loan after 5 years and to add to this the value of the $20,000 loan after 5 years. The amount $10,000 payable after 4 years will grow during the fifth year to an amount

$$A_2 = (\$10,000)(1.06)^2 = \$11,236.00$$

Also, after 5 years the $20,000 loan payable after 6 years will have the present value

$$P_4 = \frac{\$20,000}{(1.06)^2} = \$17,799.93$$

Thus a single 5-year loan should be for the amount

$$A_2 + P_4 = \$29,035.93 \qquad \square$$

EXAMPLE 11-9

Suppose that an amount P is to be invested in an account which pays 6 percent interest compounded quarterly. After 2 years $1000 is to be withdrawn and after 4 years another $1000 is to be withdrawn. After the second withdrawal the account should have value zero. What is the proper value for P?

Solution: After 2 years the principal P will have the value

$$V = P(1.015)^8$$

A withdrawal of $1000 is to be made from the amount V, and the remainder $V - 1000$ is to be invested for 2 more years. After these last 2 years the value is $(V - 1000)(1.015)^8$. This new value is to be exactly $1000 so that a second withdrawal of $1000 can be made to close the account. We have the equation (in dollars)

$$(V - 1000)(1.015)^8 = 1000$$

or
$$V = \frac{1000}{(1.015)^8} + 1000 = 887.71 + 1000 = 1887.71$$

Finally, $P = V/(1.015)^8$, and we obtain

$$P = \frac{1887.71}{(1.015)^8} = 1675.74$$

Thus the initial deposit should be $1675.74. $\qquad \square$

EXAMPLE 11-10

Nick and Peni Saver hold a winning ticket in a state lottery. This ticket will bring them payments of $100 per month for 20 years. The Savers plan to use the money to establish a fund to provide for their daughter's college education. They agree to deposit $100 on the last day of each month in an account at the Safety First Savings and Loan which earns interest at the rate of 6 percent per year compounded monthly. They made their decision on January 1 and their first payment on January 31. What is the value of the account 2 years later on December 31?

Solution: We assume that all payments (deposits) are made and that interest is credited at the end of each month. The first payment of $100 is made at the end of the first month. and of course no interest is paid at that time. However, the first payment will accumulate interest for the remaining 23 months of the 2 years. At the end of 2 years the first payment will have a value (in dollars) of $100(1 + .005)^{23}$. The term .005 appears in this expression since the interest rate is .5 percent per month. The second payment will accumulate interest for 22 months and at the end of 2 years will have value $100(1.005)^{22}$. The value after 2 years of the remaining payments can be determined similarly, and at the end of 2 years the sum of all payments and accumulated interest is

$$S = 100(1.005)^{23} + 100(1.005)^{22} + \cdots + 100(1.005) + 100$$
$$= 100[1 + (1.005) + \cdots + (1.005)^{22} + (1.005)^{23}]$$

In this formula for S the sum of the powers of the term 1.005 forms a geometric series, and there is a simple formula for the value of such a series:

Formula for the Sum of a Geometric Series

If a is any number, $a \neq 1$, then the sum $1 + a + a^2 + \cdots + a^n$ has the value

$$\frac{a^{n+1} - 1}{a - 1}$$

Using this formula in the expression for S, we have

$$S = 100 \frac{(1.005)^{24} - 1}{(1.005) - 1} = \frac{100}{.005}[(1.005)^{24} - 1] = \$2543.20$$

The situation studied in Example 11-10 is an instance of an annuity.

Definition

An *annuity* is a sequence of a specific number of equal payments made at equally spaced times. The times at which payments are made are called *payment dates* and the interval between successive payment dates is the *payment period*. In an *ordinary annuity* interest is credited on payment dates. The *amount of an annuity* is the sum of all payments made (or to be made) plus all interest accumulated (or to be accumulated).

In Example 11-10 the payment dates are the last day in each month, and the payment period is 1 month. In this example, payments are actually deposits.

The technique used in Example 11-10 to find S can be used to find the amount of any annuity. If P represents the payment made on each payment date, and if the interest rate is k per period, then after n periods the value of the first payment (which is made at the end of the first period) is $P(1 + k)^{n-1}$. After n periods the value of the second payment is $P(1 + k)^{n-2}$, ..., the value of the last payment (made at the end of the last period) is P. The amount of the annuity is

$$S = P(1 + k)^{n-1} + P(1 + k)^{n-2} + \cdots + P(1 + k) + P$$

Using the formula for the sum of a geometric series, we have

$$S = P\frac{(1 + k)^n - 1}{(1 + k) - 1} = \frac{P}{k}[(1 + k)^n - 1]$$

To summarize:

The amount of an annuity with payments P, n payment dates, and an interest rate of k per period is

$$S = \frac{P}{k}[(1 + k)^n - 1] \qquad\qquad (11\text{-}7)$$

Annuities are bought and sold (in Example 11-10, the Savers could sell their winning ticket), and to determine a fair price for an annuity it is important to know its present value. Since an annuity is a sequence of payments made on specific dates, it is reasonable to make the definition:

Definition

The *present value of an annuity* is the sum of the present values of all payments.

EXAMPLE 11-11

What is the present value of an ordinary annuity consisting of payments of $1000 per quarter for 4 years at 8 percent per year?

Solution: The first payment is made at the end of the first quarter. We use Eq. (11-6) with $k = .02$, and we conclude that the first payment has present value (in dollars) $1000(1 + .02)^{-1}$. Likewise, we conclude that the ith payment has present value $1000(1 + .02)^{-i}$. Therefore, the present value of the annuity (in dollars) is

$$V = 1000(1.02)^{-1} + 1000(1.02)^{-2} + \cdots + 1000(1.02)^{-15} + 1000(1.02)^{-16}$$

$$= \frac{1000}{(1.02)^{16}} [(1.02)^{15} + (1.02)^{14} + \cdots + (1.02) + 1]$$

$$= \frac{1000}{(1.02)^{16}} \left[\frac{(1.02)^{16} - 1}{(1.02) - 1} \right] = 728.446 \frac{.372786}{.02} = 13{,}577.71 \qquad \square$$

Again, the method introduced in this example is a general one, which can be used to compute the present value of any ordinary annuity. If the annuity consists of n payments of amount P, with interest rate k per period, the present value of the annuity is

$$V = P(1 + k)^{-1} + P(1 + k)^{-2} + \cdots + P(1 + k)^{-n+1} + P(1 + k)^{-n}$$

$$= \frac{P}{(1 + k)^n} \frac{(1 + k)^n - 1}{(1 + k) - 1} = \frac{P}{k} \left[1 - \left(\frac{1}{1 + k} \right)^n \right]$$

To summarize:

The present value of an annuity with payments P, n payment dates, and an interest rate of k per period is

$$V = \frac{P}{k} \left[1 - \left(\frac{1}{1 + k} \right)^n \right] \qquad (11\text{-}8)$$

Remark: If the amount of an annuity has already been computed, Eq. (11-8) for finding the present value an annuity can be simplified. Recall that the amount of an annuity with n payments of size P and interest rate of k per payment period is $S = (P/k)[(1 + k)^n - 1]$. Using this and (11-8), we have

$$V = \frac{P}{k} \left[1 - \left(\frac{1}{1 + k} \right)^n \right] = \frac{1}{(1 + k)^n} \left\{ \frac{P}{k} [(1 + k)^n - 1] \right\} = \frac{S}{(1 + k)^n}$$

Thus the present value of an annuity is simply the present value of the amount of the annuity.

It is common for annuities to arise as periodic payments from a sum deposited in a pension fund or with an insurance company. Although it is usual for the amounts of the payments to be determined on an actuarial basis, the idea can be illustrated through a simple example.

EXAMPLE 11-12

Nick and Peni Saver plan to establish a savings account to help pay for their daughter's education. They propose to make a single deposit and to deposit an amount sufficient to provide $1000 quarterly for 4 years with no balance remaining after that time. They intend to open an account at Safety First Savings and Loan, which pays interest at a rate of 6 percent per year, compounded quarterly. If they make the deposit on July 1, 1980, and if withdrawals begin on September 30 of that year and continue quarterly through June 30, 1984 how large does their initial deposit need to be?

Solution: Withdrawals of $1000 are to be made for 16 consecutive quarters, after which no balance is to remain. The entire amount deposited, call it D, earns interest for 1 quarter. Thus the amount in the account on September 30 (after interest is added) is $D(1.015)$. After the withdrawal of $1000 is made on October 1, there remains $D(1.015) - 1000$, and this amount earns interest for the second quarter. The amount in the account at the end of the second quarter, after interest is added but before any withdrawal is made, is

$$[D(1.015) - 1000](1.015)$$

After the withdrawal is made at the beginning of the third quarter, the amount in the account is

$$[D(1.015) - 1000](1.015) - 1000 = D(1.015)^2 - 1000(1.015) - 1000$$

Likewise, the amount in the account at the beginning of the fourth quarter is

$$D(1.015)^3 - [1000 + 1000(1.015) + 1000(1.015)^2]$$

Continuing, the amount in the account after the last withdrawal (at the end of the sixteenth quarter) is

$$D(1.015)^{16} - [1000 + 1000(1.015) + \cdots + 1000(1.015)^{15}]$$

But this amount is to be zero, so we have

$$D(1.015)^{16} = 1000[1 + (1.015) + \cdots + (1.015)^{15}]$$

or
$$D = \frac{1000}{(1.015)^{16}} [1 + (1.015) + \cdots + (1.015)^{15}]$$

In this formula for D the sum of the powers of the term 1.015 forms a geometric series, and we can use our formula to determine the value of such a series:

$$D = \frac{1000}{(1.015)^{16}} \frac{(1.015)^{16} - 1}{(1.015) - 1} = \frac{1000}{.015} \left[1 - \frac{1}{(1.015)^{16}} \right]$$

$$= 14,131.26$$

The initial deposit should be \$14,131.26. Notice that the amount of this annuity, i.e., the sequence of withdrawals, is $(1000/.015)[(1.015)^{16} - 1]$ and that the initial deposit required is just the present value of this amount. $\square$

Exercises for Sec. 11-3

1. What is the present value of \$10,000 payable in 10 years if interest is 10 percent per year, compounded yearly?

2. Find the amounts of the following annuities:
   a. \$1200 semiannually for 4 years at 12 percent per year interest.
   b. \$600 quarterly for 4 years at 12 percent per year interest.
   c. \$200 per month for 4 years at 12 percent per year interest.

3. Find the present value of each of the annuities in Exercise 2.

4. Find the amount and present value of an annuity for which one pays \$10 per week for 128 weeks with an interest rate of .1 percent per week.

5. Ms. Perkins makes a down payment of \$1000 on a new car. She then agrees to pay the balance of the car's price plus interest by paying \$100 per month for 36 months. If interest is charged at the rate of 18 percent per year and there are no other charges added to the price of the car, what is the cash price of the car?

6. Jack owes Jill \$5000 payable in 3 years and \$3000 payable in 9 years. The two loans are to be consolidated into one loan payable in 5 years. What is the amount payable in 5 years if the interest rate is 12 percent per year and interest is compounded semiannually?

7. A company has three loans outstanding. The first is for \$10,000 payable in 1 year, the second is for \$20,000 payable in 2 years, and the third is for \$30,000 payable in 3 years. The company seeks to defer payment on these loans and to consolidate them into one loan payable in 5 years. The interest rate is 10 percent, compounded annually. Assuming the lender is willing to consolidate the loans, what amount should be paid at the end of 5 years?

8. Mr. Perkins and the sales manager of Ampsonwatts Electric Automobile Co. agree on a price of $3500 for a new Glomobile. Mr. Perkins plans to make a down payment of $1000. The sales manager proposes that Mr. Perkins make an additional 36 monthly payments of $90 each. What is the effective annual interest rate implicit in this proposal?

9. (Continuation of 8) Mr. Perkins checks with the local finance company and finds that their rate on new car loans is 15 percent effective annual interest. If Mr. Perkins plans to finance $2500 over 2 years, what will his monthly payments be?

10. The Safety First Savings and Loan pays interest at a rate of 6 percent compounded quarterly. Nick Saver wants to establish an annuity for his father which will pay $500 quarterly for a period of 10 years. How much should Nick deposit so that the annuity can be paid and no balance will remain after 10 years?

11-4 AMORTIZATION AND SINKING FUNDS

It is common for both the principal and interest of a loan to be repaid by making a sequence of equal payments at regular intervals (usually monthly). In such a case the loan is said to be *amortized*. Naturally the amount of each payment depends on the number of payments and the interest rate charged for the loan. The payments are determined so that an annuity with these payments for the specified period and interest rate has a present value equal to the amount of the loan.

EXAMPLE 11-13

The cash price for a new Trimobile is $5000. If the total amount is to be paid by monthly payments, and if interest is 1.5 percent per month, what monthly payments will pay off the loan in 36 months? In 48 months?

Solution: Let the monthly payment be denoted by P. If there are 36 payments and interest is 1.5 percent per payment period, the present value of the total amount of payments is [from (11-8)]

$$V = \frac{P}{.015}\left[1 - \left(\frac{1}{1 + .015}\right)^{36}\right]$$

Since the present value is to be $5000, we can determine the payments by solving for P in the equation

$$\$5000 = \frac{P}{.015}\left[1 - (1 + .015)^{-36}\right]$$

This gives

$$P = \frac{(.015)(\$5000)}{1 - (1.015)^{-36}} = \$180.76$$

If the loan is amortized over 48 months, the payments are

$$P = \frac{(.015)(\$5000)}{1 - (1.015)^{-48}} = \$146.87$$

Thus we see that the payments are about $34 per month less when the loan is repaid over 48 months than when it is repaid over 36 months. However, we also note that the amount paid over 36 months is $6507.36, while the amount paid over 48 months is $7049.76. ☐

The method used in Example 11-13 can be used in general.

If a loan of amount L is amortized with n payments of amount P and interest rate k per payment period, then each each payment must be

$$P = \frac{kL}{1 - (1 + k)^{-n}} \qquad (11\text{-}9)$$

Sinking Funds

Frequently a company or an individual will make deposits into an account in such a way that the deposits together with the interest earned is enough to pay off a debt which comes due on a fixed date in the future. Such an account is called a *sinking fund*. Such funds are also often used by companies to finance the replacement of old equipment.

EXAMPLE 11-14

The Fast Haul Trucking Company plans to buy a new $50,000 truck in 5 years. If money is deposited in an account which pays 8 percent compounded semi-annually, how much should be deposited each half-year so that at the end of 5 years there will be enough cash to buy a new truck?

Solution: If the deposit per half year is denoted by P, the amount in the account after 5 years is [by (11-7)]

$$S = \frac{P}{.04}[(1.04)^{10} - 1]$$

Since this amount must be $50,000, P is given (in dollars) by

$$P = \frac{(50,000)(.04)}{(1.04)^{10} - 1} = 4164.55$$ ☐

EXAMPLE 11-15

The Bilow-Selhi, Inc. brokerage firm owes $100,000 payable in 2 years together with interest accumulated at a rate of 12 percent per year compounded semi-annually. It has a sinking fund which pays 9 percent per year compounded *monthly*. How much should be deposited each month in the sinking fund to amortize the debt?

Solution: The total amount to be paid is the value of $100,000 after 2 years with interest of 12 percent per year compounded semiannually, namely,

$$A = (\$100,000)(1.06)^4 = \$126,247.70$$

If deposits of amount P are made monthly into a sinking fund paying 9 percent per year compounded monthly, then after 2 years the sinking fund has the value

$$S = \frac{P}{.0075} [(1.0075)^{24} - 1]$$

Since P is to be chosen so that $S = \$126,247.70$, we see that (in dollars)

$$P = \frac{(.0075)(126,247.70)}{(1.0075)^{24} - 1} = 4820.74$$ ☐

The technique used in Examples 11-14 and 11-15 can be used in any similar problem involving a sinking fund.

If an amount S is to be accumulated in a sinking fund through n payments of amount P with interest rate k per period, then each payment must be

$$P = \frac{kS}{(1 + k)^n - 1} \qquad (11\text{-}10)$$

Exercises for Sec. 11-4

1. What monthly payment is needed to amortize a loan of $5000 payable in 1 year at 12 percent per year interest, compounded monthly?

2. What quarterly payment is needed to amortize a loan of $20,000 payable in 3 years at 12 percent per year interest, compounded quarterly?

3. What monthly payments should be made into a sinking fund which pays compound interest at 1 percent per month if the fund is to pay off the loan and accumulated interest in Exercise 1?

4. What monthly payments should be made into a sinking fund which pays interest at 6 percent per year compounded monthly in order to accumulate $5000 in 1 year?

5. What monthly payment is needed to amortize a loan of $30,000 with interest of 9 percent per year compounded quarterly for 5 years?

6. Shortshocks Electronics Inc. arranges a loan of $500,000 to expand its semiconductor division. The interest rate on the loan is 9 percent compounded yearly, and the loan together with accumulated interest is to be repaid in 3 years. A sinking fund is established in an account which pays 8 percent per year compounded quarterly. What quarterly payments into the sinking fund will give the amount required to pay off the loan plus accumulated interest in 3 years?

7. Sam and Pam Dram are setting up a college fund for their children, 6-year-old Fran and 10-year-old Sam, Jr. They are investing quarterly in an account which pays 8 percent per year, compounded quarterly. They estimate that each child will need $3000 per year for college, and they will each begin college at age 18 and go for exactly 4 years. The Drams intend to stop investing in the account when Sam, Jr. starts college. How much should they invest each quarter so that each child can receive $3000 at the start of each year of college and the account will have value zero at the time of the last withdrawal?

8. Nick and Peni Saver plan to buy a new house. They will need a mortgage of $30,000, and they have several options:

    8.5 percent per year interest rate for 17 years
    9 percent per year interest rate for 20 years
    9.5 percent per year interest rate for 25 years

    In each case the interest is compounded monthly.
    a. Which option results in the lowest monthly payments?
    b. What is the total amount of interest paid in each case?
    c. If they can afford payments of $180 per month and they would like to have the mortgage paid off in 25 years, what interest rate (to the nearest 0.1 percent) should they look for? [Hint: Assuming a monthly payment of $180, Eq. (11-9) gives an equation for the interest rate r. Find an approximate solution to this equation by systematically trying selected values of k.]

9. The Bilow-Selhi, Inc. brokerage firm can afford payments of $2000 a month to amortize a loan. The current rate of interest for companies with credit ratings similar to that of Bilow-Selhi, Inc. is 9 percent per year. How large a loan can the firm afford if it is to be amortized over 8 years? Over 12 years?

10. A company is considering buying a new computer which costs $1.2 million. As an alternative they could lease the computer for $15,000 per month. The life of the computer is estimated to be 10 years, after which the computer will have a salvage value of $100,000. If the company buys the computer, they must pay all maintenance

costs and a 10-year maintenance agreement will cost $150,000. If current interest rates are 8 percent, compounded quarterly, should the company buy or lease the computer? (*Hint:* Find the present value of the amount paid to lease the computer and compare with the present value of the money needed to buy the computer.)

11. The Fast Haul Trucking Company can arrange to make sinking fund payments of $6000 semiannually. The best interest rate it can receive on its money is 10 percent per year, compounded semiannually. How many periods are necessary for Fast Haul to accumulate $100,000?

12. Bilow-Selhi, Inc. sells $1 million of 8 percent (interest rate per year) sinking fund debentures due in 10 years with interest payable semiannually. The company can earn 9 percent per year compounded semiannually on money it deposits in a sinking fund. What amount does the company need every 6 months to meet interest and sinking-fund payments?

11-5 STOCKS AND BONDS

As an alternative to investing in savings accounts at savings institutions, one can also invest in the stock issued by corporations and in notes and bonds issued by corporations and various governmental units. In such cases there is usually an element of uncertainty about the return on the investment, which does not exist for investments in savings accounts. On the other hand there is also the potential for much greater gain. Since notes and bonds correspond more closely with savings accounts, we begin our study with them.

A *bond* is an agreement to pay a fixed sum at a certain date in the future, *the redemption date*, together with fixed interest payments at prescribed intervals up to the redemption date. Notes are very similar to bonds (the difference is in the degree of security of the investment), and for our purposes the two can be studied together. Bonds are sold on the open market, and the selling price may differ substantially from the *redemption value* (the amount to be paid on the redemption date). A person who purchases a bond obtains the rights to all *future* interest payments as well as the right to the redemption value. Normally the next interest payment is prorated between the current holder and the purchaser. At the time of purchase the purchaser must pay a fraction of the next interest payment to the current holder. The fraction is that proportion of the interest period which has passed. Thus in computing the return on a bond both the redemption value and future interest payments must be considered. If the purchase price of a bond is less than the redemption value, the bond is said to have been purchased at a *discount*. If the purchase price is more that the redemption value, then the bond is purchased at a *premium*. Finally, if the bond is purchased for the same price as the redemption value, then the bond is purchased *at par*.

EXAMPLE 11-16

Describe the payments received by a purchaser of a bond with a $1000 redemption value, an $850 purchase price, and an interest rate of 8 percent per year, based on the redemption value, and paid semiannually. The bond is purchased on January 1, 1977, the redemption date is January 1, 1980, and interest is paid every January 1 and July 1 up to and including the redemption date.

Solution: Since interest is based on the redemption value and is 4 percent per half year, each interest payment is $(.04)(\$1000) = \$40$. There are six future interest payments made to the purchaser. They are on the dates July 1, 1977, January 1 and July 1, 1978, January 1 and July 1, 1979, and January 1, 1980. Thus the total interest paid is $240. Moreover there is a payment of $1000 (the redemption value) on January 1, 1980 (the redemption date). Thus the total payments to the purchaser amount to $1240. Since the purchase price is $850, the purchaser of the bond makes a net profit of $390 over the period of ownership of the bond (January 1, 1977 to January 1, 1980). ☐

Since the decision to purchase a bond is normally based on a comparison of the returns on this and other investments, the rate of return is a significant item. The term used to describe the rate of return available to a purchaser of a bond is yield to maturity. The *yield to maturity* for a specific purchase price of a bond is the *interest rate* for which the sum of the present values of the interest payments yet to be made and the present value of the redemption value is equal to the purchase price. We illustrate this idea with an example.

EXAMPLE 11-17

Find the yield to maturity for the purchaser of the bond in Example 11-16.

Solution: The unknown in this problem is the interest rate k per period of payment (a half year). The present value of each of the interest payments and of the redemption value is shown in Table 11-4, based on the unknown interest rate k for a half year.

We see from Table 11-4 that the sum of the present values of all payments is

$$S = \frac{1000}{(1+k)^6} + \frac{40}{1+k}\left[1 + \frac{1}{1+k} + \frac{1}{(1+k)^2} + \frac{1}{(1+k)^3} + \frac{1}{(1+k)^4} + \frac{1}{(1+k)^5}\right]$$

TABLE 11-4

Date	Payment	Present Value
July 1, 1977	$ 40	$ $\dfrac{40}{1 + k}$
January 1, 1978	40	$\dfrac{40}{(1 + k)^2}$
July 1, 1978	40	$\dfrac{40}{(1 + k)^3}$
January 1, 1979	40	$\dfrac{40}{(1 + k)^4}$
July 1, 1979	40	$\dfrac{40}{(1 + k)^5}$
January 1, 1980	40	$\dfrac{40}{(1 + k)^6}$
January 1, 1980	1000	$\dfrac{1000}{(1 + k)^6}$

Using the formula for the sum of a geometric series, we have

$$S = \frac{1000}{(1 + k)^6} + \frac{40}{1 + k} \frac{1 - 1/(1 + k)^6}{1 - 1/(1 + k)}$$

or

$$S = \frac{1000}{(1 + k)^6} + \frac{40}{k} \left[1 - \frac{1}{(1 + k)^6} \right] \tag{11-11}$$

The problem reduces to one of finding k such that S, as given by (11-11), has the value $850, the purchase price of the bond. When phrased in this way, the problem is essentially one of finding a root of a sixth-degree polynomial. Although there are computational algorithms for solving such problems, we use a systematic trial-and-error approach. In proceeding by trial and error, better described as proceeding by successive approximations, it is useful to keep in mind that in Eq. (11-11) an increase in k results in a decrease in the present value S. For $k = .05$, .06, .07, and .08, the present value S in (11-11) has the following values (in dollars):

$$S = \begin{cases} 746.22 + 203.03 = 949.25 & k = .05 \\ 704.96 + 196.69 = 901.65 & k = .06 \\ 666.34 + 190.66 = 857.00 & k = .07 \\ 630.17 + 184.92 = 815.09 & k = .08 \end{cases}$$

We see from these choices of k that $k = .07$ is a little too small and $k = .08$ is quite a bit too large. At the next stage we try $k = .071$, $.072$, and $.073$ and obtain

$$S = \begin{cases} 662.62 + 190.07 = \$852.69 & k = .071 \\ 658.92 + 189.49 = \$848.41 & k = .072 \\ 655.24 + 188.91 = \$844.15 & k = .073 \end{cases}$$

Thus, $k = .071$ is too small (S is larger than \$850), and $k = .072$ is too large (S is smaller than \$850). Since the value of S for $k = .072$ is a little closer to \$850 than that for $k = .071$, we next try $k = .0716$, $k = .0717$, and $k = .0718$. We obtain $S = \$850.12$ for $k = .0716$, $S = \$849.69$ for $k = .0717$, and $S = \$849.26$ for $k = .0718$. The value of S for $k = .0716$ is quite close to \$850, and since k is now accurate to at least three decimal places, we consider the problem to be solved. We take $k = .0716$ as the proper interest rate for a half year. Thus the yield to maturity, based on interest rate per year, is $.1432$ or, equivalently, 14.32 percent. $\qquad\square$

The price of a bond is related to many factors, including the fiscal state of the corporation or municipality issuing the bond and the general state of the economy. In particular, the price of a bond is strongly influenced by the prevailing interest rate in the economy. If interest rates are high, prices of bonds tend to be low so that bonds can compete as an investment. Conversely, if interest rates are low, a bond paying a high fixed rate of interest is more valuable and the price of the bond tends to rise. Thus, in addition to buying a bond to obtain the interest payments, one can also purchase a bond as a speculative investment, hoping to buy the bond at one price and sell it at some future date at a higher price. Of course, not all such plans work, and it is always possible that at the time you wish to sell the bond its market price is the same or even less than the amount you paid for it. In such a case you at least have the consolation of having received the interest payments while you held the bond.

Another investment option in which there is substantial uncertainty about the return is the purchase of the common stock of a corporation. The ownership of shares of *common stock* (the only type we consider here) constitutes part ownership of a corporation. If a company issues 3 million shares of stock, an individual who owns 3000 shares owns one-thousandth of the company. All company decisions are, in theory, made by the stockholders (persons owning the company's stock) and by individuals hired by the stockholders to run the day-to-day affairs of the company (the management). In practice most stockholders of a company do not buy the stock to enable them to share in the process of owning and managing the company. Instead they buy the stock as an investment in the hope of obtaining a return which exceeds returns on other investments such as bonds and savings accounts.

One who invests in stocks can profit in two ways. First, the company may issue dividends to all stockholders. A dividend is a payment made by the company which issued the stock to the owners of the stock. Dividends are often issued on a regular basis (such as quarterly), and they are of a fixed amount so that a stockholder can depend on these payments as a source of income. However, these payments must come from earnings of the company, and they are not guaranteed. If the company suffers losses, it may well have to cease paying dividends. Also, even if a company has the money to pay dividends it need not do so. Instead, the management may choose to use the money to modernize the plant or equipment or for other purposes, e.g., to buy another company.

The second method by which a shareholder can gain from owning shares is by having the shares increase in value. It is said that the shares have *appreciated* when they increase in value. A shareholder with shares that have appreciated can sell the shares and realize a gain in this way. Shares of stock of most major companies are traded (bought and sold) on stock exchanges each weekday, and the system handles trades almost constantly. Thus the price of a stock often (in fact usually) varies in the course of a day and can vary greatly in periods such as a week or month. This provides the opportunity for a rapid gain (or loss) on an investment. If the stock appreciates substantially in a short period of time, the *return on investment* can be very large. Here by return on investment we mean the increase in value of the investment as a percent of the investment.

EXAMPLE 11-18

Gus buys 1000 shares of MDI (Multi-Decibel, Incorporated) at $4 per share. When, 1 month later, the company announces the development of new very-long-playing record (4 hours), the stock appreciates quickly to $5.20 a share. Gus sells his shares immediately. What is his return on investment?

Solution: Since Gus sells his shares for $1.20 per share more than he paid for them, his profit is 1000 ($1.20) = $1200. His return on investment is $1200/$4000 = .30 = 30 percent. On a yearly basis this is 360 percent.   □

Remark: In Example 11-18 and in the example which follows we ignore the commissions and fees which investors must pay a stockbroker in order to buy and sell stocks. These fees vary according to the price per share of the stock and the number of shares bought or sold. Typically, for a small investor (less than $10,000 per transaction), the commission and fees will be between 1 and 4 percent of the value of the transaction.

Values of shares of stock may decrease as well as increase, and the behavior of stock prices is very unpredictable. One method of attempting to deal with the uncertainty of the future values of a stock is to estimate the probabilities of various actions of the stock market and then to compute the expected value of the return.

EXAMPLE 11-19

Ms. Shrewd is considering buying 100 shares of a new company, FemSport, which specializes in women's sports equipment. The shares now sell for $16 a share, and the company has not yet made a substantial inroad into the market. However, they plan to introduce a new line of tennis and golf equipment, and if it sells well, the company will profit greatly. Ms. Shrewd estimates the probability that the new line will sell well at .2, she estimates the probability of it selling moderately well at .5, and the probability of it being a complete failure at .3. If the line does well, she believes the stock price will double in a year to $32. If the line does moderately well, she believes that the stock will increase in value to $20, and if the line fails to sell, she believes the stock will fall in value to $4 per share. Assuming these three possibilities are the only ones which can occur, what is the expected return on an investment in these shares?

Solution: The expected value of the price of the stock in a year's time is (see Chap. 3 for the computation of expected values)

$$E = (\$32)(.2) + (\$20)(.5) + (\$4)(.3) = \$17.60$$

The expected return on investment is

$$\frac{\$17.60 - \$16.00}{\$16.00} = \frac{\$1.60}{\$16.00} = .1 = 10\%$$

□

Remarks: As the reader is no doubt aware, the analysis in Example 11-19 is quite naïve. In an actual case the future price of the stock of FemSport will depend on many factors other than just the success of their new line of equipment. In particular, the state of the entire economy will be important in determining the price of a share of FemSport stock. It is possible that a company does well in the sense of sales and profits and yet the price of its stock falls because of factors other than the success of the company. To be realistic in an example such as 11-19, many other factors must be considered. This requires the estimation of probabilities for events which are very complex and interrelated. However, the method of computing expected return in these more realistic settings is the same as that used in Example 11-19.

Exercises for Sec. 11-5

1. Describe the payments received by a purchaser of a bond with a $10,000 redemption value, an $8700 purchase price, and an interest rate of 10 percent per year based on the redemption value and paid semiannually. The bond is purchased on July 1, 1978, the redemption date is January 1, 1981, and interest is paid every January 1 and July 1 up to and including the redemption date.

2. Describe the payments received by a purchaser of a bond with a $5000 redemption value, a $5500 purchase price, and an interest rate of 12 percent per year based on the redemption value and paid quarterly on February 1, May 1, August 1, and November 1, up to and including the redemption date. The bond is purchased on February 1, 1979, the redemption date is February 1, 1981.

3. Find the *yield to maturity* for the purchaser of the bond in Exercise 1.

4. Find the *yield to maturity* for the purchaser of the bond in Exercise 2.

5. What is the return on investment of a share of stock purchased at $33\frac{1}{8}$ and sold at $55\frac{5}{8}$? What is the rate of return per year if the stock is held for 9 months?

6. In a stock split the number of shares is increased and the price per share is decreased by an appropriate amount to keep the total value of all shares the same. Thus in a two-for-one split the number of shares is doubled, and the price per share is cut in half.

    Mr. Gamble bought shares in MDI for $12 per share. Suppose MDI has a three-for-one split, and shares which are selling at $27 per share before the split sell for $9 per share after the split. Mr. Gamble sells his stock immediately after the split. What is his return on investment? What is the rate of return per year if he held the stock for 30 months?

7. Repeat Exercise 6 for a stock which split five for four and an investor who bought shares worth $400 at $20 per share before the split and sold out at $25 per share after the split.

8. Find the expected return on investment in Example 11-19 if the possible future (1 year from now) stock prices and their probabilities are as follows:

Possible Value	Probability
$40	.1
32	.1
20	.4
16	.3
2	.1

9. A $1000 bond which pays interest semiannually at the rate of 4 percent per year is discounted so that its current yield (the yield based on its current market price) is 8 percent.
    a. What is the current market price?
    b. If the redemption date of the bond is 6 years in the future, what is its yield to maturity based on the current market price?

10. A $1000 bond which pays interest semiannually at the rate of 9 percent per year sells at a premium so that its current yield is 8 percent.
    a. What is the current market price?
    b. If the redemption date of the bond is 10 years in the future, what is its yield to maturity based on the current market price?

IMPORTANT TERMS

You should be able to describe, define, or give examples of each of the following:

Interest
Interest rate
Principal of a loan
Simple interest
Compound interest
Interest compounded continuously
Effective rate of interest
Present value
Annuity
Amount of an annuity
Present value of an annuity

Amortization of a loan
Sinking fund
Bond
Redemption value
Purchase at a discount
Purchase at a premium
Purchase at par
Yield to maturity
Common stock
Return on investment

REVIEW EXERCISES

1. An amount of $12,000 is deposited in an account on which interest is compounded quarterly at the rate of 6 percent per year. The funds are invested for 10 years. Find the effective rate of interest.

2. Mr. Saver has $2000 in the bank now. How much did he have invested in the bank 3 years ago if the interest is 5 percent per year compounded twice a year and if no deposits or withdrawals were made during the 3 years?

3. Find the amount and present value of an annuity for which one pays $15 a week for 2 years with an interest rate of .25 percent per week.

4. The Fast Haul Trucking Company needs $50,000 at the end of 6 years to purchase new equipment. How much should be invested per quarter if the interest rate is 6.5 percent per year and interest is compounded quarterly?

5. The owner of a small business takes out a loan for $8000 payable in 5 years with accumulated interest. The interest rate is 9 percent per year,

and interest is compounded semiannually. What is the total amount due at the end of 5 years?

6. What quarterly payment is needed to amortize a loan of $4500 with interest of 8 percent per year compounded quarterly for 4 years?

7. Ivan Investor buys 3500 shares of stock at $2.25 per share. There is a two-for-one stock split, after which Ivan sells his stock for $1.25 per share. What is his return on investment?

8. Ms. Saver purchases a bond for $4500. It has a redemption value of $5000 and an interest rate of 9 percent per year paid 3 times a year—January 1, May 1, and September 1. If she purchased the bond on January 1, 1977 and the redemption date is January 1, 1980, what is her profit (total income minus purchase price) over the period of ownership of the bond?

9. What is the yield to maturity of the bond purchased in Exercise 8?

10. A company is planning to purchase a new computer 3 years from now at an estimated cost of $3 million, and 1 year later they plan to buy an auxiliary memory unit at an estimated cost of $1 million. The company has established a sinking fund so it will have the cash available at the time of each purchase. The company will make payments into the fund on a quarterly basis, and interest will be paid on the fund at a rate of 8 percent per year compounded quarterly. What size payment should be made into the fund each quarter?

APPENDICES

AREAS UNDER THE STANDARD NORMAL CURVE

A-1   USE OF THE TABLE

The entries in the table below are the areas under the standard normal curve from 0 to x where the units and tenths digit of x are given at the left of the table and the hundredths digit is given at the top of the table. For example, if $x = 1.32$, we look in the row labeled 1.3 and in the column labeled .02. We find the area under the standard normal curve from 0 to 1.32 to be .4066.

A-2   INTERPOLATION IN THE TABLE

In some applications it is necessary to determine areas defined by values of x which are not listed in the table. For example, to determine $\Pr[0 \leq Z \leq 1.363]$ one cannot use the method described in Sec. A-1. For our purposes it is adequate to use linear interpolation between the values given in the table, and we now outline this technique. To obtain the value (area) corresponding to 1.363 we interpolate between the values for 1.36 and 1.37. Since 1.363 is three-tenths of the way from 1.36 to 1.37, we obtain an approximation to the value for 1.363 by taking the value which is three tenths of the way between the values for 1.36 and 1.37. The value for 1.36 is .4131, and the value for 1.37 is .4147.

Therefore, an approximate value for 1.363 is

$$.4131 + .3(.4147 - .4131) = .4131 + .3(.0016)$$
$$= .4131 + .0005 = .4136$$

We therefore have $\Pr[0 \le Z \le 1.363] = .4136$.

In general, if we wish to obtain $\Pr[0 \le Z \le z_2]$, where $z_1 \le z_2 \le z_1 + 0.01$ and the value for z_1 is given in the table, we use the formula

$$\Pr[0 \le Z \le z_2]$$
$$= \Pr[0 \le Z \le z_1] + 100(z_2 - z_1)(\Pr[0 \le Z \le z_1 + .01] - \Pr[0 \le Z \le z_1])$$

Table of Areas under the Standard Normal Curve

z	.00	.01	.02	.03	.04	.05	.06	.07	.08	.09
0.0	.0000	.0040	.0080	.0120	.0160	.0199	.0239	.0279	.0319	.0359
0.1	.0398	.0438	.0478	.0517	.0557	.0596	.0636	.0675	.0714	.0753
0.2	.0793	.0832	.0871	.0910	.0948	.0987	.1026	.1064	.1103	.1141
0.3	.1179	.1217	.1255	.1293	.1331	.1368	.1406	.1443	.1480	.1517
0.4	.1554	.1591	.1628	.1664	.1700	.1736	.1772	.1808	.1844	.1879
0.5	.1915	.1950	.1985	.2019	.2054	.2088	.2123	.2157	.2190	.2224
0.6	.2257	.2291	.2324	.2357	.2389	.2422	.2454	.2486	.2517	.2549
0.7	.2580	.2611	.2642	.2673	.2704	.2734	.2764	.2794	.2823	.2852
0.8	.2881	.2910	.2939	.2967	.2995	.3023	.3051	.3078	.3106	.3133
0.9	.3159	.3186	.3212	.3238	.3264	.3289	.3315	.3340	.3365	.3389
1.0	.3413	.3438	.3461	.3485	.3508	.3531	.3554	.3577	.3599	.3621
1.1	.3643	.3665	.3686	.3708	.3729	.3749	.3770	.3790	.3810	.3830
1.2	.3849	.3869	.3888	.3907	.3925	.3944	.3962	.3980	.3997	.4015
1.3	.4032	.4049	.4066	.4082	.4099	.4115	.4131	.4147	.4162	.4177
1.4	.4192	.4207	.4222	.4236	.4251	.4265	.4279	.4292	.4306	.4319
1.5	.4332	.4345	.4357	.4370	.4382	.4394	.4406	.4418	.4429	.4441
1.6	.4452	.4463	.4474	.4484	.4495	.4505	.4515	.4525	.4535	.4545
1.7	.4554	.4564	.4573	.4582	.4591	.4599	.4608	.4616	.4625	.4633
1.8	.4641	.4649	.4656	.4664	.4671	.4678	.4686	.4693	.4699	.4706
1.9	.4713	.4719	.4726	.4732	.4738	.4744	.4750	.4756	.4761	.4767
2.0	.4772	.4778	.4783	.4788	.4793	.4798	.4803	.4808	.4812	.4817
2.1	.4821	.4826	.4830	.4834	.4838	.4842	.4846	.4850	.4854	.4857
2.2	.4861	.4864	.4868	.4871	.4875	.4878	.4881	.4884	.4887	.4890
2.3	.4893	.4896	.4898	.4901	.4904	.4906	.4909	.4911	.4913	.4916
2.4	.4918	.4920	.4922	.4925	.4927	.4929	.4931	.4932	.4934	.4936
2.5	.4938	.4940	.4941	.4943	.4945	.4946	.4948	.4949	.4951	.4952
2.6	.4953	.4955	.4956	.4957	.4959	.4960	.4961	.4962	.4963	.4964
2.7	.4965	.4966	.4967	.4968	.4969	.4970	.4971	.4972	.4973	.4974
2.8	.4974	.4975	.4976	.4977	.4977	.4978	.4979	.4979	.4980	.4981
2.9	.4981	.4982	.4982	.4983	.4984	.4984	.4985	.4985	.4986	.4986
3.0	.4987	.4987	.4987	.4988	.4988	.4989	.4989	.4989	.4990	.4990

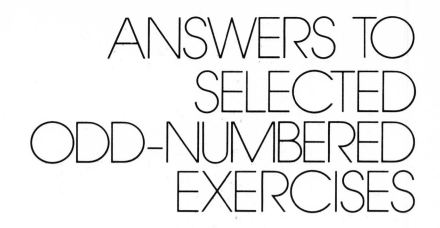

ANSWERS TO SELECTED ODD-NUMBERED EXERCISES

CHAPTER 1

Section 1-4

1. a. True   **b.** False   **c.** True   **d.** False   **e.** False   **f.** True

3. a. $\{a, b, c, 2, 3\}$ **b.** $\{2, 3\}$ **c.** $\{a, 2, 3\}$ **d.** $\{1, 2, 3\}$ **e.** $\{b, c\}$
   f. $\{a, b, c\}$

5. a. $\{GM, Ford\}, \{GM\}, \{Ford\}, \varnothing$
   b. $\{GM, Ford, Chrysler\}, \{GM, Ford\}, \{GM, Chrysler\}, \{Ford, Chrysler\}, \{GM\},$
   $\{Ford\}, \{Chrysler\}, \varnothing$

7. a. False   **b.** True   **c.** True   **d.** True

9. a. No   **b.** Yes   **c.** No   **d.** Yes

11. $A \times B = \{(x,i), (x,e), (y,i), (y,e), (z,i), (z,e)\}$

Section 1-5

1.

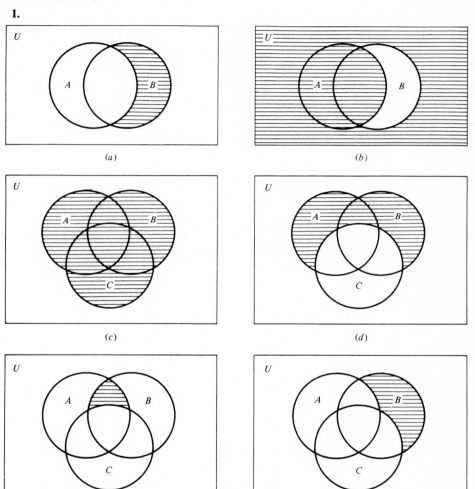

(a)

(b)

(c)

(d)

(e)

(f)

3. a. *y, z*   **b.** *z*   **c.** *x, y, z, v, w*   **d.** *y, z, v*   **e.** *y, z, v*
5. 9   **7.** 32
9. a. 456   **b.** 688   **c.** Insufficient information   **d.** Insufficient information
11. a. 10   **b.** 8

Section 1-6

1. a. No   **b.** Yes   **c.** No   **d.** Yes
3. $f(x) = \frac{1}{2}x - \frac{1}{2}$ **5.** $f(x) = 3x^3 + 5$
7. a. 3   **b.** 45   **c.** $8x + 29$

CHAPTER 2

Section 2-2

1. $\mathcal{O}_1$ long widget, $\mathcal{O}_2$ short widget, $\mathcal{O}_3$ widget of proper length,

$$w_1 = \tfrac{5}{1000} = \tfrac{1}{200}, \; w_2 = \tfrac{10}{1000} = \tfrac{1}{100}, \; w_3 = \tfrac{985}{1000} = \tfrac{197}{200}$$

3. 52; each outcome has the same weight, $\frac{1}{52}$
5. $w_H = \frac{3}{4}; \; w_T = \frac{1}{4}$
7. $\mathcal{O}_1$ has $w_1 = .1$; $\mathcal{O}_2$ has $w_2 = .09$; $\mathcal{O}_3$ has $w_3 = .81$

Section 2-3

1. $G = \{(O,O), (O,U), (O,N)\}$, $H = \{(O,O), (O,U), (O,N), (U,O), (N,O)\}$,
   $I = \{(U,U), (U,N), (N,U), (N,N)\}$
3. Each batch can be labeled *PP*, pure both times; *PI*, pure the first time and impure
   the second; or *I*, impure the first time. The outcomes are reported as ordered
   triples (label of first batch, label of second batch, label of third batch), for example
   (PP,PI,I).

   $S = \{(PP,PP,PP), (PP,PP,PI), (PP,PP,I), (PP,PI,PP), (PP,PI,PI), (PP,PI,I),$
     $(PP,I,PP), (PP,I,PI), (PP,I,I), (PI,PP,PP), (PI,PP,PI), (PI,PP,I), (PI,PI,PP),$
     $(PI,PI,PI), (PI,PI,I), (PI,I,PP), (PI,I,PI), (PI,I,I), (I,PP,PP), (I,PP,PI),$
     $(I,PP,I), (I,PI,PP), (I,PI,PI), (I,PI,I), (I,I,PP), (I,I,PI), (I,I,I)\}$
   $E = \{(PI,PI,PI), (PI,PI,I), (PI,I,PI), (PI,I,I), (I,PI,I), (I,PI,PI), (I,I,PI), (I,I,I)\}$

5. Denote outcomes by (face up on nickel, face up on quarter)

$$S = \{(H,H), (H,T), (T,H), (T,T)\}, \; E = \{(H,T), (T,T)\}$$

7. Denote outcomes by (number of red balls, number of white balls, number of blue
   balls)

   $S = \{(3,1,0), (3,0,1), (2,2,0),(2,1,1), (2,0,2), (1,2,1), (1,1,2), (1,0,3), (0,2,2), (0,1,3)\}$
   $E = \{(2,2,0), (1,2,1), (0,2,2)\}$

9. $S = \{(R,R,R), (R,R,W), (R,R,B), (R,W,R), (R,W,W), (R,B,R), (R,B,W), (R,B,B),$
     $(W,W,W), (W,W,R), (W,W,B), (W,R,W), (W,R,R), (W,R,B), (W,B,W),$
     $(W,B,R), (W,B,B), (B,B,B), (B,B,R), (B,B,W), (B,R,B), (B,R,R), (B,R,W),$
     $(B,W,B), (B,W,R), (B,W,W)\}$

Event that balls of three colors are selected = {(R,W,B), (R,B,W), (B,R,W), (B,W,R), (W,R,B), (W,B,R)}
Event of red ball selected first = {first 9 elements of S}

Section 2-4

1. E = {AAAA, AAAD, AADA, AADD, ADAA, ADAD, DAAA, DAAD, DADA}
3. S = {AAA, AADA, AADD, ADAA, ADAD, ADDA, ADDD, DAAA, DAAD, DADA, DADD, DDAA, DDAD, DDD}

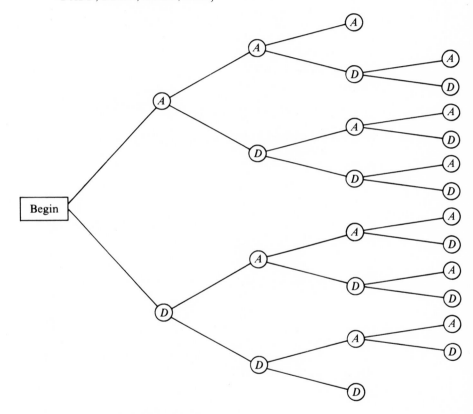

5. 27   **7.** 36   **9.** 24   **11.** 28,800

Section 2-5

1. 120 different ways
3. a. 720   **b.** 1680   **c.** 70   **d.** 270,725   **e.** 161,700
5. If team A (home) is distinguished from team B (visitors), there are 10! different game lineups. If no such distinction is made, there are 10!/2 game lineups.
7. 7200   **9.** 2^{10}

Section 2-6

1. a. 840   **b.** 126   **c.** $\frac{45}{4}$ **d.** 64
3. 30   **5.** 84   **7.** 75,287,520   **9.** 150   **11.** $2^{10} \cdot 4^5$ **13.** 21

Section 2-7

1. $\frac{1}{10}, \frac{6}{10}$ **3. a.** $\frac{1}{6}$ **b.** $\frac{1}{2}$ **c.** $\frac{1}{6}$
5. a. $\frac{1}{5}$ **b.** $\frac{4}{5}$ **c.** $\frac{4}{15}$
7. $\frac{10}{21}$ **9. a.** $\binom{15}{10} \Big/ \binom{20}{10}$ **b.** $\binom{5}{5}\binom{15}{5} \Big/ \binom{20}{10}$

11. $\frac{1}{720}$

Review Exercises

1. $S = \{HHH, HTH, HHT, HTT, TTT, TTH, THT, THH\}$
$E = \{HHH, HTH, HHT, THH\}$
3. $E = \{(H,2,A), (H,2,E), (H.2,O), (H,4,A), (H,4,E), (H,4,O), (H,6,A), (H, 6, E),$
$(H,6,O)\}$
5. 663   **7.** 15

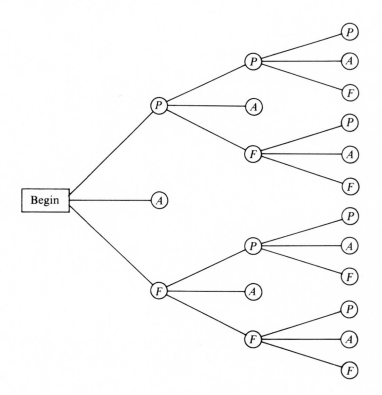

9. $4!$

11. a. $6!$ **b.** $2 \cdot 3! \cdot 3!$ **c.** $4 \cdot 3! \cdot 3!$ **d.** $8 \cdot 3!$ **e.** $4 \cdot 3! \cdot 3!$

13. a. 120 **b.** 56 **15.** $\frac{1}{27}$

17. a. $\dfrac{\binom{17}{3}}{\binom{20}{3}}$ **b.** $\dfrac{1}{\binom{20}{3}}$ **c.** $\dfrac{\binom{17}{3} + \binom{17}{2}\binom{3}{1}}{\binom{20}{3}}$ **d.** $\dfrac{\binom{3}{1}\binom{17}{2} + \binom{3}{2}\binom{17}{1}}{\binom{20}{3}}$

19. $S = \{RR, RW, RB, WW, WB\}$

CHAPTER 3

Section 3-2

1. b and d **3.** $.66$

5. a. $.18$ **b.** $.16$ **c.** $.02$ **d.** $.07$

7. $w_1 = \frac{4}{8}$, $w_2 = \frac{2}{8}$, $w_3 = \frac{1}{8}$, $w_4 = \frac{1}{8}$ **9.** E_1 and E_2

Section 3-3

1. $\Pr[A|B] = \frac{7}{10}$, $\Pr[B|A] = \frac{7}{16}$ **3. a.** $\frac{1}{3}$ **b.** $\frac{1}{5}$ **c.** $\frac{2}{11}$

5. $\frac{1}{5}$ **7.** $\frac{3}{8}$ **9.** No   **11.** $\frac{1}{3}$

13.

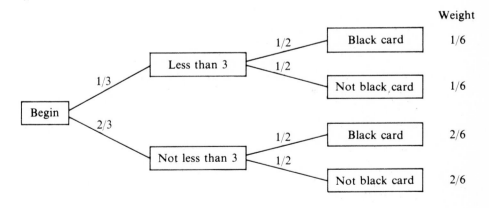

Section 3-4

1. $\Pr[A] = \frac{1}{6}$, $\Pr[B] = \frac{1}{6}$, $\Pr[A \text{ and } R] = \frac{1}{9}$, $\Pr[B \text{ and } W] = \frac{1}{24}$, $\Pr[R|A] = \frac{2}{3}$, $\Pr[W|B] = \frac{1}{4}$

3. $\frac{3}{18}$

5. Pr[either 2 heads and 1 tail or 1 head and 2 tails] = $\frac{9}{16}$

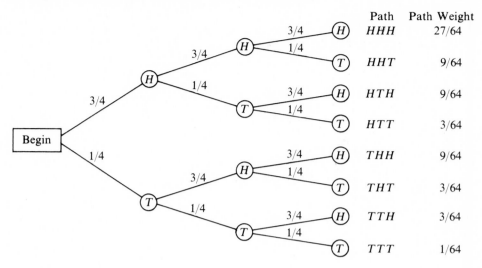

	Path	Path Weight
	HHH	27/64
	HHT	9/64
	HTH	9/64
	HTT	3/64
	THH	9/64
	THT	3/64
	TTH	3/64
	TTT	1/64

7. .1625   **9.** $\frac{1}{2}, \frac{1}{2}$

Section 3-5

1. a. $\binom{6}{3}\left(\frac{1}{4}\right)^3\left(\frac{3}{4}\right)^3$ **b.** $\binom{5}{2}(.3)^2(.7)^3$ **c.** $\binom{7}{4}(.8)^4(.2)^3$

3. $1 - \left[\binom{5}{0}\left(\frac{1}{4}\right)^0\left(\frac{3}{4}\right)^5 + \binom{5}{1}\left(\frac{1}{4}\right)^1\left(\frac{3}{4}\right)^4\right]$

5. $\binom{10}{4}(.20)^4(.80)^6$ **7.** at least 3 calls   **9.** 6 lines

11. $\left[\binom{10}{7} + \binom{10}{8} + \binom{10}{9} + \binom{10}{10}\right]\left(\frac{1}{2}\right)^{10} = 176\left(\frac{1}{2}\right)^{10}$

Section 3-6

1. From top to bottom $\Pr[C|A] = \frac{30}{59}$, $\Pr[F|A] = \frac{15}{59}$, $\Pr[P|A] = \frac{14}{59}$, $\Pr[C|N] = \frac{20}{41}$, $\Pr[F|N] = \frac{15}{41}$, $\Pr[P|N] = \frac{6}{41}$; $\frac{15}{41}$

3. $\Pr[I|(A \text{ and } a)] = \frac{9}{25}$ **5.** $\frac{16}{25}$ **7.** Type E **9.** $\frac{2}{5}$ **11.** $\frac{18}{37}$

Section 3-7

1. and **7.**

x	$\Pr[X = x]$	Product
6	$\binom{3}{3}\left(\frac{1}{2}\right)^3\left(\frac{1}{2}\right)^0 = \frac{1}{8}$	$\frac{6}{8}$
7	$\binom{3}{2}\left(\frac{1}{2}\right)^2\left(\frac{1}{2}\right) = \frac{3}{8}$	$\frac{21}{8}$
8	$\binom{3}{1}\left(\frac{1}{2}\right)\left(\frac{1}{2}\right)^2 = \frac{3}{8}$	$\frac{24}{8}$
9	$\binom{3}{0}\left(\frac{1}{2}\right)^0\left(\frac{1}{2}\right)^3 = \frac{1}{8}$	$\frac{9}{8}$
		$\frac{60}{8} = E[X]$

3. and **9.**

x	$\Pr[X = x]$	Product
0	.001	0
1	.027	.027
2	.243	.486
3	.729	2.187
		$\overline{2.7} = E[X]$

5. and **11.**

x	$\Pr[X = x]$	Product
0	$\binom{9}{5}\Big/\binom{12}{5} = \frac{126}{792}$	0
1	$\binom{9}{4}\binom{3}{1}\Big/\binom{12}{5} = \frac{378}{792}$	$\frac{378}{792}$
2	$\binom{9}{3}\binom{3}{2}\Big/\binom{12}{5} = \frac{252}{792}$	$\frac{504}{792}$
3	$\binom{9}{2}\binom{3}{3}\Big/\binom{12}{5} = \frac{36}{792}$	$\frac{108}{792}$
		$\frac{990}{792} = 1.25 = E[X]$

13. $E[X] = 40/27$ **15.** $E[X] = 13/12$

Review Exercises

1. a. .54   **b.** 1   **c.** .23   **d.** .27

3. 4   **5.** No

7. a. $\dfrac{1}{54}$ **b.** $\dbinom{8}{1}\left(\dfrac{3}{8}\right)^1\left(\dfrac{5}{8}\right)^7 + 2\dbinom{8}{2}\left(\dfrac{3}{8}\right)^2\left(\dfrac{5}{8}\right)^6 + 3\dbinom{8}{3}\left(\dfrac{3}{8}\right)^3\left(\dfrac{5}{8}\right)^0$

c. $\dfrac{30\dbinom{3}{3} + 45\dbinom{3}{2}\dbinom{5}{1} + 60\dbinom{3}{1}\dbinom{5}{2} + 75\dbinom{5}{3}}{\dbinom{8}{3}}$

9. a. $\dfrac{10}{20}\left(\dfrac{9}{19}\right)$ **b.** $1 - \left[\dfrac{10}{20}\left(\dfrac{4}{19}\right) + \dfrac{6}{20}\left(\dfrac{4}{19}\right) + \dfrac{4}{20}\right]$

c. $\dfrac{10}{20}\left(\dfrac{6}{19}\right) + \dfrac{4}{20}\left(\dfrac{6}{19}\right) + \dfrac{6}{20}$ **d.** $1 - \left[\dfrac{10}{20}\left(\dfrac{6}{19}\right) + \dfrac{4}{20}\left(\dfrac{6}{19}\right) + \dfrac{6}{20}\right]$ **e.** $\dfrac{9}{19}$

11. $\dfrac{15}{44}$ **13. a.** $\dfrac{3}{5}$ **b.** $\dfrac{3}{5}$ **c.** $\dfrac{7}{20}$ **d.** $\dfrac{7}{20}$ **15.** $\dfrac{600}{691}$

CHAPTER 4

Section 4-2

1.

x	Density function $\Pr[X = x]$	Distribution function $d(x)$
0	$\frac{25}{36}$	$\frac{25}{36}$
1	$\frac{10}{36}$	$\frac{35}{36}$
2	$\frac{1}{36}$	$\frac{36}{36} = 1$

3.

x	Density function $\Pr[X = x]$	Distribution function $d(x)$
0	$\dbinom{5}{0}\left(\dfrac{1}{5}\right)^0\left(\dfrac{4}{5}\right)^5 = \left(\dfrac{4}{5}\right)^5 = .32768$	.32768
1	$\dbinom{5}{1}\left(\dfrac{1}{5}\right)^1\left(\dfrac{4}{5}\right)^4 = \left(\dfrac{4}{5}\right)^4 = .40960$	.73728
2	$\dbinom{5}{2}\left(\dfrac{1}{5}\right)^2\left(\dfrac{4}{5}\right)^3 = \dfrac{2(4^3)}{5^4} = .20480$	.94208
3	$\dbinom{5}{3}\left(\dfrac{1}{5}\right)^3\left(\dfrac{4}{5}\right)^2 = \dfrac{2(4^2)}{5^4} = .05120$	.99328
4	$\dbinom{5}{4}\left(\dfrac{1}{5}\right)^4\left(\dfrac{4}{5}\right)^1 = \dfrac{4}{5^4} = .00640$	.99968
5	$\dbinom{5}{5}\left(\dfrac{1}{5}\right)^5\left(\dfrac{4}{5}\right)^0 = \dfrac{1}{5^5} = .00032$	1.00000

5. a. .1915   **b.** .2417   **c.** .0668   **d.** .6247

7. a. $\dbinom{5}{5}(.6)^5(.4)^0 = (.6)^5$

b. Pr[profit on at least one stock] $= 1 -$ Pr[profit on none of the stocks] $=$

$$1 - \left[\binom{5}{0}(.6)^0(.4)^5\right]$$

Section 4-3

1. a. Mean $= 3$, spread $= 4$, variance $= 2$, standard deviation $= \sqrt{2}$
   b. Mean $= 3$, spread $= 4$, variance $= 10/3$, standard deviation $= \sqrt{10/3}$
   c. Mean $= 3$, spread $= 3$, variance $= 1$, standard deviation $= 1$
5. Mean $= .9$, standard deviation $= .7$ **7.** At least 137
9. a. Pr$[(3 - \sqrt{2}) \le x \le (3 + \sqrt{2})] =$ Pr[2 or 3 or 4] $= \frac{3}{5}$
   b. Pr$[(3 - \sqrt{10/3}) \le x \le (3 + \sqrt{10/3})] =$ Pr[2 or 3 or 4] $= \frac{3}{15}$
   c. Pr$[(3 - 1) \le x \le (3 + 1)] =$ Pr[2 or 3 or 4] $= .9$
11. Mean $= 0$, standard deviation $= \sqrt{20.4}$, 100%

Section 4-4

1. $\mu = 20$, $\sigma = 4$ **a.** $Z \le 2.5$ **b.** $-1.25 \le Z \le 1.25$ **c.** $Z \ge 1.25$
   d. $-1 \le Z \le 1$
3. $n = 10$, $p = .2$, $\mu = 2$, $\sigma = \sqrt{1.6}$

x	Interval	Area under binomial density	Area under normal density
0	$(-.5,.5)$	$\binom{10}{0}(.2)^0(.8)^{10} = .107$	.095
1	$(.5,1.5)$	$\binom{10}{1}(.2)^1(.8)^9 = .268$	.229
2	$(1.5,2.5)$	$\binom{10}{2}(.2)^2(.8)^8 = .302$	.303
3	$(2.5,3.5)$	$\binom{10}{3}(.2)^3(.8)^7 = .201$	.229
4	$(3.5,4.5)$	$\binom{10}{4}(.2)^4(.8)^6 = .088$	.095
5	$(4.5,5.5)$	$\binom{10}{5}(.2)^5(.8)^5 = .026$	.021
6	$(5.5,6.5)$	$\binom{10}{6}(.2)^6(.8)^4 = .006$	.0028
7	$(6.5,7.5)$	$\binom{10}{7}(.2)^7(.8)^3 = .0008$	.0002
8	$(7.5,8.5)$	$\binom{10}{8}(.2)^8(.8)^2 = .00007$	Very small
9	$(8.5,9.5)$	$\binom{10}{9}(.2)^9(.8)^1 = .0000004$	Very small
10	$(9.5,10.5)$	$\binom{10}{10}(.2)^{10}(.8)^0 = .0000001$	Very small

5. $\mu = 50$, $\sigma = 6.7$, $\Pr[X \geq 59.5] = .0784$

7. $3\% - A$, $29\% - B$, $53\% - C$, $13\% - D$, $1\% - F$

9. $\mu = -1$, $\sigma = 9.74$, $\Pr[\mu - 2\sigma \leq X \leq \mu + 2\sigma] = 1 > .75$

11. **a.** $\mu = 0$, $\sigma = k\sqrt{2p}$ **b.** $\sigma = |k|$ **c.** $\frac{1}{8}$, $\frac{1}{18}$

Section 4-5

1. $\mu = 20$, $\sigma = 4.24$, $a = 20$, $b = (1.96)\sigma = 8.31$

3. **a.** For $n \leq 1610$ Sam can be 90% sure the claim is false.

 b. For $n \geq 2472$, there is a 90% chance that the company's claim is true, provided no other weed seeds are on the table.

5. $\Pr[\text{reject good lot}]$
 $= \Pr[\text{at least one defective is found when 5 or fewer exist}]$
 $= 1 - \Pr[\text{no defectives found when 5 or fewer exist}]$
 $\leq 1 - \Pr[\text{no defectives found when 5 defectives are in the lot}]$
 $= 1 - .067 = .933$

7. $C(n) = \dfrac{10}{n} + 13 - (11.5)\left(\dfrac{19}{20}\right)^n$: the company should pool 5 batches at a time.

Review Exercises

1.

Events	x	$\Pr[X = x]$	$d(x)$
$\{(1,1)\}$	2	$\frac{1}{36}$	$\frac{1}{36}$
$\{(1,2), (2,1)\}$	3	$\frac{2}{36}$	$\frac{3}{36}$
$\{(1,3), (3,1), (2,2)\}$	4	$\frac{3}{36}$	$\frac{6}{36}$
$\{(1,4), (4,1), (2,3), (3,2)\}$	5	$\frac{4}{36}$	$\frac{10}{36}$
$\{(1,5), (5,1), (2,4), (4,2), (3,3)\}$	6	$\frac{5}{36}$	$\frac{15}{36}$
$\{(1,6), (6,1), (2,5), (5,2), (3,4), (4,3)\}$	7	$\frac{6}{36}$	$\frac{21}{36}$
$\{(3,5), (5,3), (2,6), (6,2), (4,4)\}$	8	$\frac{5}{36}$	$\frac{26}{36}$
$\{(3,6), (6,3), (5,4), (4,5)\}$	9	$\frac{4}{36}$	$\frac{30}{36}$
$\{(4,6), (6,4), (5,5)\}$	10	$\frac{3}{36}$	$\frac{33}{36}$
$\{(5,6), (6,5)\}$	11	$\frac{2}{36}$	$\frac{35}{36}$
$\{(6,6)\}$	12	$\frac{1}{36}$	$\frac{36}{36}$

3.

x	$\Pr[X = x]$	Product	
0	$\dbinom{3}{0}\dbinom{5}{4}\Big/\dbinom{8}{4} = \dfrac{5}{70}$	0	
1	$\dbinom{3}{1}\dbinom{5}{3}\Big/\dbinom{8}{4} = \dfrac{30}{70}$	$\dfrac{30}{70}$	$\mu = 1.5$,
2	$\dbinom{3}{2}\dbinom{5}{2}\Big/\dbinom{8}{4} = \dfrac{30}{70}$	$\dfrac{60}{70}$	$v(X) = .54$
3	$\dbinom{3}{3}\dbinom{5}{1}\Big/\dbinom{8}{4} = \dfrac{5}{70}$	$\dfrac{15}{70}$	

5. a. 1.0　　**b.** 3.1　　**c.** 1.76　　**d.** .6
7. a. .7327　　**b.** .3446　　**c.** .8531
9. a. 4　　**b.** $\sqrt{3}$
11. a. $\mu = 54$, $\sigma = 6$　　**b.** $npq = 36 \geq 12$　　**c.** .9241
13. Pr[at least 50 hits] = Pr[$Z \geq 7.5$] $\leq 10^{-12}$
15. a. Less than 10^{-4}　　**b.** .0044

CHAPTER 5

Section 5-3

1. a. 1　　**b.** -1　　**c.** 2　　**d.** 0
3. a. $y = -2(x - 1)$　　**b.** $y = -1$　　**c.** $x = 2$
5. a. $y = -2x + 2$　　**b.** $y = 2x - 2$
7. 20　　**9.** 3000 miles　　**11.** No, data are not linear.
13. Since (x_1, y_1) and (x_2, y_2) both satisfy $Ax + By = C$, we have
$A(x_1 - x_2) = -B(y_1 - y_2)$, and since the points are distinct, at least one of the
inequalities $x_1 \neq x_2$, $y_1 \neq y_2$ must hold. If $x_1 = x_2$, then the fact that $B \neq 0$ means
that y_1 must equal y_2, which is impossible.

Section 5-4

1. a. Intersect in one point　　**b.** Intersect in one point　　**c.** Coincident
d. Intersect in one point
3. $100,000 building: $v = 100,000 - 5000t$, slope $= -5000$; $150,000 building:
$v = 150,000 - 12,000t$, slope $= -12,000$; the two buildings have the same value when
$t = 50/7$.
5. $\frac{3}{4}$ mile
7. $4000 in utility bonds and $6000 in a savings account
9. a. One solution　　**b.** No solutions　　**c.** Infinitely many solutions

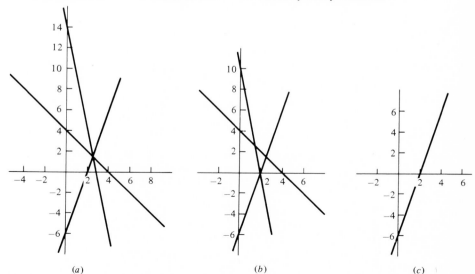

(a)　　　　　　　　(b)　　　　　　　　(c)

Section 5-5

3. a. $x \geq 0$, $y \geq 0$, $x + y \leq 3$ **b.** $y \geq -3$, $5x + 2y \leq 4$, $-5x + 2y \leq 4$
7. $x \geq 1$, $y \leq 2$, $2y - x \geq 1$
9.

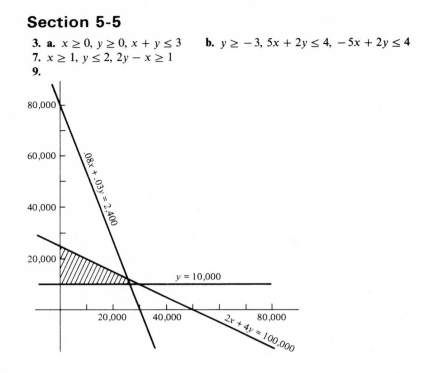

Section 5-6

1. Let x and y denote the numbers of barrels of distillates A and B, respectively. The problem is to maximize $2x + y$ subject to $x \geq 0$, $y \geq 0$ and

$$
\begin{aligned}
2x &\leq 500 \\
25x &\leq 4000 \\
2x + 3y &\leq 1200 \\
25x + 15y &\leq 7500
\end{aligned}
$$

3. Let x and y denote the numbers of desks and filing cabinets, respectively. The problem to maximize $20x + 15y$ subject to $x \geq 0$, $y \geq 0$ and

$$
\begin{aligned}
75x + 50y &\leq 18{,}000 \\
2x + 3y &\leq 750 \\
x &\leq 200
\end{aligned}
$$

5. Let x and y denote the numbers of single-family homes and condominium units, respectively. The problem is to maximize $8000x + 10{,}000y$ subject to $x \geq 0$, $y \geq 0$ and

$$
\begin{aligned}
x &\leq 30 \\
y &\leq 40 \\
3x + y &\leq 60 \\
8x + 5y &\leq 200 \\
9x + 15y &\leq 300
\end{aligned}
$$

7. Let x and y denote the numbers of 100-pound sacks of lawn and garden fertilizer, respectively. The problem is to maximize $6x + 4y$ subject to $x \geq 0$, $y \geq 0$ and

$$20x + 10y \leq 7000$$
$$5x + 15y \leq 4000$$
$$5x + 10y \leq 4000$$
$$y \geq 20$$

9. Let x and y denote the numbers of full teams and half teams, respectively. The problem is to maximize $180x + 100y$ subject to $x \geq 0$, $y \geq 0$ and

$$x + y \leq 200$$
$$3x + 2y \leq 450$$

Section 5-7

1. a. Corner points: $(0, -2)$, $(2,0)$, $(-2,2)$
   b. Corner points: $(1,1)$, $\left(-\frac{1}{2}, \frac{5}{2}\right)$

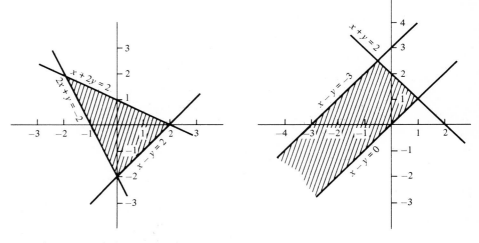

3. 150   **5.** -25 **7.** 50/3
9. 132 desks and 162 filing cabinets
11. The corner point which gives the maximum value of the objective function is $\left(\frac{50}{3}, 10\right)$, where the first coordinate denotes the number of single-family homes and the second coordinate denotes the number of condominium units. Since the number of single-family homes must be an integer, we check feasible integer values of the variables. We conclude that the solution is 16 homes and 10 condominium units.
13. 260 sacks of lawn fertilizer and 180 sacks of garden fertilizer

Review Exercises

3. $y = -3x - 5$ **5.** No solutions
9. a. Equation: $y = 6$, slope $= 0$, y intercept $= 6$

    b. Equation: $y = 1.5x + 3$, slope $= 1.5$, y intercept $= 3$, x intercept $= -2$
    c. Equation: $x = -3$, slope undefined, x intercept $= -3$
    d. Equation: $y = -2x/3$, slope $= -2/3$, y intercept $= 0$, x intercept $= 0$
    e. Equation: $y = 3x/2 + 17/2$, slope $= 3/2$

11. a. $1825 at end of 3 years, $250 at end of 7 years
    b. $v = 3400 - 525t$ for $t = 1, 2, 3, 4, 5, 6$; $v = 250$ for $t = 7$ and above
    c. -525 and 0

13. They never have the same value.

15. a.

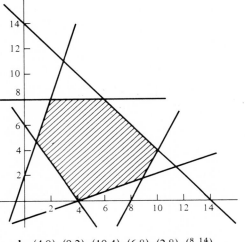

    b. $(4,0)$, $(9,2)$, $(10,4)$, $(6,8)$, $(2,8)$, $(\frac{8}{9},\frac{14}{3})$
    c. 36 at $(6,8)$ **d.** -19 at $(9,2)$

17. Maximum profit of $8750 is attained with the production of 1500 left-handed widgets and 2500 right-handed widgets. Maximum profit attained by producing one kind of widget is $7500.

CHAPTER 6

Section 6-2

1. a. $\begin{bmatrix} 17 \\ 39 \end{bmatrix}$ **b.** $\begin{bmatrix} -2 \\ 2 \\ 0 \end{bmatrix}$

3. a. $A = \begin{bmatrix} 3 & 2 & -1 \\ 1 & 1 & 1 \\ 2 & 1 & -1 \end{bmatrix}$ $b = \begin{bmatrix} 4 \\ 1 \\ \frac{5}{2} \end{bmatrix}$ **b.** $A = \begin{bmatrix} 3 & -2 & 4 & -1 \\ 2 & 1 & 5 & 2 \\ -1 & 1 & 3 & 0 \end{bmatrix}$ $b = \begin{bmatrix} 1 \\ 3 \\ 2 \end{bmatrix}$

5. a. $2x_1 + 3x_2 + x_3 = 6$ **b.** $2x_1 + 2x_2 + \frac{3}{2}x_3 = 2$
$\qquad 6x_1 - 2x_2 - 8x_3 = 7 \qquad\quad 4x_1 + \frac{5}{2}x_2 + 2x_3 = 8$
$\qquad 8x_1 + 5x_2 - 3x_3 = 17 \qquad\; 2x_1 - \frac{3}{2}x_2 - x_3 = 10$

Section 6-3

1. a. $(8/5, 6/5)$ **b.** $(8/5, 6/5)$ **3. a.** $(2, 1/2, 3/2)$ **b.** $(1, -5/2, 1)$
5. $(0, 0, -2, 3)$ and $(1, 2, 1, 1)$
7. $(7/4 - 7t/8, 2, 19/4 + 5t/8, t)$ for any number t.
9. Not all cabinets can be used. The maximum number of cabinets will be used if 3 ranch, 2 bilevel, and 5 colonial houses are built.

Section 6-4

1. a. Not defined   **b.** $\begin{bmatrix} 1 & 2 & 0 \\ 10 & 0 & 7 \end{bmatrix}$ **c.** $\begin{bmatrix} 10 & 3 & 4 \\ 10 & 4 & 5 \end{bmatrix}$

3. a. $\begin{bmatrix} 6 & 2 \\ -2 & 0 \\ 8 & 4 \end{bmatrix}$ **b.** $\begin{bmatrix} 9 & 7 \\ -2 & -2 \\ 14 & 10 \end{bmatrix}$ **c.** $\begin{bmatrix} 50 & 38 \\ 39 & 29 \\ 39 & 29 \end{bmatrix}$

5. $c = 3$ **9.** $\mathbb{A}^2 = \begin{bmatrix} 1 & 0 \\ \frac{3}{4} & \frac{1}{4} \end{bmatrix}$ $\mathbb{A}^3 = \begin{bmatrix} 1 & 0 \\ \frac{7}{8} & \frac{1}{8} \end{bmatrix}$

Section 6-5

3. a. $\begin{bmatrix} 1 & 1 & -\frac{1}{2} & \frac{1}{2} \\ -\frac{1}{2} & -\frac{1}{2} & \frac{3}{4} & -\frac{1}{4} \\ \frac{1}{2} & \frac{1}{2} & -\frac{1}{4} & \frac{1}{4} \\ \frac{1}{2} & -\frac{1}{2} & -\frac{1}{4} & \frac{3}{4} \end{bmatrix}$ **b.** $\begin{bmatrix} \frac{15}{14} & \frac{4}{14} & \frac{1}{14} \\ \frac{4}{14} & \frac{16}{14} & \frac{4}{14} \\ \frac{1}{14} & \frac{4}{14} & \frac{15}{14} \end{bmatrix}$

5. $\mathbb{A}^{-1} = \begin{bmatrix} 1 & -2 \\ -2 & 5 \end{bmatrix}$ $(\mathbb{A}^{-1})^{-1} = \begin{bmatrix} 5 & 2 \\ 2 & 1 \end{bmatrix}$

7. a. $\mathbb{B} = \begin{bmatrix} \frac{1}{5} & \frac{9}{5} \\ \frac{8}{5} & \frac{12}{5} \end{bmatrix}$ **b.** $\mathbb{B} = \begin{bmatrix} 4 & 2 \\ -\frac{8}{5} & -\frac{7}{5} \end{bmatrix}$

9. $\mathbb{A}^2 = \begin{bmatrix} 4 & 18 \\ 6 & 28 \end{bmatrix}$ $\mathbb{A}^{-1} = \begin{bmatrix} \frac{5}{2} & -\frac{3}{2} \\ -\frac{1}{2} & \frac{1}{2} \end{bmatrix}$ $\mathbb{A}^{-1}\mathbb{A}^{-1} = \begin{bmatrix} 7 & -\frac{9}{2} \\ -\frac{3}{2} & 1 \end{bmatrix}$

$(\mathbb{A}^2)^{-1} = \begin{bmatrix} 7 & -\frac{9}{2} \\ -\frac{3}{2} & 1 \end{bmatrix}$

Section 6-6

1.

$$\begin{array}{c} \\ A \\ B \\ C \\ D \\ E \end{array}\begin{array}{ccccc} A & B & C & D & E \\ \left[\begin{array}{ccccc} 0 & 1 & 1 & 0 & 0 \\ 1 & 0 & 0 & 1 & 0 \\ 1 & 0 & 0 & 0 & 1 \\ 0 & 0 & 0 & 0 & 1 \\ 0 & 0 & 1 & 0 & 0 \end{array}\right] \end{array}$$

3.

$$\begin{array}{c} \\ A \\ B \\ C \\ D \\ E \\ F \\ G \end{array}\begin{array}{ccccccc} A & B & C & D & E & F & G \\ \left[\begin{array}{ccccccc} 0 & 1 & 1 & 1 & 0 & 0 & 0 \\ 0 & 0 & 1 & 0 & 1 & 1 & 0 \\ 0 & 0 & 0 & 1 & 0 & 1 & 0 \\ 0 & 0 & 1 & 0 & 0 & 1 & 1 \\ 0 & 0 & 0 & 0 & 0 & 0 & 0 \\ 0 & 1 & 0 & 0 & 1 & 0 & 0 \\ 0 & 0 & 0 & 0 & 0 & 0 & 0 \end{array}\right] \end{array}$$

7. A **9.** The unit corresponding to the last row of the matrix

Section 6-7

1. a. $\begin{bmatrix} 55 \\ 115 \end{bmatrix}$ **b.** $\begin{bmatrix} 10{,}375/3 \\ 21{,}125/3 \end{bmatrix}$ **3.** $\begin{bmatrix} 20/3 \\ 10/3 \end{bmatrix}$

5. $\begin{bmatrix} 15 \\ 35 \\ 15 \end{bmatrix}$ **7.** $\begin{bmatrix} 144.0 \\ 114.8 \\ 81.6 \end{bmatrix}$

Review Exercises

1. a. $\begin{bmatrix} 2 & 2 \\ 6 & -1 \end{bmatrix}\begin{bmatrix} x_1 \\ x_2 \end{bmatrix} = \begin{bmatrix} 4 \\ 6 \end{bmatrix}$ **b.** $\begin{bmatrix} 2 & -2 & 2 \\ 4 & -9 & -7 \\ -11 & 1 & 1 \end{bmatrix}\begin{bmatrix} x_1 \\ x_2 \\ x_3 \end{bmatrix} = \begin{bmatrix} 7 \\ 4 \\ 15 \end{bmatrix}$

3. a. $\begin{bmatrix} 17 \\ 39 \end{bmatrix}$ **b.** $\begin{bmatrix} 1 \\ 4 \\ 7 \end{bmatrix}$ **7.** No solutions

9. a. $\begin{bmatrix} 5 & -1 & 0 \\ 0 & 2 & -5 \end{bmatrix}$ **b.** $\begin{bmatrix} 5 & 3 & -10 \\ -10 & 4 & -5 \end{bmatrix}$

11. $\begin{bmatrix} 6 & -1 & -1 \\ -2 & 1 & 0 \\ -3 & 0 & 1 \end{bmatrix}$ **13.** $\begin{bmatrix} 15 \\ 35 \end{bmatrix}$

15.

$$\begin{array}{c} \\ A \\ B \\ C \\ D \\ E \end{array}\begin{array}{ccccc} A & B & C & D & E \\ \left[\begin{array}{ccccc} 0 & 1 & 1 & 0 & 0 \\ 0 & 0 & 1 & 1 & 0 \\ 0 & 0 & 0 & 0 & 1 \\ 0 & 1 & 1 & 0 & 0 \\ 0 & 0 & 0 & 0 & 0 \end{array}\right] \end{array}$$

A, B, and D all give directives to 3 other units in one or two steps.

CHAPTER 7

Section 7-2

1. Find the maximum of $p = 1.1x + 1.3y + 2z$ for $x \geq 0$, $y \geq 0$, $z \geq 0$, and

$$5x + 3y + 10z \leq 192$$
$$2x + 4y + z \leq 55$$
$$3x + 4y + 6z \leq 45$$

3. a. No, because all constraints are not of the form: linear function $\leq$ positive constant.
   b. No, because the objective function is not to be maximized.
5. Exercises 3 and 4 are SMPs. Exercise 5 is not an SMP because the objective function is minimized.

Section 7-3

1. a. Find the maximum of $3x + 8y$ for $x \geq 0$, $y \geq 0$, $u \geq 0$, $v \geq 0$, and

$$2x + 2y + u = 10$$
$$3x + 6y + v = 18$$

Corner points: (0,0), (5,0), (4,1), (0,3)
   b. Find the maximum of $x + y$ for $x \geq 0$, $y \geq 0$, $u \geq 0$, $v \geq 0$, $w \geq 0$, and

$$2x + 8y + u = 16$$
$$3x + 3y + v = 15$$
$$10x + 3y + w = 30$$

Corner points: (0,0), (3,0), (0,2), (96/37,50/37)

3.

x	y	z	u	v	w	Type
0	0	0	160	50	50	Basic
0	0	$\frac{160}{9}$	0	$\frac{290}{9}$	$-\frac{30}{9}$	Basic
0	$\frac{160}{3}$	0	0	$-\frac{490}{3}$	$-\frac{490}{3}$	Basic
1	1	1	144	42	41	Not basic
2	2	2	128	34	32	Not basic

Section 7-4

1. $$\begin{vmatrix} 2 & 3 & 1 & 0 & 12 \\ 6 & 1 & 0 & 1 & 9 \\ -1 & -3 & 0 & 0 & 0 \end{vmatrix}$$

3. $$\begin{vmatrix} 5 & 0 & 1 & 1 & 0 & 0 & 100 \\ 0 & 1 & 3 & 0 & 1 & 0 & 300 \\ 2 & -1 & 1 & 0 & 0 & 1 & 900 \\ -2 & -3 & -1 & 0 & 0 & 0 & 0 \end{vmatrix}$$

5.

	x	y	z	u	v	Basic Solution
u	1	0	5	1	0	50
y	0	1	$\frac{1}{10}$	0	$\frac{1}{10}$	3
p	-1	0	$-\frac{3}{2}$	0	$\frac{1}{2}$	15

Basic solution:
$$y = 3$$
$$u = 50$$
$$x = z = v = 0$$
$$p = 15$$

7. Find the maximum of $x + 3y$ for $x \geq 0$, $y \geq 0$ and

$$2x + 3y \leq 12$$
$$6x + y \leq 9$$

9. Find the maximum of $2x + 3y + z$ for $x \geq 0$, $y \geq 0$, $z \geq 0$ and

$$5x + z \leq 100$$
$$y + 3z \leq 300$$
$$2x - y + z \leq 900$$

11.

	x	y	z	u	v	Basic Solution
x	1	0	5	1	0	50
y	0	1	$\frac{1}{10}$	0	$\frac{1}{10}$	3
p	0	0	$\frac{35}{10}$	1	$\frac{1}{2}$	65

Original basic solution: $z = 10$, $y = 2$, $u = v = x = 0$, $p = 30$
New basic solution: $x = 50$, $y = 3$, $u = v = z = 0$, $p = 65$
The new basic solution is better.

13.

	x	y	z	u	v	w	Basic Solution
x	1	0	0	$\frac{3}{4}$	$-\frac{5}{4}$	$\frac{3}{4}$	$\frac{27}{4}$
z	0	0	1	$-\frac{5}{36}$	$\frac{5}{12}$	$-\frac{1}{4}$	$\frac{15}{4}$
y	0	1	0	$\frac{1}{36}$	$-\frac{1}{12}$	$\frac{1}{4}$	$\frac{9}{4}$
p	0	0	0	$\frac{5}{9}$	$-\frac{2}{3}$	$\frac{35}{27}$	21

Original basic solution: $y = 2$, $z = 5$, $u = 9$, $x = v = w = 0$, $p = 16$
New basic solution: $x = 27/4$, $y = 9/4$, $z = 15/4$, $u = v = w = 0$, $p = 21$
The new basic solution is better.

Section 7-5

1. a. Entering variable is y, departing variable is u, pivot element is 4.
   b. Entering variable is x, departing variable is v, pivot element is 4.
   c. Basic solution shown is optimal.

3. Find the maximum of $x + 2y$ for $x \geq 0$, $y \geq 0$, and

$$x + 4y \leq 12$$
$$5x + 4y \leq 20$$

5. No optimal vector exists.
7. No optimal vector exists.

Section 7-6

1. Find the maximum of $\mathbf{c}^T \mathbf{x}$ for $\mathbf{x} \geq \mathbf{0}$ and $\mathbb{A}\mathbf{x} \leq \mathbf{b}$. Here $\mathbf{x}^T = (x, y)$, $\mathbf{c}^T = (1, 1)$,
$\mathbb{A} = \begin{bmatrix} 5 & 5 \\ 2 & 10 \end{bmatrix}$, $\mathbf{b}^T = (20, 40)$.

3. Find the maximum of $\mathbf{c}^T \mathbf{x}$ for $\mathbf{x} \geq \mathbf{0}$ and $\mathbb{A}\mathbf{x} \leq \mathbf{b}$. Here $\mathbf{x}^T = (x_1, x_2, x_3, x_4)$,
$\mathbf{c}^T = (3, 4, -1, 4)$.

$$\mathbb{A} = \begin{bmatrix} 1 & 2 & 3 & 4 \\ 5 & 6 & 7 & 8 \\ 10 & -1 & 1 & 10 \\ -1 & 1 & 3 & 7 \end{bmatrix} \qquad \mathbf{b}^T = (20, 30, 40, 30)$$

5. Find the maximum of $2x_1 - 2x_2 + 3x_3 + 7x_4$ for $x_1 \geq 0$, $x_2 \geq 0$, $x_3 \geq 0$, $x_4 \geq 0$ and

$$x_1 - x_3 + 5x_4 \leq 25$$
$$-x_1 + 2x_2 + 7x_3 + 4x_4 \leq 50$$
$$3x_2 - 2x_3 \leq 20$$
$$3x_2 \leq 24$$
$$-3x_2 + 6x_4 \leq 30$$

7. Find the maximum of $400r + 200s$ for $r \geq 0$, $s \geq 0$, $t \geq 0$, and

$$r - t \leq 8$$
$$r + s + 2t \leq 10$$

9. Solution to original problem: $x = 200$, $y = \frac{500}{3}$, $p = \frac{1700}{3}$. Solution to dual: $r = \frac{1}{3}$, $s = 0$, $t = \frac{1}{15}$, $c = \frac{1700}{3}$.

11. Solution to original problem: $p = 1$ when $x = 0$, $y = 0$, $z = 1$. Solution to dual: $c = 1$ when $r = 0$, $s = 1$.

13. Solution to SMP: $x = 160$, $y = \frac{700}{3}$, $p = \frac{1660}{3}$. Solution to dual: $r_1 = 0$, $r_2 = \frac{1}{15}$, $r_3 = 0$, $r_4 = \frac{1}{15}$, $c = \frac{1660}{3}$.

15. Solution to SMP: $x = 162 =$ number of file cabinets, $y = 132 =$ number of desks, $p = \$5070$. Solution of dual: $r = \frac{6}{25}$, $s = 1$, $t = 0$, $c = \$5070$.

Review Exercises

1. Not an SMP
3. Find the maximum of $10x + 5y + 15z = p$ for $x \geq 0$, $y \geq 0$, $z \geq 0$, and

$$20x + 10y + 5z \leq 2000$$
$$5x + 10y + 15z \leq 1500$$
$$5x + 10y + 10z \leq 1500$$

Initial tableau

	x	y	z	u	v	w	**Basic Solution**
u	20	10	5	1	0	0	2000
v	5	10	15	0	1	0	1500
w	5	10	10	0	0	1	1500
p	-10	-5	-15	0	0	0	0

5. $p = 75$ when $x = 15$ and $y = 0$.

7. Maximize $\begin{bmatrix} x & y \end{bmatrix} \begin{bmatrix} 1 \\ -1 \end{bmatrix}$ when $\begin{bmatrix} 5 & -10 \\ 1 & 1 \end{bmatrix} \begin{bmatrix} x \\ y \end{bmatrix} \leq \begin{bmatrix} 20 \\ 5 \end{bmatrix}$ and $\begin{bmatrix} x \\ y \end{bmatrix} \geq \mathbf{0}$.

9. Solution: $p = 4\frac{1}{3}$ when $x = 4\frac{2}{3}$ and $y = \frac{1}{3}$. Solution to dual: $c = 4\frac{1}{3}$ when $r = \frac{2}{15}$ and $s = \frac{1}{3}$.

11. $r = \frac{3}{11}$, $s = \frac{10}{11}$, $t = 0$, and $c = \frac{21000}{11}$.

CHAPTER 8

Section 8-2

3. c. $\mathbb{P}(2) = \begin{bmatrix} \frac{5}{12} & \frac{3}{12} & \frac{4}{12} \\ \frac{2}{6} & \frac{3}{6} & \frac{1}{6} \\ \frac{17}{48} & \frac{12}{48} & \frac{19}{48} \end{bmatrix}$ **d.** $\left(\frac{9}{24}, \frac{9}{24}, \frac{6}{24} \right)$

5. b. If states 1, 2, and 3 correspond to I, SB, SE respectively, then

$\mathbb{P} = \begin{bmatrix} .7 & .2 & .1 \\ .3 & .5 & .2 \\ .3 & .3 & .4 \end{bmatrix}$ **c.** .58

7. b. $\mathbb{P} = \begin{bmatrix} \frac{1}{2} & \frac{1}{2} & 0 & 0 \\ \frac{1}{3} & \frac{1}{3} & \frac{1}{3} & 0 \\ 0 & \frac{1}{3} & \frac{1}{3} & \frac{1}{3} \\ 0 & 0 & \frac{1}{2} & \frac{1}{2} \end{bmatrix}$ **c.** 7/27

Section 8-3

1. a. $\mathbf{x}_1 = \left(\frac{1}{4}, \frac{1}{4}, \frac{1}{2}, 0 \right)$, $\mathbf{x}_2 = \left(\frac{7}{48}, \frac{17}{48}, \frac{12}{48}, \frac{12}{48} \right)$
 b. $\mathbf{x}_1 = \left(\frac{16}{96}, \frac{23}{96}, \frac{39}{96}, \frac{18}{96} \right)$

3. $\left(\frac{1}{2}, \frac{1}{2} \right)$ **5.** Regular, $\left(\frac{1}{5}, \frac{1}{5}, \frac{3}{5} \right)$ **7.** Not regular

9. a. $\mathbb{P} = \begin{bmatrix} .3 & .2 & .5 \\ .2 & .5 & .3 \\ .5 & .3 & .2 \end{bmatrix}$ regular   **b.** $\left(\frac{1}{3}, \frac{1}{3}, \frac{1}{3} \right)$

c. Amount of time spent by the service specialist in each district in the long run is consistent with the fact that the service calls originate in equal numbers from all districts. Time spent travelling depends on the geography.

Section 8-4

1. **a.** Absorbing
   c. 1 transition for state 2 and 2 transitions for state 4
3. **a.** Not absorbing
   b. It is impossible to reach an absorbing state from states 3 and 4.
5. $\begin{bmatrix} 1 & 0 & 0 & 0 \\ \frac{1}{2} & \frac{1}{2} & 0 & 0 \\ 0 & \frac{2}{3} & \frac{1}{3} & 0 \\ 0 & 0 & 1 & 0 \end{bmatrix}$ $\mathbb{N} = \begin{bmatrix} 2 & 0 & 0 \\ 2 & \frac{3}{2} & 0 \\ 2 & \frac{3}{2} & 1 \end{bmatrix}$

7. $11/3$ **9.** 15

Review Exercises

1. **b.** $\begin{bmatrix} \frac{58}{144} & \frac{17}{144} & \frac{69}{144} \\ \frac{70}{192} & \frac{35}{192} & \frac{87}{192} \\ \frac{5}{12} & \frac{1}{12} & \frac{6}{12} \end{bmatrix}$ **c.** $(\frac{382}{960}, \frac{119}{960}, \frac{459}{960})$

3. **a.** Regular   **b.** $(\frac{2}{17}, \frac{3}{17}, \frac{6}{17}, \frac{6}{17})$
5. **b.** 2 transitions for initial state 2 and 3/2 transitions for initial state 4
7. 8
9. **a.**
Red	$\frac{2}{3}$	$\frac{2}{9}$	0	$\frac{1}{9}$
Blue	$\frac{1}{9}$	$\frac{2}{3}$	$\frac{2}{9}$	0
Yellow	0	$\frac{1}{9}$	$\frac{2}{3}$	$\frac{2}{9}$
Green	$\frac{2}{9}$	0	$\frac{1}{9}$	$\frac{2}{3}$

   b. $\frac{27}{2}$

CHAPTER 9

Section 9-2

1. $1.25¢$
3. Labor has two strategies: sign a contract or strike. Management has two strategies: sign a contract or do not sign. Payoffs in each case depend on the industry and union involved as well as many external factors, e.g., the size of the union's strike fund and the general condition of the economy.

5.

	H	T
H	-5	25
T	25	-5

Expected gain to Rolf $= 10\cancel{c}$ per play

Section 9-3

1. a. No saddle point
   b. Circled entries are saddle points
$$\begin{bmatrix} 1 & 2 & 3 \\ ② & ② & 3 \\ ② & ② & 2 \end{bmatrix}$$

3. a. Circled entries are saddle points
$$\begin{bmatrix} 0 & -1 & -3 & 0 \\ 1 & ⓪ & ⓪ & 3 \\ 1 & ⓪ & ⓪ & 1 \\ 0 & -2 & -1 & 1 \end{bmatrix}$$

   b. No saddle points
5. All games have saddle points except **3b**, which has the reduced form

$$\begin{array}{c} \\ r_2 \\ r_3 \end{array} \begin{matrix} c_1 & c_2 & c_3 \\ \begin{bmatrix} 0 & 3 & -3 \\ 1 & 0 & 2 \end{bmatrix} \end{matrix}$$

7. The payoff matrix is
$$\begin{array}{c} \\ S \\ B \end{array} \begin{matrix} S & B \\ \begin{bmatrix} 0 & -4 \\ 4 & ⓪ \end{bmatrix} \end{matrix},$$
where the circled entry is a saddle point.

Section 9-4

1. $s^* = \left(\frac{1}{2},\frac{1}{2}\right)$, $t^* = \left(\frac{7}{10},\frac{3}{10},0\right)$, $v = -\frac{1}{2}$
3. $s^* = \left(\frac{3}{7},\frac{4}{7},0\right)$, $t^* = \left(\frac{4}{7},\frac{3}{7}\right)$, $v = \frac{5}{7}$
5. $s^* = \left(\frac{7}{12},0,\frac{5}{12},0\right)$, $t^* = \left(\frac{1}{2},\frac{1}{2}\right)$, $v = -\frac{1}{2}$
7. $s^* = \left(\frac{1}{2},0,\frac{1}{2}\right)$, $t^* = \left(\frac{1}{2},\frac{1}{2}\right)$, $v = \frac{1}{2}$
9. $s^* = (0,1,0,0)$, $t^* = (0,0,1,0)$, $v = -1$

Section 9-5

1. $s^* = \left(\frac{1}{2},\frac{1}{2}\right)$, $t^* = \left(0,\frac{9}{16},\frac{7}{16}\right)$, $v = -\frac{1}{2}$ **3.** $s^* = \left(\frac{4}{81},\frac{70}{81},\frac{7}{81}\right)$, $t^* = \left(\frac{44}{81},0,\frac{2}{81},\frac{35}{81}\right)$, $v = -\frac{83}{81}$
5. If a payoff matrix is multiplied by a positive constant k, the optimal strategies for the game are unchanged and the value of the game is multiplied by the same constant k.
7. $s^* = \left(\frac{9}{14},\frac{5}{14},0\right)$, $t^* = \left(0,\frac{4}{7},\frac{3}{7}\right)$, $v = -\frac{1}{7}$

Review Exercises

1. The saddle points are circled
$$\begin{bmatrix} ① & 2 & 3 & ① \\ -1 & 5 & -3 & 0 \\ 0 & -2 & 6 & -2 \\ ① & 4 & 1 & ① \end{bmatrix}$$

3. $\mathbf{s}^* = \left(\frac{4}{9},\frac{5}{9}\right)$, $\mathbf{t}^* = \left(\frac{2}{3},0,0,\frac{1}{3}\right)$, $v = \frac{2}{3}$

5. $\mathbf{s}^* = \left(0,\frac{7}{13},\frac{6}{13}\right)$, $\mathbf{t}^* = \left(0,\frac{9}{13},\frac{4}{13}\right)$, $v = \frac{2}{13}$

7. $\mathbf{s}^* = (0,0,1)$, $\mathbf{t}^* = (0,1,0)$, $v = -1$ and $\mathbf{s}^* = (0,1,0)$, $\mathbf{t}^* = (0,1,0)$, $v = -1$.

CHAPTER 10

Section 10-2

1.
$$\begin{bmatrix} 0 & 1 & 0 & 0 \\ 1 & 0 & 1 & 0 \\ 0 & 0 & 0 & 0 \\ 1 & 1 & 1 & 0 \end{bmatrix}$$

3.
$$\begin{bmatrix} 0 & 1 & 0 & 0 \\ 0 & 0 & 1 & 0 \\ 0 & 0 & 0 & 1 \\ 0 & 1 & 0 & 0 \end{bmatrix}$$

5. b.
$$\begin{bmatrix} 0 & 1 & 0 & 0 & 0 & 1 \\ 0 & 0 & 0 & 0 & 1 & 0 \\ 0 & 1 & 0 & 0 & 0 & 0 \\ 0 & 0 & 1 & 0 & 0 & 0 \\ 0 & 0 & 1 & 0 & 0 & 0 \\ 1 & 1 & 0 & 0 & 0 & 0 \end{bmatrix}$$

c.
$$\begin{bmatrix} 0 & 1 & 3 & \infty & 2 & 1 \\ \infty & 0 & 2 & \infty & 1 & \infty \\ \infty & 1 & 0 & \infty & 2 & \infty \\ \infty & 2 & 1 & 0 & 3 & \infty \\ \infty & 2 & 1 & \infty & 0 & \infty \\ 1 & 1 & 3 & \infty & 2 & 0 \end{bmatrix}$$

7.
$$\begin{bmatrix} 0 & 1 & 2 \\ 1 & 0 & 1 \\ 2 & 1 & 0 \end{bmatrix}$$

9. F   **11.** $\frac{15}{3}$ **13. b.**
$$\begin{bmatrix} 0 & 1 & 0 & 0 \\ 0 & 0 & 1 & 0 \\ 0 & 0 & 0 & 0 \\ 0 & 1 & 0 & 0 \end{bmatrix}$$

15. $S(P) = 1(3) + 2(2) = 7$, $S(VPF) = 1(2) + 2(2) = 6$, $S(VPP) = 1(3) + 2(1) = 5$, $S(T) = 1(2) = 2$, $S(DR) = S(DLR) = 0$

Section 10-3

1. Geodesic: $ACFG$, cost: 11

3. Geodesic: $ABEFHI$, cost: 18   **5.** 7000 pounds

Review Exercises

1. $\begin{bmatrix} 0 & 1 & 1 & 1 \\ 0 & 0 & 0 & 0 \\ 0 & 1 & 0 & 0 \\ 0 & 0 & 1 & 0 \end{bmatrix}$

5.

	P	VPI	VPII	VPIII	T	C	DR	DLR
P	0	1	1	1	2	2	2	3
VPI	∞	0	∞	∞	1	1	2	2
VPII	∞	∞	0	1	∞	1	2	3
VPIII	∞	∞	1	0	∞	2	1	2
T	∞	∞	∞	∞	0	1	2	1
C	∞	∞	∞	∞	∞	0	1	2
DR	∞	∞	∞	∞	∞	∞	0	1
DLR	∞	∞	∞	∞	∞	∞	∞	0

7. $S(P) = 12$, $S(T) = 4$, $S(VPI) = 6$, $S(C) = 3$, $S(VPII) = 7$, $S(DR) = 1$, $S(VPIII) = 6$, $S(DLR) = 0$

9. Geodesic: $BHGLMP$; cost: 17

CHAPTER 11

Section 11-2

1. a. $11,596.93   **b.** $6,092.01   **c.** $1822.12   **3.** 21% **5.** $1977.73
7. $2219.28   **9.** 5.09% and 5.61%

Section 11-3

1. $3855.43   **3. a.** $7451.75   **b.** $7536.66   **c.** $7594.79
5. $3766.07   **7.** $77,561.00   **9.** $120.74

Section 11-4

1. $444.24   **3.** $444.24   **5.** $1879.26   **7.** $386.23
9. $136,516.88 and $175,742.18   **11.** 13 payments are needed

Section 11-5

1. Payments of $500 on each of January 1 and July 1 for 1979 and 1980 and on January 1, 1981. A payment of $10,000 on January 1, 1981.
3. 16.56% per year   **5.** 67.9% return on investment, 90.5% per year
7. 56.25% return on investment, 22.5% per year
9. a. $500   **b.** 17.94% per year

Review Exercises

1. 6.14% per year   **3.** $1779.06 and $1372.09   **5.** $12,423.76   **7.** 11.11%
9. 13.10% per year

INDEX

Absorbing Markov chain, 329
Absorbing state, 329
Adjacency matrix, 370
Agriculture, 40 (Ex. 7), 94 (Ex. 6), 112 (Ex. 11),
    142, 166 (Ex. 3)
Amortization of a loan, 406
    formula for, 407
Amount of an annuity, 402
Annuity, 402
    amount of, 402
Appreciated value of stock, 414
Assignment of weights, 77
    measure defined by, 78
Asymptotic behavior of a Markov chain, 320
Augmented matrix, 233
Axions of probability theory, 72

Back substitution, 232
Basic solution, 275
Bayes' formula, 105
Bayes' probabilities, 102, 105
Bell-shaped curve, 129
Bernoulli process, 95
Bernoulli trial, 95
Binomial density function, 125
Binomial distribution, 124, 128
Binomial process, 95
Binomial random variable, 125
Binomial trial, 95
Biology, 52, 72, 200 (Ex. 10)
Bond, 410
    purchased at a discount, 410
    purchased at par, 410
    redemption date, 410
    redemption value, 410
Bounded set, 212
Branch weights, 90
Business:
    decision problems, 159
    depreciation, 194 (Ex. 3), 218 (Exs. 11 and
        12)
    finance, 406, 410
    product design, 69 (Ex. 12)
    production scheduling, 186, 200 (Ex. 9), 201,
        206–208 (Exs. 1–7), 220, 269, 302 (Exs. 2

Business:
    production scheduling:
        and 3)
    purchasing, 203
    quality control, 11, 40 (Ex. 1), 47, 53 (Ex. 3),
        60, 66, 68, 76, 80 (Ex. 5), 87, 99, 102,
        117, 120 (Exs. 3 and 5), 124, 162
    routing, 53 (Exs. 7 and 8), 368, 380
    sales projections, 2

Canonical form of a Markov chain, 330
Cartesian coordinate system, 172
Cartesian product, 13
Chebychev's theorem, 147, 158 (Ex. 9)
Coefficient matrix, 223
Combination principle, 59
Committee formation, 62, 64 (Ex. 10), 67
Common stock, 413
Communication matrix, 256
Complement of A, 13
    with respect to B, 13
Compound interest, 391
    formula for, 393
Conditional probability of A given B, 82
Consistent system, 228
Constraints of a linear programming problem,
    202
Consumerism, 318 (Ex. 8), 338 (Ex. 10), 390ff.
Continuous interest, 396
Coordinates:
    of a point, 173
    of a vector, 222
Corner point, 212

Deductive method of assigning weights, 38
Demography, 174, 194 (Ex. 8)
Density function of a random variable, 116
Departing variable, 282
Digraph, 369
Directed graph, 369
Discount, bond purchased at, 410
Disjoint, 12
Distance from u to v in a digraph, 372
Distance matrix, 372

Distance sum:
  of a digraph, 378
  of a vertex, 376
Distribution function, 126
Domain, 26
Dominance, simplifications in a game due to, 349
Dual linear programming problems, 295
  interpretation of, 299
  properties of, 297

Echelon form, 230
  reduction to, 232
Economics, 176 (Ex. 4), 260
Edges of a digraph, 369
Effective rate of interest, 396
Element, 9
Empty set, 11
Entering variable, 282
Equality:
  of matrices, 242
  of sets, 10
  of vectors, 225
Equally likely outcomes, 39, 65
Equation of a line, 177
Equiprobable measure, 65
Event, 42
Expected value of a random variable, 117
Experiment, 41
  sample space of, 41
External demand vector, 262

Fair die, 38
Feasible set of a linear programming problem, 202
Feasible vector, 271
Fork in a tree diagram, 46–52
Function, 26
  binomial density, 125
  binomial distribution, 128
  density, 116
  distribution, 126
  linear, 270
  objective, 202
  probability, 116
Fundamental matrix, 332

Game, 339
  two-person zero-sum, 343
Gaussian elimination, 248
Geodesic, 380
Geometric series, sum of, 401

Graph, 369

Identity matrix, 245
  properties of, 246
Included in (for sets), 10
Inconsistent system, 228
Independent events, 84
Independent trials, 95
Index of relative centrality of a vertex, 378 (Ex. 10)
Input-output matrix, 262
Intercept, 178
Interest, 390
Interest rate, 390
  compound, 391
  continuous, 396
  simple, 391
Intersection of sets, 10
Intersection digraph, 379 (Ex. 13)
Inverse of a matrix, 247
  method of determining, 249
Invertible matrix, 247
Investments, 64 (Ex. 6), 176 (Ex. 10), 199 (Ex. 5), 200 (Ex. 11), 240 (Ex. 8), 340, 363, 389

kth power of a matrix, 257
k-step transition:
  matrix, 316
  probabilities, 314

Length of a path, 372
Leontief, Wassily, 260
Leontief model, 260
Line, 177
  general equation of, 178
  point-slope form of, 183
  slope-intercept form of, 179
Linear function of n variables, 270
Linear inequalities, 195
Linear model, 4
Linear programming problem, 200, 201

Markov, A. A., 308
Markov chain, 308
  absorbing, 329
  regular, 321
Mathematical models, 2
Matrix, 222
  adjacency, 370
  augmented, 233
  coefficient, 223

Matrix:
  communication, 256
  distance, 372
  i,j entry of, 222
  invertible, 247
  payoff, 344
  size of, 222
  technology, 262
  transition, 313
Matrix addition, 242
Matrix product, 244
Matrix-vector product, 223
Max-min reasoning in games, 354
Mean, 136
  of binomial random variable, 137, 138
  of normal random variable, 138, 139
Measure:
  defined by weights, 78
  probability, 65, 72
  tree, 90
Medicine, 6, 21, 40 (Ex. 8), 45 (Exs. 3 and 4),
    108, 160, 164
Method:
  of approximating binomial random variables,
    154
  of finding a geodesic in a network, 384
  of finding saddle points, 347
  of inverting matrices, 249
  of solving games: using graphs, 356
    using simplex method, 362
  of solving linear programming problems, 213
  of solving systems of linear equations, 230
Model:
  linear, 4
  mathematical, 2
Multiplication principle, 51
Mutually exclusive events, 79

Negative axis, 172
Network, 379
Normal random variable, 129–134
  as an approximation to binomial random
    variable, 147, 152, 154
Nutrition, 195, 273 (Ex. 2)

Objective function, 202, 271
Optimal strategy, 352
Optimization problem, 200
Ordinary annuity, 402
Origin of a line, 172

Pairwise disjoint, 12

Par, bond purchased at, 410
Parallel lines, 182
Parameter, 237
Partition, 12
Path:
  in a digraph, 371
  in a tree diagram, 52
Path weights, 90
Payoff matrix, 344
Payoffs in a game, 343
Performance of an experiment, 41
Permutation principle, 55
Pivot element, 282
Pivot operation, 281
Point-slope form of equation of a line, 183
Political science, 7, 102 (Ex. 11), 113 (Ex. 12),
    136 (Ex. 3), 204
Positive axis, 172
Premium, bond purchased at, 410
Present value, 398
  of an annuity, 402, 403
Principal of a loan, 390
Probability function of a random variable, 116
Probability measure, 65, 72
Production schedule, 262
Properties:
  of matrix addition and scalar multiplication,
    243
  of probability measures, 74
Psychology, 63, 64 (Ex. 9), 308, 313, 314
Pure strategy, 343

Random variable, 114
Range, 26
Regular Markov chain, 321
Return on investment, 414
Row operations, 230
Row vector, 222

Saddle point, 346
  how to find a, 347
Sample-frequency method of assigning weights,
    37
Sample space, 41
Scalar multiplication, 242
Scheduling, 8, 58 (Ex. 6), 309
Selected at random, 65
Selection:
  with replacement, 53
  without replacement, 53
Set, 9
  equality, 10
  inclusion, 10

Set:
  intersection, 10
  union, 10
Simplex method:
  for linear programming, 268
    finding optimal vectors via, 285
    summary of, 290
  for solving games, 359
    summary of, 362
Sinking fund, 407
  formula for, 408
Slack variable, 274
Slope-intercept form of a line, 179
Slope of a line, 179, 180
SMP, 271
Sociology, 100, 174, 255, 260 (Exs. 10 and 11),
    308, 367, 375, 378, 379 (Ex. 14)
Solution:
  of a game, 352
    by graphing, 356
    by simplex method, 362
  of a system of equations, 222
Spread of a random variable, 140
Stable probabilities for a Markov chain, 321
Stable vector, 321
Stage of a Markov chain, 310
Standard deviation, 141
Standard maximum problem, 271
Standard normal curve, 129
Standard normal random variable, 130
State vector, 319
States of a Markov chain, 310
Status, 376
Stochastic process, 88
Stock split, 416
Strictly determined game, 346
Subset, 10
System of m linear equations and n variables,
    222
  solution of, 222

Tableau, 278
  formation of initial, 279
Technology matrix, 262
Transition diagram, 312
Transition matrix, 313
Transitions between states, 310
Tree diagram, 46
Tree measure, 90
Trial of a Markov chain, 310
Two-person zero-sum game, 343
Two-point form of equation of a line, 183

Unbounded set, 212
Union of sets, 10
Universal set, 13

Value:
  of a game, 356
  of a path, 380
Variance of a random variable, 141
Vector, 222
Vector inequality, 292
Venn diagram, 15
Vertices. 369

Weights, 39
  assignment of, 77
  branch, 90
  path, 90

x axis, 173
x intercept, 178

y axis, 173
y intercept, 178
Yield to maturity of a bond, 411

Zero vector, 226